T0211190

Communications in Computer and Information Science 641

Commenced Publication in 2007
Founding and Former Series Editors:
Alfredo Cuzzocrea, Dominik Ślęzak, and Xiaokang Yang

More information about this series at http://www.springer.com/series/7899

Harald Sack · Stefan Dietze
Anna Tordai · Christoph Lange (Eds.)

Semantic Web Challenges

Third SemWebEval Challenge at ESWC 2016
Heraklion, Crete, Greece, May 29 – June 2, 2016
Revised Selected Papers

 Springer

Editors

Harald Sack
IT Systems Engineering
Hasso-Plattner Institute
Potsdam
Germany

Stefan Dietze
Leibniz Universität Hannover
Hannover
Germany

Anna Tordai
Elsevier B.V.
Amsterdam
The Netherlands

Christoph Lange
Universität Bonn
Bonn
Germany

ISSN 1865-0929   ISSN 1865-0937 (electronic)
Communications in Computer and Information Science
ISBN 978-3-319-46564-7   ISBN 978-3-319-46565-4 (eBook)
DOI 10.1007/978-3-319-46565-4

Library of Congress Control Number: 2016951976

Printed on acid-free paper

This Springer imprint is published by Springer Nature
The registered company is Springer International Publishing AG Switzerland

Preface

Common benchmarks, established evaluation procedures, comparable tasks, and public datasets are vital to ensure reproducible, evaluable, and comparable scientific results. To assess the current state of the art and foster the systematic comparison of contributions to the Semantic Web community, open challenges are now a key scientific element of the Semantic Web conferences. Following the success of the previous years, reflected by the high number of high-quality submissions, we organized the third edition of the "Semantic Web Challenges" as an official track of the ESWC 2016 conference (held in Anissaras, Crete, Greece, from May 30 to June 2, 2016), one of the most important international scientific events for the Semantic Web research community. The purpose of challenges is to validate the maturity of the state of the art in tasks common to the Semantic Web community and adjacent academic communities in a controlled setting of rigorous evaluation, thereby providing sound benchmarks, datasets, and evaluation approaches, which contribute to the advancement of the state of the art. This third edition included five challenges: Open Knowledge Extraction (OKE 2016), Semantic Publishing (SemPub 2016), Semantic Sentiment Analysis (SSA 2016), Question Answering over Linked Data (QALD 6) and Top-K Shortest Paths in RDF Datasets. A total of 31 teams competed in the different challenges. The event attracted attendees from across the conference, with a high attendance for all challenge-related activities during ESWC 2016. This included the dedicated conference track and participation of challenge candidates during the ESWC poster and demo session. The very positive feedback and resonance suggests that the ESWC challenges provided a central contribution to the ESWC 2016 program.

This book includes the descriptions of all methods and tools that competed at the "Semantic Web Challenges 2016," together with a detailed description of the tasks, evaluation procedures, and datasets, offering to the community a snapshot of the advancement in those areas at that moment in time and material for replications of the results. The editors have divided the book content into five chapters, each dedicated to one area, or challenge. Each chapter includes an introductory section by the challenge chairs providing a detailed description of the challenge tasks, the evaluation procedure, and associated datasets and peer-reviewed descriptions of the participants' methods, tools, and results.

We would like to thank all challenge chairs, whose hard work during the organization of the 2016 edition of the Semantic Web Challenges helped in making it a fruitful event. Thanks to their work, we experienced a successful and inspiring scientific event, and we are now able to deliver this book to the community.

June 2016

Harald Sack
Stefan Dietze
Anna Tordai

Preface

Common benchmarks, established evaluation procedures, comparable tasks, and public datasets are vital to ensure reproducible, evaluable, and comparable scientific results. To assess the current state of the art and foster the systematic comparison of contributions to the Semantic Web community, open challenges are now a key scientific element of the Semantic Web conferences. Following the success of the previous years, reflected by the high number of high-quality submissions, we organized the third edition of the "Semantic Web Challenges" as an official track of the ESWC 2016 conference (held in Anissaras, Crete, Greece, from May 30 to June 2, 2016), one of the most important international scientific events for the Semantic Web research community. The purpose of challenges is to validate the maturity of the state of the art in tasks common to the Semantic Web community and adjacent academic communities in a controlled setting of rigorous evaluation, thereby providing sound benchmarks, datasets, and evaluation approaches, which contribute to the advancement of the state of the art. This third edition included five challenges: Open Knowledge Extraction (OKE 2016), Semantic Publishing (SemPub 2016), Semantic Sentiment Analysis (SSA 2016), Question Answering over Linked Data (QALD 6) and Top-K Shortest Paths in RDF Datasets. A total of 31 teams competed in the different challenges. The event attracted attendees from across the conference, with a high attendance for all challenge-related activities during ESWC 2016. This included the dedicated conference track and participation of challenge candidates during the ESWC poster and demo session. The very positive feedback and resonance suggests that the ESWC challenges provided a central contribution to the ESWC 2016 program.

This book includes the descriptions of all methods and tools that competed at the "Semantic Web Challenges 2016," together with a detailed description of the tasks, evaluation procedures, and datasets, offering to the community a snapshot of the advancement in those areas at that moment in time and material for replications of the results. The editors have divided the book content into five chapters, each dedicated to one area, or challenge. Each chapter includes an introductory section by the challenge chairs providing a detailed description of the challenge tasks, the evaluation procedure, and associated datasets and peer-reviewed descriptions of the participants' methods, tools, and results.

We would like to thank all challenge chairs, whose hard work during the organization of the 2016 edition of the Semantic Web Challenges helped in making it a fruitful event. Thanks to their work, we experienced a successful and inspiring scientific event, and we are now able to deliver this book to the community.

June 2016

Harald Sack
Stefan Dietze
Anna Tordai

Organization

Organizing Committee (ESWC)

General Chair
Harald Sack   Hasso Plattner Institute (HPI), Germany

Program Chairs
Mathieu d'Aquin   Knowledge Media Institute KMI, UK
Eva Blomqvist   Linköping University, Sweden

Workshops Chairs
Dunja Mladenic   Jožef Stefan Institute, Slovenia
Sören Auer   University of Bonn and Fraunhofer IAIS, Germany

Poster and Demo Chairs
Giuseppe Rizzo   Istituto Superiore Mario Boella, Italy
Nadine Steinmetz   Technische Universität Ilmenau, Germany

Tutorials Chairs
Tommaso di Noia   Politecnico di Bari, Italy
H. Sofia Pinto   INESC-ID, Instituto Superior Técnico, Universidade
                                     de Lisboa, Portugal

PhD Symposium Chairs
Simone Paolo Ponzetto   Universität Mannheim, Germany
Chiara Ghidini   Fondazione Bruno Kessler, Italy

Challenge Chairs
Stefan Dietze   L3S, Germany
Anna Tordai   Elsevier, The Netherlands

Semantic Technologies Coordinators
Andrea Giovanni Nuzzolese   University of Bologna/STLab ISTC-CNR, Italy
Anna Lisa Gentile   University of Mannheim, Germany

EU Project Networking Session Chairs
Erik Mannens   Data Science Lab – iMinds – Ghent University,
                                     Belgium

Mauro Dragoni   Fondazione Bruno Kessler, Italy
Lyndon Nixon   Modul Universität Vienna, Austria
Oscar Corcho   Universidad Politécnica de Madrid, Spain

Publicity Chair

Heiko Paulheim   University of Mannheim, Germany

Sponsor Chairs

Steffen Lohmann   Fraunhofer IAIS, Germany
Freddie Lecue   IBM, Ireland

Web Presence

Venislav Georgiev   STI International, Austria

Proceedings Chair

Christoph Lange   University of Bonn and Fraunhofer IAIS, Germany

Treasurer

Ioan Toma   STI International, Austria

Local Organization and Conference Administration

Katharina Haas   YouVivo GmbH, Germany

Challenges Organization

Challenge Chairs

Stefan Dietze   L3S, Germany
Anna Tordai   Elsevier, The Netherlands

Open Knowledge Extraction Challenge

Andrea Giovanni Nuzzolese STLab-CNR, Italy
Anna Lisa Gentile   University of Mannheim, Germany
Valentina Presutti   STLab-CNR, Italy
Robert Meusel   University of Mannheim, Germany
Heiko Paulheim   University of Mannheim, Germany
Aldo Gangemi   Université Paris 13, France

ESWC-16 Challenge on Semantic Sentiment Analysis

Mauro Dragoni   Fondazione Bruno Kessler, Italy
Diego Reforgiato Recupero University of Cagliari, Italy

6th Open Challenge on Question Answering over Linked Data

Christina Unger	Universität Bielefeld, Germany
Axel-Cyrille Ngonga Ngomo	University of Leipzig, Germany
Elena Cabrio	University of Nice Sophia Antipolis, France

Top-K Shortest Path in Large Typed RDF Graphs Challenge

Ioannis Papadakis	Ionian University, Greece
Michalis Stefanidakis	Ionian University, Greece
Phivos Mylonas	Ionian University, Greece
Brigitte Endres-Niggemeyer	Hannover University for Applied Sciences, Germany
Spyridon Kazanas	Ionian University, Greece

Semantic Publishing Challenge 2016

Angelo Di Iorio	University of Bologna, Italy
Anastasia Dimou	Ghent University, Belgium
Christoph Lange	University of Bonn/Fraunhofer IAIS, Germany
Sahar Vahdati	University of Bonn, Germany

Sponsoring Institutions

Gold Sponsors

http://www.eutravelproject.eu/

Silver Sponsors

http://byte-project.eu/   http://entropy-project.eu/

http://project-hobbit.eu/   http://www.springer.com/lncs

Sponsors of Specific Challenges

http://www.blazegraph.com/

Contents

Challenge on Question Answering over Linked Data

Top-k Shortest Paths in Large Typed RDF Datasets Challenge

Semantic Publishing Challenge

Open Knowledge Extraction Challenge

The Second Open Knowledge Extraction Challenge

Andrea Giovanni Nuzzolese[1(✉)], Anna Lisa Gentile[2], Valentina Presutti[1],
Aldo Gangemi[1,3], Robert Meusel[2], and Heiko Paulheim[2]

[1] Semantic Technology Laboratory, ISTC-CNR, Rome, Italy
`andrea.nuzzolese@istc.cnr.it`
[2] Data and Web Science Group, University of Mannheim, Mannheim, Germany
[3] LIPN, UMR CNRS, Université Paris 13, Sorbone Cité, Paris, France

Abstract. The Open Knowledge Extraction (OKE) challenge, at its
second edition, has the ambition to provide a reference framework for
research on Knowledge Extraction from text for the Semantic Web by
re-defining a number of tasks (typically from information and knowledge
extraction), taking into account specific SW requirements. The OKE
challenge defines two tasks: (1) Entity Recognition, Linking and Typing
for Knowledge Base population; (2) Class Induction and entity typing
for Vocabulary and Knowledge Base enrichment. Task 1 consists of identifying Entities in a sentence and create an OWL individual representing
it, link to a reference KB (DBpedia) when possible and assigning a type
to such individual. Task 2 consists in producing rdf:type statements,
given definition texts. The participants will be given a dataset of sentences, each defining an entity (known a priori). The following systems
participated to the challenge: WestLab to both Task 1 and 2, ADEL and
Mannheim to Task 2 only. In this paper we describe the OKE challenge,
the tasks, the datasets used for training and evaluating the systems, the
evaluation method, and obtained results.

1 Introduction

The vision of the Semantic Web (SW) is to populate the Web with machine
understandable data so as to make intelligent agents able to automatically interpret its content - just like humans do by inspecting Web content - and assist users
in performing a significant number of tasks, relieving them of cognitive overload.
The Linked Data movement [1] kicked-off the vision by realising a key bootstrap
in publishing machine understandable information mainly taken from structured
data (typically databases) or semi-structured data (e.g. Wikipedia infoboxes).
However, most of the Web content consists of natural language text, e.g., Web
sites, news, blogs, micro-posts, etc., hence a main challenge is to extract as much
relevant knowledge as possible from this content, and publish it in the form of
Semantic Web triples.

© Springer International Publishing Switzerland 2016
H. Sack et al. (Eds.): SemWebEval 2016, CCIS 641, pp. 3–16, 2016.
DOI: 10.1007/978-3-319-46565-4_1

There is huge work on knowledge extraction (KE) and knowledge discovery contributing to address this problem, and several contests addressing the evaluation of Information Extraction systems. Hereafter we shortly list some of the most popular initiatives which have contributed to the advancement of research on automatic content extraction:

MUC-6: The Message Understanding Conferences is a series of conferences designed to evaluate research in information extraction. MUC-6 [7] was the first to define the "named entity" task, where the participants had to identify the names of all the people, organizations, and geographic locations in a collection of textual documents in English.

HUB-4: The Hub-4 Broadcast News Evaluation[1] included a MUC-style evaluation for Named Entity Recognition, but with the focus on speech input in the domain of broadcast news.

MUC-7 and MET-2: The main difference between MUC-7/MET-2[2] to previous MUC is the introduction of multilingual NE evaluation, using training and test articles from comparable domains for all languages.

CONLL: The CoNLL-2002 and CoNLL-2003 shared task focused on language independent named entity recognition. The evaluation focused on entities of four types: persons (PER), organizations (ORG), locations (LOC) and miscellaneous (MISC) and the task was performed on Spanish and Dutch for CoNLL-2002 and German and English for CoNLL-2003 [15,16].

ACE: The Automatic Content Extraction program evaluates methods to extract (i) entities, (iii) relations among these entities and (iii) the events in which these entities participate. In the first edition extraction tasks were available in English, Arabic and Chinese. In the entity detection and tracking (EDT) task, all mentions of an entity, whether a name, a description, or a pronoun, are to be found. ACE defines seven types of entities: Person, Organization, Location, Facility, Weapon, Vehicle and Geo-Political Entity (GPEs). Each type is further divided into subtypes (for instance, Organization subtypes include Government, Commercial, Educational, Non-profit, Other) [4]. ACE started in 2004 with following successful editions[3].

TAC: The Text Analysis Conference[4] is a series of evaluation workshops on Natural Language Processing, with several specific tasks (known as "tracks"). The Knowledge Base Population (KBP) task[5] is present since 2009 and has the goal to populate knowledge bases (KBs) from unstructured text. The current KB schema consists of named entities that can be a person (PER), organization (ORG), or geopolitical entity (GPE) and predefined attributes (or slots) to fill for those named entities.

---

[1] http://www.itl.nist.gov/iad/mig/publications/proceedings/darpa99/html/ie5/ie5.htm.

[2] http://www.itl.nist.gov/iaui/894.02/related_projects/muc/proceedings/muc_7_proceedings/overview.html.

[3] https://www.ldc.upenn.edu/collaborations/past-projects/ace/annotation-tasks-and-specifications.

[4] http://www.nist.gov/tac/tracks/index.html.

[5] http://www.nist.gov/tac/2015/KBP.

TREC-KBA: The Knowledge Base Acceleration (KBA) track[6] ran in TREC 2012, 2013 and 2014. It evaluates systems that filter a time-ordered corpus for documents and slot fills that would change an entity profile in a predefined list of entities. The focus is therefore on spotting novelty and changes for predefined entities.

SemEval-2015 Task 13: The Multilingual All-Words Sense Disambiguation (WSD) and Entity Linking (EL) are tasks that address the lexical ambiguity of language, but they use different meaning inventories: EL uses encyclopedic knowledge, while WSD uses lexicographic information. The main goal of this combined task is to treat the two problems holistically using a resource that integrates both kinds of inventories (i.e., BabelNet 2.5.1).

Despite the numerous initiatives for benchmarking KE systems, there is lack of a "genuine" SW reference evaluation framework for helping researchers and the whole community to assess the state of the art in this domain. In fact, results of Knowledge Extraction systems are usually evaluated against tasks that do not focus on specific Semantic Web goals. For example, tasks such as named Entity Recognition, Relation Extraction, Frame Detection, etc. are certainly of importance for the SW, but in most cases such tasks are designed without considering the output design and formalisation in the form of Linked Data and OWL ontologies. This makes results of existing methods often not directly reusable for populating the SW, until a translation from linguistic semantics to formal semantics is performed.

The OKE challenge, originally inspired by [11] and launched as first edition at last year Extended Semantic Web Conference [10] has the ambition to provide a reference framework for research on *Knowledge Extraction from text for the Semantic Web* by re-defining a number of tasks (typically from information and knowledge extraction), taking into account specific SW requirements.

2 Tasks

2.1 Task 1: Entity Recognition, Linking and Typing for Knowledge Base population

This task consists of (i) identifying Entities in a sentence and create an OWL individual (`owl:Individual` statement) representing it, (ii) link (`owl:sameAs` statement) such individual, when possible, to a reference Knowledge Base (i.e., DBpedia [2]) and (iii) assigning a type to such individual (`rdf:type` statement) selected from a set of given types. In this task by Entity we mean any discourse referent (the actors and objects around which a story unfolds), either named or anonymous that is an individual of one of the following DOLCE Ultra Lite classes[7] [6], i.e., `dul:Person`[8], `dul:Place`, `dul:Organization`, and `dul:Role`.

---

[6] http://trec-kba.org/.

[7] http://stlab.istc.cnr.it/stlab/WikipediaOntology/.

[8] The prefix `dul:` stands for the namespace http://www.ontologydesignpatterns.org/ont/dul/DUL.owl#.

By entities we also refer to anaphorically related discourse referents. Hence, anaphora resolution is part of the requirements for the identification of entities. As an example, for the sentence:

Florence May Harding studied at a school in Sydney, and with Douglas Robert Dundas, but in effect had no formal training in either botany or art.

we want to recognize the entities reported in Table 1.

Table 1. Task 1: example.

Recognized entity	Generated URI	Type	SameAs
Florence May Harding	oke:Florence_May_Harding	dul:Person	dbpedia:Florence_May_Harding
school	oke:School	dul:Organization	
Sydney	oke:Sydney	dul:Place	dbpedia:Sydney
Douglas Robert Dundas	oke:Douglas_Robert_Dundas	dul:Person	

Sentences were provided in input to systems as RDF by using the NIF notation[9] [9]. The following is an example of input for the previous sentence.

```
oke:task-1/sentence-1#char=0,146
        a                 nif:RFC5147String , nif:String , nif:Context ;
    nif:beginIndex        "0"^^xsd:nonNegativeInteger ;
    nif:endIndex          "146"^^xsd:nonNegativeInteger ;
    nif:isString          "Florence May Harding studied at a school in Sydney, and with
                          Douglas Robert Dundas , but in effect had no
                          formal training in either botany or art."@en .
```

System were asked to provide recognised entities by using a NIF-compliant output as shown in the following example.

```
... oke:Florence_May_Harding
        a                 owl:Individual, dul:Person ;
    rdfs:label            "Florence May Harding"@en ;
    owl:sameAs            dbpedia:Florence_May_Harding .

oke:task-1/sentence-1#char=0,20
        a                 nif:RFC5147String , nif:String ;
    nif:anchorOf          "Florence May Harding"@en ;
    nif:beginIndex        "0"^^xsd:nonNegativeInteger ;
    nif:endIndex          "20"^^xsd:nonNegativeInteger ;
    nif:referenceContext  oke:task-1/sentence-1#char=0,146 ;
    itsrdf:taIdentRef     oke:Florence_May_Harding .
```

---

[9] http://persistence.uni-leipzig.org/nlp2rdf/.

The RDF above[10] is an example of possible output for annotating the string that represents the entity *Florence May Harding* in the original sentence. This string is typed as a `nif:RFC5147String` and is related to a reference context (cf., property `nif:referenceContext`), which identifies the input sentence, and to an `owl:Individual` (cf., property `itsrdf:taIdentRef`), which represents the entity within the dataset. This entity is further typed as `dul:Person` and linked to its corresponding entity in DBpedia (cf. property `owl:sameAs`). The namespace prefix `oke:` is used to identify the URIs of recognised entities. There is not a given rule for generating these URI, thus any system can implement its own algorithm for generating URIs. The linking to DBpedia can be omitted in case a system is not able to identify a corresponding entity in such a dataset. This means that it might be possible to have entities that cannot be linked to any DBpedia entities. This is always the case occurring when dealing with anonymous entities. For example, given the sentence:

She was appointed as Senator for Life in Italy by the President Carlo Azeglio Ciampi.

We want to recognise the term *She* an as anonymous entity within our dataset and to type it as `owl:Individual` and `dul:Person`. However, we do not want any linking to DBpedia because it would introduce an error.

2.2   Task 2: Class Induction and Entity Typing for Vocabulary and Knowledge Base Enrichment

This task was designed for producing `rdf:type` statements for an entity, given its definition as natural language text. The participants were provided with a dataset of sentences, each defining an entity (known a priori). More in detail the task required the participants to (i) identify the type(s) of the given entity as they are expressed in the given definition, (ii) create a `owl:Class` statement for defining each of them as a new class in the target knowledge base, (iii) create a `rdf:type` statement between the given entity and the new created classes, and (iv) align the identified types, if a correct alignment is available, to a set of given types from a subset of DOLCE+DnS Ultra Lite classes. Table 2 shows the complete list of these types[11]

For example, given the entity `dbpedia:Skara_Cathedral` and its definition

Skara Cathedral is a church in the Swedish city of Skara.

the types that the systems were asked to recognise are reported in Table 3.

---

[10] The prefixes `nif:`, `itsrdf:`, `dul:`, and `dbpedia:` identify the namespaces http://persistence.uni-leipzig.org/nlp2rdf/ontologies/nif-core#, http://www.w3.org/2005/11/its/rdf#, http://www.ontologydesignpatterns.org/ont/dul/DUL.owl#, and http://dbpedia.org/resource/ respectively.

[11] Prefixes `d0:` and `dul:` stand for namespaces http://ontologydesignpatterns.org/ont/wikipedia/d0.owl# and http://www.ontologydesignpatterns.org/ont/dul/DUL.owl# respectively.

Table 2. The subset of DOLCE+DnS Ultra Lite classes used for typing entities in the Task 2.

Class	Description
dul:Abstract	Anything that cannot be located in space-time
d0:Activity	Any action or task planned or executed by an agent intentionally causing and participating in it
dul:Amount	Any quantity, independently from how it is measured, computed, etc.
d0:Characteristic	An aspect or quality of a thing
dul:Collection	A container or group of things (or agents) that share one or more common properties
d0:CognitiveEntity	Attitudes, cognitive abilities, ideologies, psychological phenomena, mind, etc.
dul:Description	A descriptive context that creates a relational view on a set of data or observations
d0:Event	Any natural event, independently of its possible causes
dul:Goal	The description of a situation that is desired by an agent
dul:InformationEntity	A piece of information, be it concretely realized or not: linguistic expressions, works of art, knowledge objects
d0:Location	A location, in a very generic sense e.g. geo-political entities, or physical object that are inherently located
dul:Organism	A physical object with biological characteristics, typically able to self-reproduce
dul:Organization	An internally structured, conventionally created social entity such as enterprises, bands, political parties, etc.
dul:Person	Persons in commonsense intuition
dul:Personification	A social entity with agentive features, invented or conceived through a cultural process
dul:PhysicalObject	Any object that has a proper space region, and an associated mass: natural bodies, artifacts, substances
dul:Process	Any natural process, independently of its possible causes
dul:Process	Any natural process, independently of its possible causes
dul:Role	A concept that classifies some entity: social positions, roles, statuses
dul:Situation	A unified view on a set of entities, e.g. physical or social facts or conditions, configurations, etc.
d0:System	Physical, social, political systems
dul:TimeInterval	A time span
d0:Topic	Any area, discipline, subject of knowledge

Table 3. Task 2: example.

Recognized string for the type	Generated type	Subclass of
fictional villain	oke:FictionalVillain	dul:Personification
villain	oke:Villain	oke:FictionalVillain, dul:Person

Target entities, i.e. the entities to type, along with their definition in natural language were provided as RDF by using the NIF notation. The following is an example of input for the previous example.

```
oke:task-2/sentence-1#char=0,150>
    a               nif:RFC5147String , nif:String , nif:Context ;
    nif:isString        "Brian Banner is a fictional villain from the Marvel Comics Universe
                        created by Bill Mantlo and Mike Mignola and first
                        appearing in print in late 1985." ;
    nif:beginIndex      "0"^^xsd:int ;
    nif:endIndex        "150"^^xsd:int .

oke:task-2/sentence-1#char=0,12
    a               nif:RFC5147String , nif:String ;
    nif:anchorOf        "Brian Banner"@en ;
    nif:referenceContext oke:task-2/sentence-1#char=0,150 ;
    nif:beginIndex      "0"^^xsd:int ;
    nif:endIndex        "12"^^xsd:int ;
    itsrdf:taIdentRef    dbpedia:Brian_Banner .

dbpedia:Brian_Banner
    rdfs:label          "Brian Banner"@en .
```

Participants were asked to complete the RDF snippet above with the following information about typing by using the NIF notation:

```
... oke:FictionalVillain
    a               owl:Class ;
    rdfs:label          "fictional villain"@en ;
    rdfs:subClassOf     dul:Personification .

oke:Villain
    a               owl:Class ;
    rdfs:label          "villain"@en ;
    rdfs:subClassOf     oke:FictionalVillain, dul:Person .

oke:sentence-1#char=18,35
    a               nif:RFC5147String , nif:String ;
    nif:anchorOf        "fictional villain"@en ;
    nif:referenceContext oke:task-2/sentence-1#char=0,150 ;
    nif:beginIndex      "18"^^xsd:int ;
    nif:endIndex        "35"^^xsd:int ;
    itsrdf:taIdentRef    oke:FictionalVillain .

oke:sentence-1#char=28,35>
    a               nif:RFC5147String , nif:String ;
    nif:anchorOf        "villain"@en ;
    nif:referenceContext oke:task-2/sentence-1#char=0,150 ;
    nif:beginIndex      "28"^^xsd:int ;
    nif:endIndex        "35"^^xsd:int ;
    itsrdf:taIdentRef    oke:Villain .
```

We designed the task in order to ask participants to report as `rdfs:label` the string recognised within a definition as a valid type for a given entity.

Additionally, we asked participants to record span indexes for such strings with respect to the original definition by using `nif:beginIndex` and `nif:endIndex`. Namely, `nif:beginIndex` and `nif:endIndex` were used to identify the initial and final span index respectively.

3   Training and Evaluation Datasets

In the following sections we describe how we constructed the datasets used for Task 1 and 2 (cf. Sect. 3.1 and 3.2 respectively).

3.1   Task 1

The training set used for Task 1 consists of the training and the evaluation set used for the Task 1 at the last OKE challenge (i.e., OKE 2015). Hence, training set counts of 196 sentences selected from Wikipedia articles reporting biographies. This choice comes from the observation that biographies typically contain entities about people (e.g., the person that is subject of the given Wikipedia article, her colleagues, her relatives, etc.), locations (e.g., the places the person lived in), organisations (e.g., the organisations the person worked for) and roles (e.g., the roles held by the person during her career). The evaluation set counts of 55 sentences selected from Wikipedia by following the same rationale as for the training set and by taking care to have:

– no overlap of sentences between the two training and evaluation datasets;
– as much as possible an equal distribution of DOLCE entity types (i.e., Person, Place, Organization, and Role) within the datasets.

The evaluation set was constructed by configuring a job[12] on CrowdFlower[13]. Basically, we asked participants to annotate each sentence with (i) entities of four specific types (i.e., Persons, Places, Organizations, and Roles) and (ii) to associate, whenever possible, each identified Entity with a link to English Wikipedia.

Table 4 shows the details about the two datasets in terms of the number of sentences, the overall number of annotated entities, the average number of annotated entities per sentence, and the number of entities linked to DBpedia.

Table 4. Figures about the training and the evaluation datasets for Task 1.

Parameter	Training dataset	Evaluation dataset
# of sentences	196	55
# of annotated entities	722	218
Avg # of annotated entities per sentence	4.56	5.31
# of entities linked to DBpedia	534	172

---

[12] A preview of the job can be found at https://tasks.crowdflower.com/channels/cf_internal/jobs/913913/editor_preview.

[13] https://crowdflower.com.

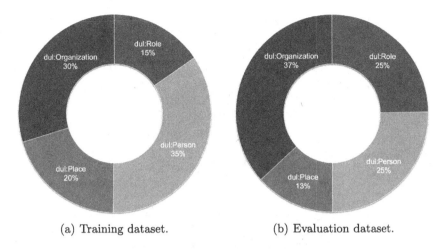

(a) Training dataset.   (b) Evaluation dataset.

Fig. 1. Distribution of entities according to their DOLCE type.

Figure 1 shows the distribution of entities with respect to the four DOLCE types used for typing in Task 1. More in detail, Fig. 1a and b show the distribution in the training dataset and the evaluation dataset, respectively.

Both datasets are available on-line as TURTLE for download.[14]

3.2   Task 2

For Task 2, similarly to Task 1, we defined a training and an evaluation dataset. The training dataset consists of both training and evaluation datasets from the previous OKE edition, for a total of 198 sentences, manually annotated using the NIF notation. Each sentence provides a definition of a DBpedia entity expressed as natural language. The evaluation dataset was manually constructed by an annotator with high expertise on DOLCE and consists of 50 sentences. Table 5 reports the details about the training and the evaluation datasets for Task 2 in terms of (i) number of sentences annotated, (ii) number of `rdfs:subClassOf`

Table 5. Figures about the training and the evaluation datasets for Task 2.

Parameter	Training dataset	Evaluation dataset
# of sentences	198	50
# of `rdfs:subClassOf` axioms	425	139
# of annotated classes	520	142

---

[14] The training dataset is available at https://github.com/anuzzolese/oke-challenge-2016/blob/master/GoldStandard_sampleData/task1/dataset_task_1.ttl.   Similarly, the evaluation dataset is available at https://github.com/anuzzolese/oke-challenge-2016/blob/master/evaluation-data/task1/evaluation-dataset-task1.ttl.

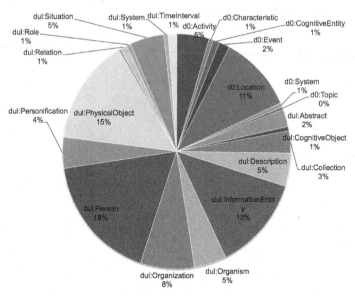

(a) Training dataset.

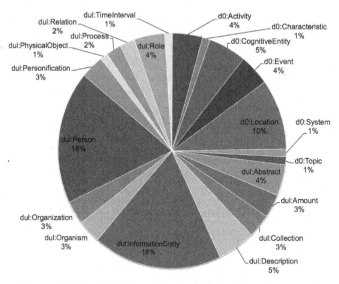

(b) Evaluation dataset

Fig. 2. Distribution of entities over the subset of DOLCE Ultra Lite classes used for Task 2.

axioms used within the datasets and (iii) number of classes extracted from the natural language and used for typing the DBpedia entities. It is worth remarking that each sentence provided a definition for a single DBpedia entity only, meaning that the number of sentences and the number of DBpedia entities in the datasets were the same.

Figures 2a and b show the distribution of entities over the subset of DOLCE Ultra Lite classes used for Task 2, for the training and the evaluation datasets respectively.

Finally, Fig. 3 shows the comparison between the training and evaluation datasets in terms of the number of resources available for each DOLCE Ultra Lite class.

Both datasets are available on-line as TURTLE for download[15].

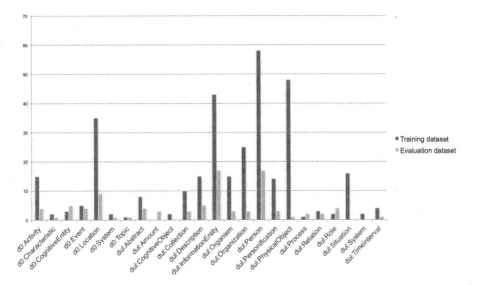

Fig. 3. Comparison between the number of resources grouped by class available in the training and evaluation datasets used for Task 2.

4   Results

The evaluation of the challenge was enabled by designing a dedicated version of GERBIL [17], which was used as benchmarking system for evaluating precision, recall and F-measure for both tasks. For Task 1 GERBIL was designed in order

---

[15] The training dataset is available at https://github.com/anuzzolese/oke-challenge-2016/blob/master/GoldStandard_sampleData/task2/dataset_task_2.ttl. Similarly, the evaluation dataset is available at https://github.com/anuzzolese/oke-challenge-2016/blob/master/evaluation-data/task2/evaluation-dataset-task2.ttl.

to evaluate systems with respect to (i) their ability to recognize entities using the NIF offsets returned by the systems (only full matches were counted as correct, e.g., if the system returned "Art School" instead of "National Art School", this was counted as a miss), (ii) their ability to assign the correct type among the 4 target DOLCE types (cf. Sect. 2.1), and (iii) their ability to link individuals to DBpedia 2014. Instead, for Task 2 GERBIL was designed in order to evaluate systems with respect to (i) their ability to recognize strings (i.e., linguistic evidences) in the definition that identify the type of a target entity (i) their ability to align identified types to the defined subset of DOLCE Ultra Lite classes (cf. Table 2 in Sect. 2.2).

We received four submissions. Table 6 lists the systems participating to the challenge by providing the names and the short descriptions of the systems, and the tasks they are involved in.

Table 6. Systems participating to the challenge.

System	Description	Participating to task
Mannheim [5]	Open Knowledge Extraction Challenge (2016) a Hearst-like Pattern-Based approach to Hypernym Extraction and Class Induction	2
WestLab-Task1 [3]	Collective disambiguation and Semantic Annotation for Entity Linking and Typing	1
Adel [12]	A Enhancing Entity Linking by Combining Models	1
WestLab-Task2 [8]	Entity Typing and Linking using SPARQL Patterns and DBpedia	2

The winner for Task 1 was WestLab-Task1, which obtained micro F1 and macro F1 of 0.6676 and 0.6516 respectively. The exhaustive results for all the systems involved in Task 1 is reported in Table 7.

Table 7. Task 1 results

Annotator	Micro F1	Micro Precision	Micro Recall	Macro F1	Macro Precision	Macro Recall
WestLab-Task1	0.6676	0.6964	0.6509	0.6516	0.6902	0.6439
Adel	0.6249	0.6689	0.5942	0.6064	0.6606	0.5846

The winner for Task 2 was WestLab-Task1, which obtained micro F1 and macro F1 of 0.4664 and 0.465 respectively. The exhaustive results for all the systems involved in Task 2 are reported in Table 8. Results also include figures obtained by running Cetus [13] as baseline. Cetus was the winning system for the task at last year edition of OKE and the service is still up and running.

Table 8. Task 2 results

Annotator	Micro F1	Micro Precision	Micro Recall	Macro F1	Macro Precision	Macro Recall
Mannheim	0.3617	0.3864	0.34	0.3233	0.315	0.34
WestLab-Task2	0.4664	0.47	0.4633	0.465	0.47	0.4633
(BASELINE from OKE2015) Cetus	0.5023	0.4546	0.5624	0.4861	0.4708	0.5624

5   Conclusions

The Open Knowledge Extraction challenge attracted research groups coming from the Knowledge Extraction (KE) and the Semantic Web (SW) communities. Indeed, the challenge proposal was aimed at attracting research groups from these two communities in order to further investigate existing overlaps between KE and the SW. Additionally, one of the goals of the challenge was to foster the collaboration between the two communities, to the aim of growing further the SW community. To achieve this goal we defined a SW reference evaluation framework, which is composed of (i) two tasks, (ii) a training and evaluation dataset for each task, and (iii) an evaluation framework to measure the accuracy of the systems.

Although the participation in terms of number of competing systems remained quite limited, we believe that the challenge is a breakthrough in the hybridisation of Semantic Web technologies with Knowledge Extraction methods. As a matter of fact, the evaluation framework is available on-line[16] and can be reused by the community and for next editions of the challenge.

References

1. Bizer, C., Heath, T., Berners-Lee, T.: Linked data - the story so far. Int. J. Semant. Web Inf. Syst. **5**(3), 1–22 (2009)
2. Bizer, C., Lehmann, J., Kobilarov, G., Auer, S., Becker, C., Cyganiak, R., Hellmann, S.: DBpedia - a crystallization point for the web of data. J. Web Semant. **7**(3), 154–165 (2009)

---

[16] https://github.com/anuzzolese/oke-challenge-2016.

3. Chabchoub, M., Gagnon, M., Zouaq, A.: Collective disambiguation and semantic annotation for entity linking and typing. In: Sack et al. [14]
4. Doddington, G.R., Mitchell, A., Przybocki, M.A., Ramshaw, L.A., Strassel, S., Weischedel, R.M.: The automatic content extraction (ACE) program-tasks, data, and evaluation. In: LREC (2004)
5. Faralli, S., Ponzetto, S.P.: Open knowledge extraction challenge a hearst- like pattern-based approach to hypernym extraction and class induction. In: Sack et al. [14] (2016)
6. Gangemi, A., Guarino, N., Masolo, C., Oltramari, A., Schneider, L.: Sweetening ontologies with DOLCE. In: Gómez-Pérez, A., Benjamins, V.R. (eds.) EKAW 2002. LNCS (LNAI), vol. 2473, pp. 166–181. Springer, Heidelberg (2002)
7. Grishman, R., Sundheim, B.: Message understanding conference-6: a brief history. In: Proceedings of 16th Conference on Computational Linguistics - COLING 1996, vol. 1, pp. 466–471. Association for Computational Linguistics, Stroudsburg (1996)
8. Haidar-Ahmad, L., Font, L., Zouaq, A., Gagnon, M.: Entity typing and linking using sparql patterns and DBpedia. In: Sack et al. [14]
9. Hellmann, S., Lehmann, J., Auer, S., Brümmer, M.: Integrating NLP using linked data. In: Alani, H., et al. (eds.) ISWC 2013, Part II. LNCS, vol. 8219, pp. 98–113. Springer, Heidelberg (2013)
10. Nuzzolese, A.G., Gentile, A.L., Presutti, V., Gangemi, A., Garigliotti, D., Navigli, R.: Open knowledge extraction challenge. In: Gandon, F., Cabrio, E., Stankovic, M., Zimmermann, A. (eds.) SemWebEval 2015. CCIS, vol. 548, pp. 3–15. Springer, Heidelberg (2015). doi:10.1007/978-3-319-25518-7_1
11. Petasis, G., Karkaletsis, V., Paliouras, G., Krithara, A., Zavitsanos, E.: Ontology population and enrichment: state of the art. In: Paliouras, G., Spyropoulos, C.D., Tsatsaronis, G. (eds.) Bridging the Semantic Gap. LNCS, vol. 6050, pp. 134–166. Springer, Heidelberg (2011)
12. Plu, J., Rizzo, G., Troncy, R.: Enhancing entity linking by combining models. In: Sack et al. [14]
13. Röder, M., Usbeck, R., Speck, R., Ngomo, A.-C.N.: CETUS – a baseline approach to type extraction. In: Gandon, F., Cabrio, E., Stankovic, M., Zimmermann, A. (eds.) SemWebEval 2015. CCIS, vol. 548, pp. 16–27. Springer, Heidelberg (2015). doi:10.1007/978-3-319-25518-7_2
14. Sack, H., Dietze, S., Tordai, A., Lange, C. (eds.): The Semantic Web: ESWC Challenges, Communications in Computer and Information Science. Springer, Berlin (2016)
15. Tjong Kim Sang, E.F., Introduction to the CoNLL- shared task: language-independent named entity recognition. In: Proceedings of 6th Conference on Natural Language Learning - COLING-2002, vol. 20, pp. 1–4. Association for Computational Linguistics, Stroudsburg (2002)
16. Iordache, O.: Introduction. In: Iordache, O. (ed.) Polystochastic Models for Complexity. UCS, vol. 4, pp. 1–16. Springer, Heidelberg (2010)
17. Usbeck, R., Röder, M., Ngomo, A.N., Baron, C., Both, A., Brümmer, M., Ceccarelli, D., Cornolti, M., Cherix, D., Eickmann, B., Ferragina, P., Lemke, C., Moro, A., Navigli, R., Piccinno, F., Rizzo, G., Sack, H., Speck, R., Troncy, R., Waitelonis, J., Wesemann, L.: GERBIL: general entity annotator benchmarking framework. In: Gangemi, A., Leonardi, S., Panconesi, A. (eds.) Proceedings of 24th International Conference on World Wide Web, WWW 2015, pp. 1133–1143. ACM (2015)

Enhancing Entity Linking by Combining NER Models

Julien Plu[1], Giuseppe Rizzo[2], and Raphaël Troncy[1(✉)]

[1] EURECOM, Sophia Antipolis, France
{julien.plu,raphael.troncy}@eurecom.fr
[2] ISMB, Turin, Italy
giuseppe.rizzo@ismb.it

Abstract. Numerous entity linking systems are addressing the entity recognition problem by using off-the-shelf NER systems. It is, however, a difficult task to select which specific model to use for these systems, since it requires to judge the level of similarity between the datasets which have been used to train models and the dataset at hand to be processed in which we aim to properly recognize entities. In this paper, we present the newest version of ADEL, our adaptive entity recognition and linking framework, where we experiment with an hybrid approach mixing a model combination method to improve the recognition level and to increase the efficiency of the linking step by applying a filter over the types. We obtain promising results when performing a 4-fold cross validation experiment on the OKE 2016 challenge training dataset. We also demonstrate that we achieve better results that in our previous participation on the OKE 2015 test set. We finally report the results of ADEL on the OKE 2016 test set and we present an error analysis highlighting the main difficulties of this challenge.

Keywords: Entity recognition · Entity linking · Entity filtering · Model combination · OKE challenge · ADEL

1 Introduction

The 2016 Open Knowledge Extraction challenge (OKE2016) aims to: *(i)* identify entities in a sentence, *(ii)* assign a type to these entities selected from a set of given types and *(iii)* link such entities, when possible, to a DBpedia resource. In this paper, we present our participation to this challenge using a newer version of the ADEL framework that extends our approach presented last year at OKE2015 [10]. A first improvement concerns the framework architecture which has became more modular: external NLP systems are used via a REST API and the knowledge base index uses more sophisticated tools (Elastic and Couchbase) while being built from additional data format (e.g. TSV). A second improvement concerns the way our Stanford-based named entity recognition module can be used, in particular, using multiple models.

© Springer International Publishing Switzerland 2016
H. Sack et al. (Eds.): SemWebEval 2016, CCIS 641, pp. 17–32, 2016.
DOI: 10.1007/978-3-319-46565-4_2

This paper mainly focuses on entity recognition, which refers to jointly performing the appropriate extraction and the typing of mentions. *Extraction* is the task of spotting mentions that can be entities in the text while *Typing* refers to the task of assigning them a proper type. Linking is the last step of our approach, and it refers to the disambiguation of mentions in a targeted knowledge base. It is also often composed of two subtasks: generating candidates and ranking them accordingly to various scoring functions. Following the challenge requirements, we make use of the 2015-04 snapshot of DBpedia as the targeted knowledge base.

The remainder of this paper is structured as follows. We first describe some recent related work, emphasizing the usage of external NLP systems when performing entity recognition (Sect. 2). Next, we detail the newest architecture of ADEL (Sect. 3). We propose two experiment settings among many variants in order to highlight the importance of a pre-processing step and to demonstrate the added-value of combining CRF models for improving the performance of the extraction step (Sect. 4). We first detail our results on the OKE2016 challenge training dataset using a 4-fold cross-validation setup and we also measure the improvements of ADEL on the 2015 test set (Sect. 5). Next, we provide the results of ADEL on the 2016 test set following an adjudication phase (Sect. 6). Finally, we conclude and outline some future work in Sect. 7.

2   Related Work

Several entity linking systems use an external named entity recognition tool such as Stanford NER [2] or the Apache OpenNLP Name Finder[1]. For example, the popular AIDA[2] system makes use of Stanford NER trained on the CoNLL2003 dataset [4]. In [7], the authors also use Stanford NER but without saying which specific model is being used. In [4], the authors use Stanford NER in a similar way than AIDA. Finally, the FOX tool proposes an ensemble learning method over Stanford NER and other NER classifiers (such as OpenNLP, Illinois Named Entity Tagger and Ottawa Baseline Information Extraction). The authors use a model trained again on the CoNLL2003 dataset for each sub-classifier [12]. In terms of architecture, while all these systems use external NER systems, they integrate them only internally, using the provided Java API directly in their source code. This kind of integration makes difficult the possibility of re-configuring those external NLP systems, or switching between different ones.

We first hypothesize that the CoNLL2003-based model for recognizing entities for a type of text than differs from a newswire article will not be necessary optimal [9]. Therefore, we propose an architecture that enables to use multiple models. We also promote a flexible way of interacting with NER components via a standard API.

Several entity linking systems advocate a so-called *E2E* (End-to-End) approach. This method uses only a semantic network of an entity catalogue to extract mentions from the text and to generate candidate links. The limitation with this

---

[1] https://opennlp.apache.org/.
[2] https://github.com/yago-naga/aida.

method is therefore its ability to extract emerging entities, since entities that are not present in the catalogue will not be extracted and disambiguated. Our hybrid approach overcomes this problem since we do not only use a catalogue of entities but also a POS and a NER tagger for extracting entities, thus mixing semantic and NLP-based methods. Let's take an example coming from the OKE2016 dataset with the sentence: *James Tobin married Elizabeth Fay Ringo, a former M.I.T. student of Paul Samuelson, on September 14, 1946.* In this sentence, the entities to be extracted according to the challenge annotation rules are: *James Tobin, Elizabeth Fay Ringo, M.I.T., student* and *Paul Samuelson.* Most of those entities can be extracted and linked to DBpedia via an E2E approach except *Elizabeth Fay Ringo* which does not exist in DBpedia yet. TagME [1] is a popular system implementing an *E2E* approach that provides a public API[3]. TagME will not effectively extract the entity *Elizabeth Fay Ringo* from this sentence, contrarily to our ADEL framework.

3   ADEL Architecture

The goal of our system is to link all the mentions occurring in a text to their counterparts in a knowledge base. Emerging entities, *i.e.* entities that are not present in a knowledge base, will be linked to *NIL*. ADEL comes with a brand new architecture compared to the version we have proposed in the previous edition of this challenge [10,11]. This architecture is composed of multiple modules spread into two main parts (Fig. 1). The first part (*Entity Extraction*) contains the modules *Extractors* and *Overlap Resolution*. The second part (*Entity Linking*) contains the modules *Indexing, Candidate Generation, NIL Clustering* and *Linkers.* We detail those modules in the reminder of this section.

3.1   Entity Extraction

In this section, we describe how we extract mentions from texts that are likely to be selected as entities with the *Extractor Module.* After having identified candidate mentions, we resolve their potential overlaps using the *Overlap Resolution Module.*

Extractors Module. We make use of three kinds of extractors: *(i)* Dictionary, *(ii)* POS Tagger and *(iii)* NER. Each of these extractors run in parallel. At this stage, an entity dictionary reinforces the extraction by bringing a robust spotting for well-known proper nouns or mentions that are too difficult to be extracted for the other extractors (e.g. *Role-type* mentions). The two other extractors use an external NLP system based on Stanford CoreNLP [8] and particularly the POS [14] and NER taggers.

We have developed a generic *NLP System REST API* wrapper to use the Stanford CoreNLP system. This wrapper has been designed while keeping in mind the core idea of ADEL, namely *adaptivity.* Hence, this module gives the

---
[3] http://tagme.di.unipi.it.

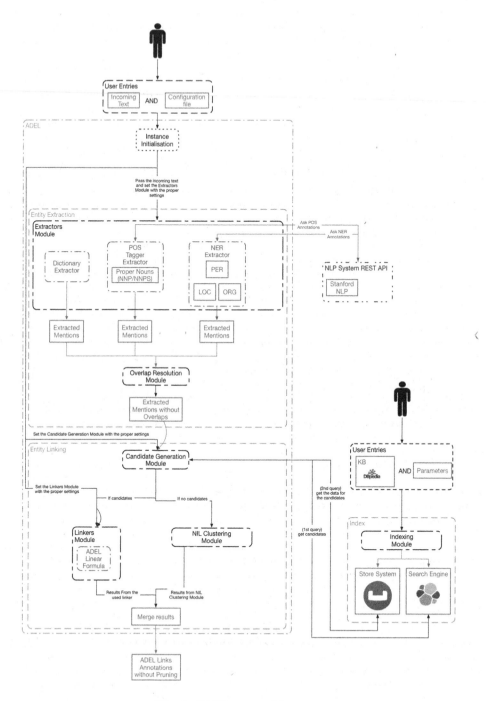

Fig. 1. ADEL new architecture

possibility to use any other NLP system such as the ones used in [12] or even systems tailored for other languages than English. The REST API provides annotations in the NIF format [3]. Therefore, by using this module, one can switch from one NLP system to another one without changing anything in the code or can combine different systems. This module enables as well to save computing time since all models being used are loaded only once at startup. A configuration file enables to parametrize how to use Stanford CoreNLP. During our tests on the OKE2016 dataset, we used the *english-bidirectional-distsim* model that provides a better accuracy but for a higher computing time for the POS tagger[4]. Contrarily to [4,7,12], we use a model combination method that aims to jointly make use of different CRF models as described in the Algorithm 1.

This algorithm shows that the order in which the models are applied is important. Hence, if a token is badly labeled by the first model, the second model cannot correct it even if it would have given the correct label in the first place. This algorithm in Stanford NER is called *NER Classifier Combiner*. An implementation of the Stanford CoreNLP as provided by our *NLP System REST API* is available on Github at https://github.com/jplu/stanfordNLPRESTAPI.

Algorithm 1. Algorithm used in ADEL to combine multiple CRF models

---

Result: Annotated tokens
Input : (Txt, M) with Txt the text to be annotated and M a list of CRF models
Output: $A = List(\{token, label\})$ a list of tuples $\{token, label\}$
1 **begin**
2    $finalTuples \leftarrow EmptyList()$;
3   **foreach** *model in M* **do**
      /* $tmpTuples$ contains the tuples $\{token, label\}$ got from *model*   */
4       $tmpTuples \leftarrow$ apply *model* over Txt;
5   **foreach** $\{token, label\}$ *in tmpTuples* **do**
6   **if** *token from* $\{token, label\}$ *not in finalTuples* **then**
7   add $\{token, label\}$ in $finalTuples$;
8   **end**
9   **end**
10   **end**
11 **end**

Overlap Resolution Module. The extractors can extract mentions that have a partial or a full overlap with others. To resolve this ambiguity, we implement an *overlap resolution* module that takes the output of each component of the extractors module and gives one output without overlaps. The logic of this module is as follows: given two overlapping mentions, *e.g.* States of America from Stanford NER and United States from Stanford POS tagger, we only take the union of the two phrases. We obtain the mention United States of America

---

[4] http://nlp.stanford.edu/software/pos-tagger-faq.shtml#h.

and the type provided by Stanford NER is selected. We have also implemented other heuristics for resolving overlaps but the choice of the proper heuristics to use is still left to be manually configured.

3.2 Entity Linking

In this section, we describe how we disambiguate candidate entities coming from the extraction step (Sect. 3.1). First, we create an index over a targeted knowledge base, *e.g.* the April 2015 DBpedia snapshot, using the *Indexing Module*. This index is used to select possible candidates with the *Candidate Generation Module*. If no candidates are provided, this entity is passed to the *NIL Clustering Module*, while if candidates are retrieved, they are given to the *Linkers Module*.

Indexing Module. Previously, we were using an index stored in Lucene. We have, however, observed unexpected behavior from Lucene such as not retrieving resources that match partially a query even when not bounding the number of results. The index is now built using *Elastic* as a search engine and *Couchbase* as data storage. First, we query *Elastic* to get the possible candidates. Second, we query *Couchbase* to get the data associated with these possible candidates. The index is built on top of both DBpedia2015-04[5] and a dump of the Wikipedia articles[6] dated from February 2015.

Candidate Generation Module. This module is querying *Elastic* and *Couchbase* to get possible candidates for the entities coming from the extraction module. If this module gets candidates for an entity, they are given to the *Linkers Module*; if not, they are given to the *NIL Clustering Module*.

NIL Clustering Module. We propose to group the *NIL* entities (emerging entities) that may identify the same real-world thing. The role of this module is to attach the same *NIL* value within and across documents. For example, if we take two different documents that share the same emerging entity, this entity will be linked to the same *NIL* value. We can then imagine different *NIL* values, such as *NIL_1*, *NIL_2*, etc.

Linkers Module. This module implements an empirically assessed function that ranks all possible candidates given by the *Candidate Generation Module*:

$$r(l) = (a \cdot L(m, title) + b \cdot max(L(m, R)) + c \cdot max(L(m, D))) \cdot PR(l) \quad (1)$$

The function $r(l)$ is using the Levenshtein distance L between the mention m and the title, the maximum distance between the mention m and every element (title) in the set of Wikipedia redirect pages R and the maximum distance between the mention m and every element (title) in the set of Wikipedia disambiguation pages D, weighted by the PageRank PR, for every entity candidate l. The weights a, b and c are a convex combination that must satisfy: $a + b + c = 1$ and $a > b > c > 0$. We take the assumption that the string distance measure

---

[5] http://wiki.dbpedia.org/services-resources/datasets/datasets2015-04.
[6] https://dumps.wikimedia.org/enwiki/.

between a mention and a title is more important than the distance measure with a redirect page which is itself more important than the distance measure with a disambiguation page.

4 Demonstrating the Added-Value of Using Multiple CRF Models

In this section, we aim to demonstrate the added-value of combining NER models for improving the named entity recognition performance. We have set up two distinct experiments, using either a single CRF model (Sect. 4.1) or multiple ones (Sect. 4.2), and that also highlight the importance of doing a proper data pre-processing and training. We performed a lightweight error analysis in the Sect. 4.4 that justifies this pre-processing step.

4.1 Experiment 1: No Role, One Single CRF Model

In the first experiment, we *(i)* pre-process the training set by removing all the occurrences of the Role type, and *(ii)* train a single CRF model with Stanford NER using the OKE 2016 training set, that will be used via the *CRF Classifier* feature. We discard the Role type on purpose as explained in the Sect. 4.4.

4.2 Experiment 2: No Role, Multiple CRF Models

In the second experiment, we perform the same pre-processing step as in the first experiment, but we make use of the *NER Classifier Combiner* feature instead of *CRF Classifier*. This feature allows to combine multiple CRF models to annotate a text where combining *model1* and *model2* means that *model1* will first be applied, followed by *model2*. Two combination options are available: if the option *ner.combinationMode* is set to *NORMAL* (the default option), any label applied by *model1* cannot be applied by subsequent models. For instance, if *model1* provides the LOCATION tag, no other model's LOCATION tag will be generated (the tag is case sensitive). If *ner.combinationMode* is set to *HIGH_RECALL*, this limitation will be deactivated.

In our experiments, we use the *HIGH_RECALL* combination mode with two CRF models (in this order): *english.all.3class.distsim.crf.ser.gz* trained over the PERSON, LOCATION and ORGANIZATION types from the CoNLL 2003 [13], MUC6, MUC7 and ACE 2002 datasets which is provided by default in the Stanford NER package, and a specific model trained from the OKE2016 dataset.

In order to illustrate this model combination functionality, let's take the following sentence from the OKE2016 dataset: *Martin Luther King then began doctoral studies in systematic theology at Boston University and received his Ph.D. degree on June 5, 1955, with a dissertation on "A Comparison of the Conceptions of God in the Thinking of Paul Tillich and Henry Nelson Wieman".* First, the *english.all.3class.distsim.crf.ser.gz* model is applied, yielding to the extraction of the candidate entities *Martin Luther King (PERSON), Boston University*

(LOCATION), *Paul Tillich (PERSON)* and *Henry Nelson Wieman (PERSON)*. Next, the OKE2016 model is applied. The combination mode prevents to re-label those candidate entities, but the second model will extract *God (PERSON)*. The entities coming from the two models are then merged without possible conflicts.

4.3 NER Results of Experiment 1 and Experiment 2 on the Training Set

Table 1-a shows the results of the NER extractor following the Experiment 1 setup (Sect. 4.1) computed with the conlleval scorer. We observe a higher recognition score for the type *Person* than for the two other types. This can be explained by the fact that there are twice more entities of type *Person* than *Place* and *Organization* in the training dataset.

Table 1. NER Results following the Experiment 1 (a) and Experiment 2 (b) corresponding to the usage of one or multiple CRF models

Type	Precision	Recall	F-measure
Organization	67.05	58.58	62.42
Person	88.09	85.73	86.87
Place	69.75	68.69	69.16
Total	78.86	74.75	76.75

(a)

Type	Precision	Recall	F-measure
Organization	88.35	80.10	83.97
Person	92.11	93.50	92.73
Place	78.03	78.58	77.83
Total	88.23	85.37	86.75

(b)

Table 1-b shows the results of the NER extractor following the Experiment 2 setup (Sect. 4.2) computed with the conlleval scorer. We observe a significant improvement in terms of recognition compared to the previous experiment. Nevertheless, the results for the type *Place* is still lower than for the other types. This can be explained by the fact that the datasets used to train the *english.all.3class.distsim.crf.ser.gz* model and the dataset from OKE2016 do not share the same definition of what is a *Place*. For example, the mention *Poughkeepsie, New York* is a single entity in the OKE2016 dataset, but correspond to two entities in the datasets used to train the *english.all.3class.distsim.crf.ser.gz* model.

4.4 Error Analysis

We have performed three other variants of the previous two experiments as follows:

1. keep the Role type annotations from the training set and use either one CRF model or multiple ones;
2. vary the experiment 2 setup using the *NORMAL* combination mode;
3. vary the experiment 2 setup using all models provided by default in the Stanford NER package.

These variants consistently yield to a drop of performance in terms of recognition results (or in terms of computing time with no gain in terms of recognition).

The first variant provides worst extraction and recognition results. This can be explained by the lack of sufficient occurrences of the Role type in the training dataset which is too small. Consequently, the results of the first variant have a high precision, but a lower recall for the Role recognition. The second variant provides a performance drop in terms of computing time, without giving more annotations compared to the *HIGH_RECALL* mode. This can be explained by the fact that there is no overlap between the types provided by the OKE model and the ones from the *english.all.3class.distsim.crf.ser.gz* model. Actually, the *NER Classifier Combiner* feature is case sensitive which means that the PERSON type (from CoNLL) is different from the Person type (from OKE). The third variant provides also a drop of performance in terms of computing time, since we obtain the same results while needing a much larger computation time.

Consequently, we pre-process the training dataset by discarding the Role type when training the NER module and we rely on a dictionary for extracting the Role type entities. This dictionary is built by using a list of the names of all the jobs existing in Wikipedia. A comparison of the pure linguistic approach (Stanford NER) and our system in terms of recognition is shown in Sect. 5.2 to demonstrate the advantage of using multiple extractors.

5   ADEL Results on the Training Set

In this section, we provide preliminary results of our ADEL framework on a 4-fold cross validation experiment using the 2016 OKE challenge training dataset. We use two different scorers: *conlleval*[7] and *neleval*[8]. We did not use the official scorer of the challenge (GERBIL [15]) since it cannot yet provide breakdown figures per entity type, like the conlleval scorer, or per sub-task (extraction, recognition, linking), like the neleval scorer. We have computed the NER results using the conlleval scorer to get the breakdown results per entity type (Person, Role, Organization and Place). We have computed the NEL results using the neleval scorer to get the breakdown results per sub-task.

The differences in terms of figures between the conlleval and the neleval scorers can be explained by the fact that conlleval does not count the entities without a type (entities coming from the POS extractor and linked to NIL). These entities are considered either as false negative or true negative by the conlleval scorer while being counted as false positive in recognition for the neleval scorer. This explains why conlleval will provide a higher precision score, while neleval will provide a higher recall score.

---

[7] http://bulba.sdsu.edu/~malouf/ling681/conlleval.
[8] https://github.com/wikilinks/neleval.

Table 2. Statistics of the OKE 2016 training dataset

Type	nb mentions	nb entities	nb mentions disambiguated (%)	nb entities disambiguated (%)
dul:Place	182	145	171 (93.96 %)	134 (92.41 %)
dul:Person	458	253	350 (76.42 %)	164 (64.82 %)
dul:Organization	237	212	198 (83.54 %)	177 (83.49 %)
dul:Role	166	109	145 (87.35 %)	90 (82.57 %)
Total	1043	719	864 (82.84 %)	565 (78.58 %)

5.1 Statistics of the Training Dataset

The training dataset provided by the OKE2016 challenge organizers is composed of a set of 196 annotated sentences using the NIF ontology[9]. The average length of the sentences is 155 chars. In total, the dataset contains 1043 mentions corresponding to 719 distinct entities that belong to one of the four types: dul:Place, dul:Person, dul:Organization and dul:Role. 565 entities (78.58 %) are linked within DBpedia, while 153 (21.28 %) are not. The breakdown of those annotations per type is provided in Table 2.

We applied a 4-fold cross validation on the training set. In each fold of the cross validation, a train and a test sets are generated and respectively used for building the supervised learning models and for benchmarking the output of the model with the expected results of the test set.

5.2 NER Results of ADEL on the Training Dataset

As we have described in the Sect. 3.1, ADEL makes use of multiple extractors (dictionary-based, POS and NER based) followed by an overlap resolution module. This hybrid approach provides the final results presented in the Table 3. We observe a general improvement and a particular high recognition of the type *Role* due to the specialized dictionary extractor. The results for the type *Person* are a little bit lower than in Experiment 2. This is due to some false positive extracted by the POS tagger extractor.

Table 3. Final ADEL results at the recognition stage on the training dataset

Type	Precision	Recall	F-measure
Organization	85.90	82.72	84.22
Person	91.27	93.27	92.10
Place	77.13	81.42	78.82
Role	95.23	98.65	96.84
Total	87.85	88.91	88.36

---

[9] http://persistence.uni-leipzig.org/nlp2rdf/ontologies/nif-core#.

5.3   NEL Results of ADEL on the Training Dataset

We use the *neleval* scorer for computing results at the linking stage. More precisely, we consider the *strong_mention_match*, *strong_typed_mention_match* and *strong_link_match* scores. The first score corresponds to a strict mention extraction. The second score corresponds to a strict mention extraction with the good type. The third score corresponds to a strict mention extraction with the good link. Considering that ADEL performs relatively well for extracting and recognizing entities, we assume that the candidate links that do not have a type corresponding to the one assigned by ADEL are likely to not be good candidates and should therefore be filtered out. Applying this simple heuristic improves the results (Table 4).

Table 4. ADEL results at the linking stage on the training dataset: (a) without using a type filter and (b) using a type filter

Level	Precision	Recall	F-measure	Precision	Recall	F-measure
strong_mention_match	81.0	88.7	84.7	81.0	88.7	84.7
strong_typed_mention_ match	78.1	85.4	81.6	78.1	85.4	81.6
strong_link_match	45.2	57.4	50.5	57.4	55.7	56.5
	No filter used			A type filter is being used		

We finally compare the previous version of ADEL (v1) used in 2015 with the newer version of ADEL (v2) presented in this paper. For both versions, we use the OKE 2015 training set for training the NER extractors and we use the OKE 2015 test set for evaluation (Table 5). We observe a relative gain of 18.75 % which is largely due to the novel model combination feature detailed in this paper.

Table 5. Comparison between ADEL-v1 and ADEL-v2 over the OKE 2015 test set

| Level | ADEL-v1 | | | ADEL-v2 | | |
	Precision	Recall	F-measure	Precision	Recall	F-measure
strong_mention_match	78.2	65.4	71.2	85.1	89.7	87.3
strong_typed_mention_ match	65.8	54.8	59.8	75.3	59.0	66.2
strong_link_match	49.4	46.6	48	85.4	42.7.	57.0

6   ADEL Results on the Test Set

In this section, we provide the official results of the ADEL framework running on the test set provided by the challenge. We used again the *conlleval* and *neleval* scorers as reported in the Sect. 5 and we also use the *GERBIL* scorer which provides the official results. We observe differences among the results provided by these scorers: GERBIL does not take into account in the evaluation the mentions retrieved by a system that do not belong to the gold standard, which means that they are not counted as false positive, contrarily to the behavior of the two other scorers. Consequently, the figures reported by the GERBIL scorer are higher. See Sect. 5 for the difference between the *conlleval* and the *neleval*. We performed two evaluations: one with the initial test set used to compute the official figures released during the conference and another one after performing an adjudication of the test set. We have indeed proposed to modify numerous annotations in the test set that were inconsistent with the rules used in the training set. The organizers have approved those corrections and merged them into the now official test set.

6.1   Statistics of the Test Set

The test dataset provided by the OKE2016 challenge organizers after adjudication is composed of a set of 55 annotated sentences using the NIF ontology. The average length of the sentences is 187 chars. In total, the test set contains 340 mentions corresponding to 218 distinct entities that belong to one of the four types: `dul:Place`, `dul:Person`, `dul:Organization` and `dul:Role`. 173 entities (79.36 %) are linked within DBpedia, while 45 (20.64 %) are not. The breakdown of those annotations per type is provided in Table 6.

Table 6. Statistics of the OKE 2016 test dataset

Type	nb mentions	nb entities	nb mentions disambiguated (%)	nb entities disambiguated (%)
dul:Place	44	29	44 (100 %)	29 (100 %)
dul:Person	105	55	82 (78.10 %)	37 (67.27 %)
dul:Organization	105	80	91 (86.67 %)	67 (83.75 %)
dul:Role	86	54	71 (82.56 %)	40 (74.07 %)
Total	340	218	288 (84.71 %)	173 (79.36 %)

The Tables 2 and 6 show that the distribution of mentions per type is dissimilar between the training and the test set. For example, there is the exact same number of mentions of type *PERSON* and *ORGANIZATION* in the test set while there is twice more mentions of type *PERSON* than *ORGANIZATION* in the training dataset. There is also twice more mentions of type *ROLE* than *PLACE*. There is a similar number of NIL entities in the two datasets. However, they are not distributed in the same way among the different types. Hence, there are more NIL entities for *ROLE* in the training set but none for *PLACE* in the test set. The percentage of disambiguated entities and mentions are similar in the two datasets.

6.2   NER and NEL Results

The Tables 7, 8 and 9 show the results of ADEL when using respectively the conlleval, neleval and GERBIL scorers on the test set after the adjudication phase. Consequently, the figures are slightly different than the ones who have been presented during the conference.

We provide some guidelines to better interpret the results shown in the Tables 7, 9 and 8. Modulo the differences in what each scorer actually evaluates (see Sect. 5), we consider that the line *Total* from the conlleval script roughly corresponds to the line *strong_typed_mention_match* from the neleval scorer and to the line *Entity Typing* from the GERBIL scorer. Similarly, the task *strong_mention_match* in neleval roughly corresponds to the task *Entity Recognition* in GERBIL. On the test set, ADEL has an overall F1 score of 72 % (or even 80 %) for properly extracting and recognizing entities of type Organization, Person, Place and Role. The task *strong_link_match* in neleval roughly corresponds to the task *D2KB* in GERBIL. For both scorers, we observe a significant loss of performance in ADEL that ultimately can only properly disambiguate 50 % of the entities.

6.3   Lessons Learned

We perform an error analysis in order to better understand what are the entities that ADEL did not recognize, wrongly typed or badly disambiguated. During the extraction process, the entities that ADEL miss-recognized are either due to the

Table 7. ADEL results with the CONLLEVAL scorer on the OKE test set after adjudication

Type	Precision	Recall	F-measure
Organization	80.90	65.45	72.36
Person	76.24	54.40	64.17
Place	75.56	59.65	66.67
Role	93.67	81.32	87.06
Total	81.60	64.92	72.31

Table 8. ADEL results with the NELEVAL scorer on the OKE test set after adjudication

Level	Precision	Recall	F-measure
strong_mention_match	83.1	73.8	78.2
strong_typed_mention_match	76.5	67.9	72.0
strong_link_match	52.8	45.8	49.1

Table 9. ADEL results with the GERBIL scorer on the OKE test set after adjudication

Level	Precision	Recall	F-measure
Entity recognition	80.78	80.56	80.06
Entity typing	80.56	80.56	80.56
D2KB	59.21	49.32	53.12

role dictionary that does not cover all possible roles or to a missing co-reference module. Hence, we observe that the performance at the extraction level (*Entity Recognition* in GERBIL or *strong mention match* in NELEVAL) drops between the training set and the test set where more co-references and new roles are present. In addition, we observe that the POS tagger brings some false positive mentions. For example, in the sentence 22 from the test set: *This was a new chair, one of the first three in theoretical physics in Italy* ..., the POS tagger extracts *physics* as a mention.

At the recognition stage, we observe some weaknesses in our model combination method which also plays a role in the extraction process. We have identified three different type of errors: *(i)* entities that should not have been extracted in the first place (e.g. *Fiat* is tagged as an Organization while the text is talking about a Product); *(ii)* entities that are simply not extracted (e.g. *dictatorship* should be extracted as an Organization according to the challenge rules but neither the POS nor the NER module is able to extract this mention); *(iii)* entities that are wrongly typed (e.g. *Cornell* should be tagged as Organization, denoting a university, but ADEL considers it to be a Place). Miss-typing organizations and places is a well-known issue in the field. Finally, ADEL makes errors on so-called *nested entities*. For example, the surface form *Stockholm, Sweden* can either be extracted as two different mention *Stockholm* and *Sweden* or as one single mention. The challenge organizers consider that this surface form corresponds to two entities while our model extracts a single entity.

At the linking stage, ADEL suffers from a disambiguation formula that gives too much importance to the absolute popularity of an entity (due to the PageRank factor). For example, the most popular entity for the mention *author* is the brand `db:Author_(bicycles)`[10] and not `db:Author`. A second problem in our ranking formula concerns the string distance measure being used, Levenshtein, which tends to give a better score with shorter strings. For example, the string distance score over the title, the redirect and the disambiguation pages between the entity mention *GM* and the entity candidate `db:Germany` (0.32879817) is higher than with the entity candidate `db:General_Motors` (0.21995464). One possibility to overcome those problems is to rely on a re-ranking module capable of better use the context surrounding the mentions [6].

---

[10] PREFIX db: <http://dbpedia.org/resource/>.

7 Conclusion and Future Work

In this paper, we have shown the benefit of combining different CRF models to improve the entity recognition, and to use it as a filter to also improve the linking. We demonstrate that the challenge dataset is not similarly distributed in terms of types accordingly to the Tables 2 and 6. The type distribution problem is revealed by the difficulty to properly recognize the *Place* or *Role* types by using only the OKE2016 dataset. In [16], we have conducted a thorough study that reveals common issues from well-known datasets that are traditionally used to evaluate the entity linking task.

This dataset is also complex since it contains mentions and disambiguation of co-references and anaphora. A co-reference denotes a situation where two or more expressions refer to the same entity in a same text, for example *Look at that man over there; he is wearing a funny hat*. An anaphora is when pronouns are used to link to an antecedent but this antecedent does not refer to any entity in the same text, for example *No man said he was hungry*. We can argue that anaphora is the generic term and co-reference is a specific kind of anaphora. Our system does not take into account this syntactic particularity, and a possible future work would be to include a module to extract and link such syntactic particularity.

We finally aim to improve our disambiguation method by better taking into account the context surrounding each mentions. One promising research direction is to take inspiration from the DSRM approach [5] that uses deep learning techniques in order to compute pairwise similarity between entities coming from the same knowledge base.

Acknowledgments. This work was partially supported by the innovation activity 3cixty (14523) of EIT Digital and by the European Union's H2020 Framework Programme via the FREME Project (644771).

References

1. Ferragina, P., Scaiella, U.: TAGME: on-the-fly annotation of short text fragments (by wikipedia entities). In: 19th ACM Conference on Information and Knowledge Management (CIKM) (2010)
2. Finkel, J., Grenager, T., Manning, C.: Incorporating non-local information into information extraction systems by Gibbs sampling. In: 43rd Annual Meeting on Association for Computational Linguistics (2005)
3. Hellmann, S., Lehmann, J., Auer, S., Brümmer, M.: Integrating NLP using linked data. In: Alani, H., et al. (eds.) ISWC 2013, Part II. LNCS, vol. 8219, pp. 98–113. Springer, Heidelberg (2013)
4. Hoffart, J., Altun, Y., Weikum, G.: Discovering emerging entities with ambiguous names. In: 23rd World Wide Web Conference (WWW) (2014)
5. Huang, H., Heck, L., Ji, H.: Leveraging deep neural networks and knowledge graphs for entity disambiguation. CoRR (2015)
6. Ilievski, F., Rizzo, G., van Erp, M., Plu, J., Troncy, R.: Context-enhanced adaptive entity linking. In: 10th International Conference on Language Resources and Evaluation (LREC) (2016)

7. Ling, X., Singh, S., Weld, D.S.: Design challenges for entity linking. Trans. Assoc. Comput. Linguist. (TACL) **3**, 315–328 (2015)
8. Manning, C.D., Surdeanu, M., Bauer, J., Finkel, J., Bethard, S.J., McClosky, D.: The stanford CoreNLP natural language processing toolkit. In: Association for Computational Linguistics (ACL) System Demonstrations (2014)
9. Plu, J.: Knowledge extraction in web media: at the frontier of NLP, machine learning and semantics. In: 25th World Wide Web Conference (WWW), Ph.D. Symposium (2016)
10. Plu, J., Rizzo, G., Troncy, R.: A hybrid approach for entity recognition and linking. In: Gandon, F., Cabrio, E., Stankovic, M., Zimmermann, A. (eds.) Semantic Web Evaluation Challenges. Communications in Computer and Information Science, vol. 548, pp. 28–39. Springer, Switzerland (2015)
11. Plu, J., Rizzo, G., Troncy, R.: Revealing entities from textual documents using a hybrid approach. In: 3rd NLP&DBpedia International Workshop (2015)
12. Speck, R., Ngonga Ngomo, A.-C.: Ensemble learning for named entity recognition. In: Mika, P., et al. (eds.) ISWC 2014, Part I. LNCS, vol. 8796, pp. 519–534. Springer, Heidelberg (2014)
13. Tjong Kim Sang, E.F., Meulder, F.D.: Introduction to the CoNLL-2003 shared task: language-independent named entity recognition. In: 17th Conference on Computational Natural Language Learning (CoNLL) (2003)
14. Toutanova, K., Klein, D., Manning, C.D., Singer, Y.: Feature-rich part-of-speech tagging with a cyclic dependency network. In: Conference of the North American Chapter of the Association for Computational Linguistics on Human Language Technology (2003)
15. Usbeck, R., Röder, M., Ngonga Ngomo, A.-C., Baron, C., Both, A., Brümmer, M., Ceccarelli, D., Cornolti, M., Cherix, D., Eickmann, B., Ferragina, P., Lemke, C., Moro, A., Navigli, R., Piccinno, F., Rizzo, G., Sack, H., Speck, R., Troncy, R., Waitelonis, J., Wesemann, L.: GERBIL: general entity annotator benchmarking framework. In: 24th World Wide Web Conference (WWW) (2015)
16. van Erp, M., Mendes, P.N., Paulheim, H., Ilievski, F., Plu, J., Rizzo, G., Waitelonis, J., Linking, E.E.: An analysis of current benchmark datasets and a roadmap for NG a better job. In: 10th International Conference on Language Resources and Evaluation (LREC) (2016)

Collective Disambiguation and Semantic Annotation for Entity Linking and Typing

Mohamed Chabchoub[1], Michel Gagnon[1(✉)], and Amal Zouaq[1,2]

[1] École Polytechnique de Montréal, Montréal, Canada
{mohamed.chabchoub,michel.gagnon}@polymtl.ca
[2] University of Ottawa, Ottawa, Canada
azouaq@uottawa.ca

Abstract. In this paper we present the WESTLAB system, the winner of the 2016 OKE challenge Task 1. Our approach combines the output of a semantic annotator with the output of a named entity recognizer, and applies some heuristics for merging and filtering the detected mentions. The approach also applies a collective disambiguation method that relies on all the previously linked entities to choose between multiple candidate entities for a given mention. Using this approach, we greatly improve the performance of all the semantic annotators that are used as baselines in our experiments and also outperform the best system of the OKE Challenge 2015.

Keywords: OKE challenge · Entity recognition · Entity typing · Entity linking

1 Introduction

The first task of the Open Knowledge Extraction (OKE) challenge is divided into three subtasks: entity recognition, entity linking and entity typing with four DUL classes (Person, Organization, Place and Role). In this paper, we show that by combining the outputs of a semantic annotator and a named entity recognizer, we can obtain very good performances for all three subtasks[1]. WESTLAB, our system, relies on the principle of collective disambiguation [5,7], where the selection of the entity to be linked to a mention takes into account the entities previously associated to other mentions in the text. To implement our approach, we propose a pipeline architecture. In the first step, the detected mentions in the text are adjusted and filtered, such that only the most relevant ones are transmitted to the entity linking module. The linking module identifies the corresponding entity in DBpedia for each mention and applies the collective disambiguation process if there are several candidates. Finally, for entity typing, our approach uses some manually defined mapping rules.

---

[1] Our service is available at the following URL: http://westlab.polymtl.ca/OkeTask1/rest/annotate/post.

H. Sack et al. (Eds.): SemWebEval 2016, CCIS 641, pp. 33–47, 2016.
DOI: 10.1007/978-3-319-46565-4_3

This paper is organized as follows. Next section discusses related works. Section 3 describes our system in more details. In Sect. 4 we present the results of our system on the training and test datasets provided by the 2016 OKE Challenge and we compare them with several baselines, using different configurations. In Sect. 5 we conclude and make some suggestions to improve WESTLAB's performance.

2   Related Work

In this section we present some state-of-the-art approaches that perform well for named entity recognition and semantic annotation, with a particular focus on systems used in our evaluation for comparison purposes. We also present the best performing systems in the 2015 OKE challenge.

2.1   Named Entity Recognition

Extracting named entities has been tackled by numerous Natural Language Processing studies in the last decade. OpenNLP[2] uses machine learning and maximum entropy models. LingPipe[3] uses n-gram character language models. OpenCalais[4] is a commercial service that uses machine learning techniques to recognize named entities and uses a proprietary taxonomy to type them. Stanford NER [4] is based on Conditional Random Fields (CRF) models and is widely used in the development of NLP applications. In an evaluation of named entity recognizers on bibliographical texts [1], Stanford NER was identified as the best system. Another evaluation on microposts [2] shows that Stanford NER is the second best after OpenCalais. This last result is especially relevant for our context, since we also have very short texts as input. Since OpenCalais is a commercial product, Stanford NER was thus our best choice. Note that this tool not only detects named entities mentions in a text, but also disambiguates them according to one of the following classes: PERSON, ORGANIZATION and LOCATION.

2.2   Semantic Annotation

Most of the semantic annotators available as online services are commercial products. For our experiments, we decided to use non-commercial systems only.

DBpedia Spotlight [6] is a semantic annotator that uses a two-step process: first, entity mentions are spotted in text and then linked to some entity in DBpedia. The spotting phase relies on a set of surface forms extracted from DBpedia (titles, redirects) and anchors of Wikipedia links. To achieve entity linking, DBpedia Spotlight pre-ranks entity candidates for each surface form spotted in

---

[2] https://opennlp.apache.org/.

[3] http://alias-i.com/lingpipe/.

[4] http://www.opencalais.com/.

the text. It combines a prior score and a contextual score to determine which candidate entity is the most relevant. The prior score represents an estimation of how often the surface form is used as an anchor in a Wikipedia hyperlink that points to the entity page. Formally it corresponds to the probability $P(e|s)$, where e and s are the entity and the surface form, respectively. The contextual score takes into account the context of the phrase (a window of words around the phrase) and the context of each candidate entity. Note that DBpedia Spotlight has a confidence parameter that can be set whose value is between 0 and 1 (default value is 0.5). The effect of this parameter is to remove some annotations. The highest the confidence value, the highest is the probability of eliminating an annotation if it has more than one candidate for entity linking.

Similarly to Spotlight, Tagme [3] uses a list of surface forms extracted from Wikipedia anchors. For entity linking, it is based on a collective disambiguation process where, for each candidate entity of a surface form, a score of relatedness with the candidates of other surface forms is computed. The selected candidate is the one that maximizes a final score that combines all these relatedness scores. Tagme also uses a pruning method that retains only annotations that have a high link probability (defined as the number of Wikipedia articles that use it as an anchor, divided by the number of articles that mention it) or that have a high coherence score, which is computed by comparing it to all other annotations (by averaging over relatedness scores). Note that Tagme has been developed to be efficient with short texts, such as micro-blog posts.

Another semantic annotator that has been proposed recently is Babelfy [7], which uses a graph-based approach. Surface forms in text are associated to one or more vertices in BabelNet [8], a semantic network built from Wikipedia and Wordnet. Disambiguation is then achieved by a process that identifies the densest sub-graph according to some coherence metric.

Finally, AIDA [5] is a semantic annotator that assigns, to each spotted surface form, a value that corresponds to the prior score in DBpedia Spotlight, and a context similarity score. These scores are weighted using links and information extracted from DBpedia and YAGO. AIDA uses an additional score that estimates the coherence between two entities. This score is calculated using the Wikipedia inlinks of each entity. The main contribution of this system is a graph-based algorithm. The graph is composed of text mentions nodes and entities nodes. AIDA extracts the sub-graph which has the highest density, where the density is a combination of the three computed scores for each annotation.

2.3   Entity Typing

None of the previous systems achieves the full task of the OKE Challenge, which also includes entity typing according to some predefined types. Stanford NER does not consider the ROLE type. Semantic annotators only identify named entities in DBpedia, without typing them according to the OKE challenge type nomenclature. The best performing system for this task in the 2015 OKE Challenge [9] is ADEL [10], whose F-score is 0.60. It uses a hybrid approach that combines linguistic and semantic features. Linguistic resources, such as POS tagger, gazetteer and named entity recognizer, are used for the entity recognition task. For entity linking, some filtering is made, based on inbound and outbound links in Wikipedia. To select the most relevant entity among candidates, a graph-based approach similar to Babelfy is used. For typing with `dul:Role` class, ADEL mainly relies on a gazetteer (containing lists of occupations and nationalities). For other types, a manual alignment is built. Finally, a classifier is trained to filter out irrelevant entities. The second best-performing system, FOX [12], has a F-score of about 0.5. Typing is achieved by using a pattern-based process that identifies the segment that expresses the type of the entity in the input. One important advantage of FOX is that it is available through the GERBIL platform, thus facilitating performance comparison.

3   System Implementation

3.1   System Architecture

WESTLAB is composed of three modules, as illustrated in Fig. 1: (1) entity spotting, which consists of identifying relevant mentions in the text, (2) entity linking, where mentions are associated to some entity in DBpedia, when possible, and (3) entity typing, where the system tries to find, for each entity, the corresponding type in the DUL ontology. We will now describe each module.

3.2   Entity Spotting

The goal of this module is to extract all relevant mentions that can be found in the input text. We relied on semantic annotators freely available as web services and one named entity recognizer, the Stanford NER [4]. We experimented with four semantic annotators: DBpedia Spotlight, Babelfy, AIDA and Tagme. These semantic annotators not only detect relevant mentions in text, but also identify a corresponding entity in DBpedia. Stanford NER detects named entities that are tagged according to one of the following classes: *person*, *organization* and *location*. Each of these types can be aligned to a type in the OKE challenge: *Person*, *Organization* and *Place*. However, the type *role*, which is also an OKE challenge type, is not recognized by the Stanford NER.

After running these semantic annotators and Stanford NER, we noticed a frequent overlap between mentions detected by semantic annotators and named entities extracted by Stanford NER. In this case, we keep the longest mention. Consider the following example, where two mentions found by DBpedia Spotlight are indicated in boldface:

> *John Stigall received a Bachelor of arts from the* **State University** *of* **New York** *at Cortland*

In this case Spotlight annotates *State University* and *New York* separately, whereas Stanford NER recognizes *State University of New York* as a single

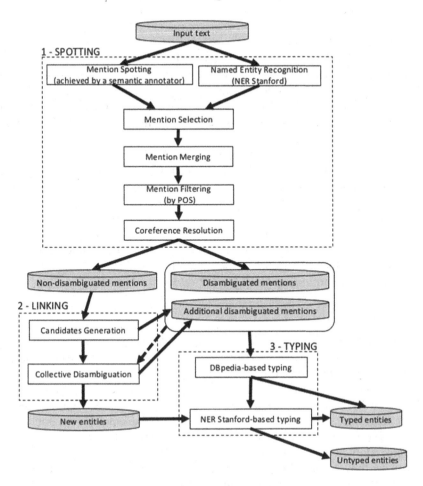

Fig. 1. System architecture

named entity. We select this last mention and discard the two separate mentions returned by DBpedia Spotlight. This is done at the *Mention selection* step.

At this stage, we still do not always have the correct mention. In fact, in this example, the correct mention is *State University of New York at Cortland*. To deal with this problem, we developed a *mention merging algorithm*. Given a mention, the algorithm attempts to expand it to cover the next mention. Such expansion is permitted only if the second mention immediately follows the first one or if the mentions are separated by one of these patterns: a word, a comma, a comma followed by a word, a period or a period followed by a word. Following this rule, in the previous example the mention *State University of New York* will be expanded into *State University of New York at Cortland*. This process is repeated until we reach a state where the mention cannot be further expanded. The expanded mention obtained at each step is memorized, such that at the end we obtain a list of mentions that are ordered from the longest one to the shortest one. We then select the first mention in this list for which there is an entity in DBpedia whose label corresponds to this mention.

The next step is *mention filtering*, where the POS of each word in a mention is identified by using the Stanford POS tagger. Every mention that contains a verb is removed from the list. Finally, we use Stanford Coreference resolution to find, for each pronoun, the coreferent mention. The mention and its co-referents are thus linked to the same entity.

The output of this module consists in two sets. One set contains the mentions detected by the semantic annotator. These mentions are already disambiguated with their corresponding entities in DBpedia. The other set contains the named entities returned by Stanford NER and the new mentions that resulted from the merging algorithm. Remember that every mention that overlaps another mention is removed if the other one is longer. The second set does not contain any disambiguated mention and must be processed by the following module, which performs entity linking for these mentions. Note that mentions already detected by the semantic annotator are not added to this set.

3.3 Entity Linking

For every non-disambiguated mention, we query DBpedia to extract the entity whose label corresponds to the mention. If there is no such entity, we create a new URI resource as described in the OKE challenge. If there is one result, we check if it corresponds to a disambiguation page. If so, all entities that are referred by this disambiguation page are taken as candidates, and are given as input to the next step, that is, the disambiguation process. If the entity corresponds to a normal page (i.e. not a disambiguation page), the entity is linked to the mention, and immediately added to the set of disambiguated entities.

When we have more than one candidate, the selection of the entity that will be linked to the mention is achieved by taking into account the other mentions that have already been disambiguated in the text. In this case, non-disambiguated mentions are processed sequentially. For each mention we compute a score for each candidate c associated to this mention. This score is based on the "outlinks" of the corresponding Wikipedia article of the entity c and is defined as follows:

$$Score(c) = \frac{\sum_{e_i \in G} outlink(e_i) \cap outlink(c)}{outlink(c) \cup outlink(G)}$$

where G is the set of all distinct entities that are already disambiguated.

We keep the candidate with the highest score value, and thus obtain a new linked mention that is added to the set of disambiguated entities. The process iterates until every mention has been disambiguated.

3.4 Entity Typing

The goal of this step is to align the extracted entities with one of the following classes in the DUL ontology: DUL:Person, DUL:Organization, DUL:Place and

DUL:Role. If a mention is not linked to a DBpedia entity, it is necessarily a mention that was recognized by Stanford NER. In this case, we simply apply the following mapping:

- Stanford:ORGANIZATION → DUL:Organization
- Stanford:LOCATION → DUL:Place
- Stanford:PERSON → DUL:Person

For mentions associated with a DBpedia URI, we try to find the type of the linked entity by executing the following SPARQL query:

`SELECT ?type WHERE {<entity> rdf:type ?type}`

If the query returns with success, the following mapping is applied, according to the instantiation of the variable `?type`:

- dbo:organisation → DUL:Organization
- yago:Organization108008335 → DUL:Organization
- dbo:Place → DUL:Place
- dbo:Location → DUL:Place
- dbo:EthnicGroup → DUL:Person
- dbo:Person → DUL:Person
- foaf:person → DUL:Person

If no type can be identified with this first query, we execute a second SPARQL query, where we check if the relevant information is represented by a predicate:

`SELECT ?predicate WHERE {[] ?predicate <entity>}`

According to the value extracted for the predicate, the following mapping is used:

- dbo:affiliation → DUL:Organization
- dbp:owner → DUL:Organization
- dbp:office → DUL:Organization
- dbo:birthPlace → DUL:Place
- dbo:location → DUL:Place
- dbo:deathPlace → DUL:Place
- dbo:occupation → DUL:Role
- dbp:occupation → DUL:Role

A conflict can appear with this method, when predicates are extracted with different types. This kind of conflict especially occurs between types *DUL:Role* and *DUL:Organization*, as for *State_University_of_New_York*, for which both predicates *dbo:affiliation* and *dbo:occupation* are extracted. In this case we count the number of occurrences of each predicate, and select the one with the highest number of occurrences. If no type can be identified with this second query, we execute a third SPARQL query, where we check if the relevant information is represented by a predicate, with the entity as subject:

```
SELECT ?predicate WHERE {<entity> ?predicate []}
```

According to the value extracted for the predicate, the following mapping is used:

- geo:geometry → DUL:Place

If this third query doesn't help to identify the type, we check if the mention was detected by Stanford NER in the spotting phase, and if so we use the mapping given at the beginning of this section. If not, the entity remains untyped.

4   Experiments and Results

4.1   Training Set Description

We conducted our experiments against the training set distributed by the OKE challenge organizers, which is composed of 195 manually annotated sentences. It contains 1030 mentions annotated with 683 distinct entities, among which 530 are linked to DBpedia and 153 are defined as NIL mentions (URIs that do not exist on DBpedia and are relevant). These new entities are used to populate knowledge bases with new resources. Table 1 provides detailed statistics on this dataset, for each type of DUL entity.

Table 1. Statistics on the training set per DUL type

Type	# mentions	# disamb. mentions	# NIL mentions
Dul:Role	165	144	21
Dul:Organization	237	198	39
Dul:Person	446	342	104
Dul:Place	182	171	11
Total	1030	855	175

4.2   Results on the OKE Challenge Training Dataset

To evaluate the potential of our approach, we applied it on the training set, using four publicly available semantic annotators: DBpedia Spotlight, Babelfy, Tagme and AIDA. We used the appropriate configuration of each system in order to obtain the highest number of possible mentions. For DBpedia Spotlight, we set the confidence parameter at 0.0, while for Babelfy we use the default configuration, which does not limit the spotting phase to named entities only. We also rely on the default configuration for Tagme. For the purpose of our evaluation, we developed our own evaluation script following the OKE challenge criteria mentioned on the homepage of the competition. Only exact matches are counted. The macro-averaged precision-recall results of the WESTLAB system coupled with each semantic annotator are shown in Table 2. As we can see, the best performance was obtained by our framework coupled with Babelfy, in the three subtasks. However the difference with the F-scores obtained using the other semantic annotators is very low.

Table 2. Results of the entity recognition, linking and typing tasks on the training dataset

	Entity recognition			Entity linking			Entity typing			Average F
	P	R	F	P	R	F	P	R	F	
DBpedia spot.	0.7388	**0.8430**	0.7875	0.5923	0.7224	0.6510	0.6538	**0.7498**	0.6985	0.7123
Babelfy	0.7742	0.8050	**0.7894**	0.6460	0.7091	0.6760	0.7031	0.7393	**0.7207**	**0.7287**
Tagme	0.7146	0.8153	0.7616	0.5912	**0.7317**	0.6540	0.6377	0.7315	0.6814	0.6990
AIDA	**0.8652**	0.6890	0.7671	**0.7670**	0.6273	**0.6901**	**0.7981**	0.6404	0.7106	0.7226

4.3 Baseline Comparisons

We compared WESTLAB with Stanford NER and each semantic annotator used in our evaluation as a baseline. To avoid improper comparison, we conducted two different evaluations. The first one compares WESTLAB with Stanford NER to evaluate its capabilities for the subtasks of entity recognition and typing. In the second evaluation, we compare WESTLAB with each of the employed semantic annotators to assess their initial performance in extracting and disambiguating mentions. Since the best performances are obtained when WESTLAB is coupled with Babelfy, as shown in Table 2, this annotator has been chosen in our following experiments.

Comparison with Stanford NER. Since Stanford NER uses 3 classes only (Organization, Person and Place) we removed Role mentions from the training dataset for this particular evaluation. The results are listed in Table 3. We can note that WESTLAB highly outperforms Stanford NER.

Comparison with Semantic Annotators. We compared WESTLAB with the four semantic annotators used individually, under various configurations. For DBpedia Spotlight, we experimented with two confidence scores: 0.5 and 0.0.

Table 3. Comparison with Stanford NER

	Entity recognition			Entity typing			Average F
	P	R	F	P	R	F	
WESTLAB	0.8395	0.8183	0.8288	0.7923	0.7754	0.7837	**0.8063**
NER Stanford	0.8120	0.6740	0.7360	0.7360	0.6130	0.6690	0.6745

For Babelfy, we tested a configuration where only named entities are extracted. Finally, for Tagme, we also experimented a configuration that extracts mentions not linked to any entity (Tagme all extractions). The results are given in Table 4 and clearly show that our approach outperforms all the others.

Table 4. Comparison with semantic annotators

	Entity recognition			Entity linking			Average F
	P	R	F	P	R	F	
WESTLAB	0.7742	0.8050	0.7894	0.6460	0.7091	0.6760	**0.7327**
DBpedia Spotlight 0.0	0.3620	0.6370	0.4620	0.2650	0.5470	0.3570	0.4095
DBpedia Spotlight 0.5	0.6570	0.4450	0.5310	0.5760	0.4630	0.5130	0.5220
AIDA	0.8245	0.5530	0.6620	0.7230	0.5160	0.6020	0.6320
Babelfy	0.3100	0.7340	0.4360	0.3250	0.6360	0.4330	0.4345
Babelfy NED	0.6500	0.4990	0.5640	0.6070	0.5470	0.5750	0.5695
Tagme	0.5110	0.7540	0.6090	0.4120	0.7060	0.5200	0.5645
Tagme (all extractions)	0.5080	0.7570	0.6080	0.4120	0.7060	0.5200	0.5640

Compared to semantic annotators, the increase in performance is at least 20 % for the F-score. Note that the improvement is more evident for the subtask of entity recognition.

4.4   Result on the OKE Challenge Evaluation Dataset

For the OKE challenge 2016, the systems were evaluated against a new dataset, composed of 55 hand-annotated sentences. Table 5 provides details about the dataset per Dul type. Note that the ratio of disambiguated mentions does not significantly differ from the training dataset. Regarding the distribution of types, the situation is different: type *Dul:Person* is more dominant in the training dataset (43 % instead of 31 %) and there is no instance of NIL mentions for type *Dul:Place*.

Table 5. Statistics on the evaluation set per DUL type

Type	# mentions	# disamb. mentions	# NIL mentions
Dul:Role	86	71	15
Dul:Organization	105	91	14
Dul:Person	105	82	23
Dul:Place	44	44	0
Total	340	288	52

The results of WESTLAB, coupled with each individual semantic annotator, are given in Table 6. These results confirm that the best performance for our system is obtained using Babelfy.

To compare WESTLAB with the semantic annotators taken separately, we used GERBIL [13]. We tested the performances for the entity recognition and linking subtasks only, since the existing systems do not provide a DUL type. The results, presented in Table 7, show that WESTLAB, coupled with Babelfy, still outperforms these baseline systems.

Table 6. Results of the entity recognition, linking and typing tasks on the evaluation dataset

	Entity recognition			Entity linking			Entity typing			Average F
	P	R	F	P	R	F	P	R	F	
DBpedia Spot	0.7737	0.6235	0.6906	**0.8255**	0.5147	0.5784	0.5853	0.5853	0.5853	0.6366
Babelfy	**0.744**	**0.7353**	**0.7396**	0.764	**0.5618**	**0.6475**	**0.6441**	**0.6441**	**0.6441**	**0.6771**
Tagme	0.6606	0.75	0.7025	0.749	0.5618	0.642	0.6412	0.6412	0.6412	0.6619
AIDA	0.8204	0.5912	0.6872	0.801	0.4735	0.5952	0.55	0.55	0.55	0.6108

Table 7. Comparison with semantic annotators

	Entity recognition			Entity linking			Average F
	P	R	F	P	R	F	
WESTLAB	0.7448	0.7382	0.7415	0.761	0.5618	0.6464	**0.6939**
Dbpedia spotlight	0.6042	0.3412	0.4361	0.931	0.3176	0.4737	0.4549
AIDA	0.8706	0.4353	0.6804	0.5803	0.3676	0.5133	0.5968
Babelfy	0.463	0.4235	0.4424	0.7972	0.3353	0.472	0.4572
Tagme	0.6192	0.55	0.5826	0.8492	0.4971	0.6271	0.6048

There were two competing systems in the OKE challenge 2016: WESTLAB and a new version of Adel [11], the winner of the 2015 OKE challenge. The results are shown in Table 8. Note that we report the results against the original dataset used during the challenge (original gold standard), and against a corrected gold standard modified by the competition organizers to fix some annotations issues after the challenge (corrected gold standard). On this corrected gold standard, the results of Adel are missing since this system does not offer a public web service to test it. The complete results can also be found at the challenge home-page[5]. Note that the evaluation uses a weak match, that is, an annotation is true if and only if it overlaps the annotation of the gold standard. Thus a weak annotation match does not require an exact match. As we can notice, WESTLAB outperforms Adel in all micro and macro measures.

---

[5] https://github.com/anuzzolese/oke-challenge-2016.

Table 8. Results on the evaluation dataset at the OKE challenge 2016, with weak match

	Original gold standard					
	Micro F1	Micro P	Micro R	Macro F1	Macro P	Macro R
WESTLAB (using Babelfy)	**0.6676**	**0.6964**	**0.6509**	**0.6516**	**0.6902**	**0.6439**
Adel	0.6249	0.6689	0.5942	0.6064	0.6606	0.5846
	Corrected gold standard					
	Micro F1	Micro P	Micro R	Macro F1	Macro P	Macro R
WESTLAB (using Babelfy)	**0.6998**	**0.7402**	**0.6696**	**0.6926**	**0.7449**	**0.6753**

4.5 Analysis of Errors

We analyze the errors generated by our system and discuss them in this section.

Spotting Errors. Table 9 shows some statistics on the evaluation dataset. 71 % of the mentions are returned by Babelfy (240 out of 337). 27 mentions among the incorrect ones should not be annotated and are due to the limits of our approach. We judge that 26 mentions are missing in the gold standard and should not be considered as errors made by WESTLAB (for example, WESTLAB spots *Scotland* in sentence *In October 1850 ... Maxwell left Scotland for the University of Cambridge*, and this mention is not in the gold standard). 8 errors are incomplete mentions (e.g. *Minister* instead of *Minister of Education*) or mentions that should be splitted (e.g. *bishop of Bologna* should be splitted into two mentions, *bishop* and *Bologna*). In fact, as we explained in Sect. 3.2, when there are overlapping mentions, we always keep the longest one. For example, we may obtain the mention *Paris, France*, whereas the gold standard contains two separate mentions, *Paris* and *France*. To avoid this kind of errors, some heuristics must be added to determine when a mention must be splitted. However, after a careful analysis of the training dataset, we did not find any clear rule governing the decision of splitting a mention or not.

Table 9. Spotting errors, according to their origin.

Annotator (240)		Stanford (54)		Coreference (43)		Total (337)
Correct	Wrong	Correct	Wrong	Correct	Wrong	
179	61	36	18	36	7	

For incorrect mentions originating from Stanford NER, which represent 16 % of the total, 9 are named entities involving a modifier, as in ***Zionist*** *affairs* and ***American*** *high schools*, whereas the gold standard does not contain modifiers, i.e. *high schools* is annotated individually. As a solution, these mentions could be filtered out using some heuristics based on syntactic parsing. Additionally,

4 mentions returned by the Stanford NER should be splitted according to the gold standard (e.g. *Principal of Marischal* should be splitted into *Principal* and *Marischal* separately).

13 % of the mentions spotted by Stanford NER are coreference mentions (personal pronouns). 7 of these mentions are not annotated in the evaluation dataset, due to wrong offsets.

Regarding the missing mentions (89 out of 340), which decrease WESTLAB's recall performance, about half of them are due to the weakness of WESTLAB in detecting roles. Roles are usually expressed by common names that are not easily detected by semantic annotators. We also note that the problem about splitted mentions discussed earlier also affects recall: 23 % of the missing mentions are caused by this problem. Taking the example given previously, the two mentions *Principal* and *Marischal* would be considered as missing, since WESTLAB only detects *Principal of Marischal*.

Table 10. Disambiguation errors

Annotator (190)		Collective disambiguation (29)		Stanford NIL (32)		Total (251)
Correct	Wrong	Correct	Wrong	Correct	Wrong	
145	45	23	6	22	10	

Disambiguation Errors. Table 10 shows the disambiguation errors made by our system. The results are calculated only for correctly spotted mentions. "Stanford NIL" represents the mentions annotated as NIL mentions. Considering the erroneous mentions generated by the annotator, we found that most of them (74 %) are real errors made by Babelfy. The remaining ones are correctly disambiguated mentions but either annotated as NIL mentions in the gold standard (for example, *author* is annotated by WESTLAB with http://dbpedia.org/resource/Author while it is annotated as NIL in the gold standard) or not found in the gold standard. Now considering the collective disambiguation process, most of the mistakes (4 out of 6) are made because the right entity is missing in the list of candidates. Note that the candidates generator is based on an exact match of the label associated to the DBpedia entity. A partial match could be more appropriate but we must determine the threshold of similarity, and this will have a cost in terms of performance. Another source of error is the fact that our algorithm is based on the outlinks of each entity. Sometimes there are not enough mentions already disambiguated by the semantic annotator, and the number of outlinks is not sufficient to identify the right candidate.

For the incorrect Stanford NIL mentions, we found two sources of errors. In some cases, our candidate generator does not return any candidate for the target mention. The other cause is the limits of the Stanford coreference resolution which sometimes does not identify correctly a co-referent.

Table 11. Typing errors

DBpedia (202)		Stanford (49)	
Correct	Wrong	Correct	Wrong
74	28	46	3

Typing Errors. Table 11 shows some statistics about the errors made when typing the annotated mentions according to DUL categories. Only 12 % among the mentions that are correctly spotted are incorrectly typed. For mentions that come from the annotator, 14 mistakes are due to the priority rule used in our implementation (i.e. Person>Place>Organization>Role). In fact, DBpedia entities might be typed with more than one of the four types used in this task. For example, *National_Academy_of_Sciences* is typed as a *dbo:place* and a *yago:Organization108008335*. To select the type, our current system simply gives priority to *dbo:Place*. In the other 14 cases, we got the wrong type simply because the entity selected during the disambiguation process is the wrong one. For the mentions that come from Stanford, the 3 errors are due to the Stanford NER, since the type it provides is directly returned by WESTLAB.

5   Conclusion

In this paper, we have shown that combining the outputs of a named entity recognizer and a semantic annotator, and applying a collective disambiguation approach for selecting an entity among candidates, outperforms these two systems taken separately. We have also shown that by querying DBpedia to extract explicit and implicit types of entities, and using some manually defined mapping to the restricted set of four types used in the challenge, we also outperform the systems that participated to the 2016 OKE Challenge. Our approach is generic and does not depend on the dataset. In terms of efficiency, the most time-consuming step is querying DBpedia, which is used for typing.

For future work, we plan to improve the linking process and define a new approach to combine multiple annotators. We also plan to develop a new typing approach which uses the hierarchy of the DBpedia ontology instead of our current manual alignment with the DUL ontology.

Aknowledgement. This research has been funded by the NSERC Discovery Grant Program.

References

1. Atdag, S., Labatut, V.: A comparison of named entity recognition tools applied to biographical texts. CoRR, abs/1308.0661 (2013)
2. Dlugolinsky, S., Ciglan, M., Laclavik, M.: Evaluation of named entity recognition tools on microposts. In: IEEE 17th International Conference on Intelligent Engineering Systems (INES), pp. 197–202. IEEE (2013)
3. Ferragina, P., Scaiella, U.: Tagme: on-the-fly annotation of short text fragments (by Wikipedia entities). In: 19th ACM Conference on Information and Knowledge Management (CIKM)
4. Finkel, J.R., Grenager, T., Manning, C.: Incorporating non-local information into information extraction systems by Gibbs sampling. In: Proceedings of 43rd Annual Meeting of the Association for Computational Linguistics (ACL), pp. 363–370 (2005)
5. Hoffart, J., Yosef, M.A., Bordino, I., Furstenau, H., Pinkal, M., Spaniol, M., Taneva, B., Thater, S., Weikum, G.: Robust disambiguation of named entities in text. In: 8th Conference on Empirical Methods in Natural Language Processing (EMNLP), Stroudsburg, PA, USA, pp. 782–792 (2011)
6. Mendes, P.N., Jakob, M., García-Silva, A., Bizer, C.: DBpedia spotlight: shedding light on the web of documents. In: Proceedings of 7th International Conference on Semantic Systems, I-Semantics 2011, pp. 1–8. ACM, New York (2011)
7. Moro, A., Raganato, A., Navigli, R.: Entity linking meets word sense disambiguation: a unified approach. In: TACL, pp. 231–244 (2014)
8. Navigli, R., Ponzetto, S.P.: BabelNet: the automatic construction, evaluation and application of a wide-coverage multilingual semantic network. Artif. Intell. **193**, 217–250 (2012)
9. Nuzzolese, A.G., Gentile, A.L., Presutti, V., Gangemi, A., Garigliotti, D., Navigli, R.: Open knowledge extraction challenge. In: Gandon, F., Cabrio, E., Stankovic, M., Zimmermann, A. (eds.) ESWC 2015. CCIS, vol. 548, pp. 3–15. Springer, Berlin (2015)
10. Plu, J., Rizzo, G., Troncy, R.: A hybrid approach for entity recognition and linking. In: Gandon, F., Cabrio, E., Stankovic, M., Zimmermann, A. (eds.) ESWC 2015. CCIS, vol. 548, pp. 28–39. Springer, Berlin (2015)
11. Plu, J., Rizzo, G., Troncy, R.: Enhancing entity linking by combining NER models. In: Open Knowledge Extraction challenge at ESWC 2016 (2016)
12. Röder, M., Usbeck, R., Speck, R., Ngomo, A.-C.N.: CETUS - a baseline approach to type extraction. In: Gandon, F., Cabrio, E., Stankovic, M., Zimmermann, A. (eds.) ESWC 2015. CCIS, vol. 548, pp. 16–27. Springer, Berlin (2015)
13. Usbeck, R., Röder, M., Ngomo, A.-C.N., Baron, C., Both, A., Brümmer, M., Ceccarelli, D., Cornolti, M., Cherix, D., Bernd Eickmann, G., et al.: General entity annotator benchmarking framework. In: Proceedings of 24th International Conference on World Wide Web, pp. 1133–1143. ACM (2015)

DWS at the 2016 Open Knowledge Extraction Challenge: A Hearst-Like Pattern-Based Approach to Hypernym Extraction and Class Induction

Stefano Faralli[(⊠)] and Simone Paolo Ponzetto

Research Group Data and Web Science,
University of Mannheim, Mannheim, Germany
{stefano,simone}@informatik.uni-mannheim.de

Abstract. In this paper we present a system for the 2016 edition of the Open Knowledge Extraction (OKE) Challenge. The OKE challenge promotes research in automatic extraction of structured content from textual data and its representation and publication as Linked Data. The proposed system addresses the second task of the challenge, namely "Class Induction and entity typing for Vocabulary and Knowledge Base enrichment" and combines state-of-the-art lexically-based Natural Language Processing (NLP) techniques with lexical and semantic knowledge bases to first extract hypernyms from definitional sentences and second select the most suitable class of the extracted hypernyms from those available in the DOLCE foundational ontology.

Keywords: Linked Open Data · Hearst patterns · Hypernym extraction · Class induction

1 Introduction

Open Knowledge Extraction (OKE) is a recently introduced paradigm [26] that focuses on developing algorithms for Open Information Extraction (OIE) and automated taxonomy learning that have been shown suitable for extracting information on a Web scale basis in an unsupervised manner. In OKE, lexicalized extractions are usually induced into a semantic model and linked to the Linked Open Data (LOD) cloud to serve as enriched resource for state-of-the-art Natural Language Processing (NLP) applications. In the context of the SemWebEval 2015 at ESWC 2015 [5] Nuzzolese et al. [15] organized in 2015 the first OKE challenge. The OKE challenge promotes the research in automatic extraction of structured content from textual data and its representation and publication as Linked Data. In this paper we present our system for the second edition of the challenge, in particular for the task no. 2, namely "Class Induction and entity typing for Vocabulary and Knowledge Base enrichment". Our approach combines three main aspects:

© Springer International Publishing Switzerland 2016
H. Sack et al. (Eds.): SemWebEval 2016, CCIS 641, pp. 48–60, 2016.
DOI: 10.1007/978-3-319-46565-4_4

1. State-of-the-art NLP methodology for hypernym extraction and class induction;
2. A fresh new dataset[1] [21] of lexical "isa" relations extracted using Hearst-like patterns from the widest publicly available crawl of the Web [13];
3. Existing Linked Data resources such as WordNet [4] and OntoWordnet [6] to target the Descriptive Ontology for Linguistic and Cognitive Engineering (DOLCE) [12].

The rest of the paper is organized as follows: in Sect. 2 we give an overview of the literature on taxonomic relation extraction and induction. In Sect. 3 we describe our approaches to hypernym extraction (see Sect. 3.1) and class induction (see Sect. 3.2). In Sect. 4 we describe the results of a preliminary evaluation of the proposed algorithms on a gold standard dataset available from the challenge official Web-site; Finally we draw some conclusions on the presented work in Sect. 5.

2   Related Work

The extraction of taxonomic relations (also known as "isa" or hypernym relations) from text is a long-standing challenge of NLP and is a basic step for higher-end knowledge acquisition tasks such as ontology learning, see e.g. Biemann [2] for a survey. The manual construction of taxonomies is a very demanding task, requiring a large amount of time and effort. Ontology learning aims at reducing such cost of manual annotation by automatically or semi-automatically creating a lexicalized taxonomies using textual data from corpora or the Web [2,3,11,17,18]. As a result, the heavy requirements of manual ontology construction are drastically reduced.

The specific methods from a large body of work on hypernym extraction range from simple lexical patterns [7,16], similar to those used in our method, all the way through more complex statistical techniques [1,19].

Snow et al. [23] perform the hypernym extraction by first collecting sentences that contain two terms that are known to be hypernyms. They parse the sentences, and extract patterns from the parse trees. Finally, they train a hypernym classifier based on these features and apply it to text corpora.

Yang and Callan [28] presented a semi-supervised taxonomy induction framework that integrates co-occurrence, syntactic dependencies, lexico-syntactic patterns and other features to learn an ontology metric, calculated in terms of the semantic distance for each pair of terms in a taxonomy. Terms are incrementally clustered on the basis of their ontology metric scores.

Snow et al. [22] perform incremental construction of taxonomies using a probabilistic model. They combine evidences from multiple supervised classifiers trained on large training datasets of hyponymy and co-hyponymy relations. The taxonomy learning task is defined as the problem of finding the taxonomy that maximizes the probability of individual relations extracted by the classifiers.

---

[1] http://webdatacommons.org/isadb/.

Kozareva and Hovy [9] start from a set of root terms and use Hearst-like lexico-syntactic patterns to harvest new terms from the Web, which results in a set of hypernym relations. Next, to induce taxonomic relations between intermediate concepts, the Web is searched again with these patterns. The last step consists of pruning of the resulting graph.

Velardi et al. [27] proposed a graph-based algorithm to learn a taxonomy from textual definitions, extracted from a corpus and the Web in an iterative fashion. The hypernym extraction is based on a set of Word-Class Lattices (WCLs) [14] classifiers trained on a domain-independent set of definitions from Wikipedia. An optimal branching algorithm is used to induce a taxonomy.

In this work we combine some of the above mentioned state-of-the-art NLP techniques of taxonomy induction, i.e., we use a collection of around 50 Hearst-like "isa" patterns collected and evaluated in [21] and similar scoring techniques to [27] - and also involve a large dataset namely the WebIsADb [21], which consists of around 40 Million matches between these patterns and a corpus built from the Common Crawl [13].[2] Finally, in our lexical-based approach, we also involve WordNet [4] and the OntoWordnet resource [6] to target the Descriptive Ontology for Linguistic and Cognitive Engineering (DOLCE) [12].

3   Approach

In Fig. 1 we represent the workflow of our system and how it interfaces with the specifications of the second task of the 2016 OKE Challenge, namely "Class Induction and entity typing for Vocabulary and Knowledge Base enrichment". The task input consists of a fragment of text s which is meant to be a definitional sentence for a given term d (in the rest of this paper we will refer to d as the *definiendum*) and the corresponding DBpedia node n for d. The expected output is a triple (s, h, c) where h is the hypernym of d, h is a substring of s and c is the most suitable class of d from the DOLCE target reference ontology.

Our methodology can be divided in two subsequent steps:

(1) **Entity Recognition**: where we extract with Hearst-like patterns from the sentence s the most suitable hypernym h in s for the *definiendum* d. In this phase, if no matches can be found in s then candidates hypernyms are obtained by consulting a large database of "isa" relations extracted from the Web, namely the WebIsADb [21] (see Sect. 2 for more details about the involved patterns and WebIsADb); The WebIsADb is a freely available resource http://webdatacommons.org/isadb/, containing more then 400 million hypernymy relations.

(2) **Entity Typing**: where we collect different chains of "isa" relations rooted on the *definiendum* d and the extracted hypernym h (i.e. the pair (d,h)) and aimed at ending with a pair (i,c) in the OntoWordnet mapping and where i belongs to WordNet lexicon and c is a label for a DOLCE class.

In the following paragraphs we describe these two steps in more details.

---

[2] The largest publicly available crawl of the Web.

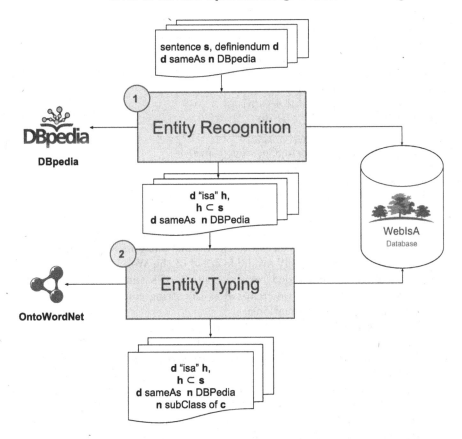

Fig. 1. Overview of our method for hypernym extraction and class induction.

3.1 Entity Recognition

With reference to Algorithm 1, initially we are given as input a triple (s,d,n) where s is a definitional sentence for the *definiendum* d and n is the DBpedia node for d. To extract the most suitable hypernym h for d which occurs in the sentence s we proceed as follows:

1. we create the set H of tuples representing the "isa" relation extracted from s and from the Web for the definiendum d (line 1). Each tuple in H includes also the number f of matches and the concatenation *Context* of a sample from all the matching sentences. The set H is created by (i) first searching for pattern matches between the sentence s and the set of Hearst-like patterns defined in [21][3] and then, (ii) in case of no matches in s, by collecting all the "isa" relations for the *definiendum* d from the WebIsADb[4] which already

---

[3] In this case $f = 1$ and $Context = s$.
[4] In this case $f = |matchingsentences|$ and $Context$ is the result of concatenation of a random sample of at most five matching sentences.

Algorithm 1. Entity Recognition

Require: - (s, d, n): s the definitional sentence for d, n the corresponding DBpedia node for d and $h \subset s$ the hypernym of d; OWN: the OntoWordnet mapping

Ensure: - (s, d, n, h)

1: $H = extractIsaTuple(s, d)$
2: $s' = s + additionalContextWords(n)$
3: **for** $t_i = \{d, h, Context, f\} \in H$ **do**
4:      $rank(t_i) = \frac{|getBoW(s') \cup getBoW(t_i.Context)|}{|getBoW(t_i.Context)|}$
5:      $rank(t_i) = rank(t_i) * stringCoverage(t_i.h, s)$
6:      $rank(t_i) = rank(t_i) * t_i.f$
7: $top = \text{argmax}_{t_i} rank(t_i)$
8: **return** $(s, d, n, top.h)$

---

holds all the matches between above Hearst-like patterns and a corpus made on top of the largest publicly available crawl of the Web [13];

2. we create s' as the expanded textual context of s (line 2). To extend the extraction context s for d we create s' as the string concatenation of s with the following textual informations:
   - the string label associated to the DBpedia node n;
   - the string label associated to all the DBpedia ontology node n_h, where n_h is a direct superclass of n in the DBpedia ontology;
   - the glosses associated to all the above DBpedia ontology node n_h.

3. we rank all the tuples t_i of H (lines 3–6) with a score that combines the following metrics:
   - BoW overlap: the normalized number of common content words[5] between the Bag of Words of s' and the extraction context of $t_i.h$ (line 4);
   - Hypernym String Coverage: is the length of the string $t_i.h$ if $t_i.h \subset s$, 0 otherwise (line 5);
   - Tuple Frequency: the number of pattern matches.

4. we select the tuple *top* with the highest score in the rank (line 7);

5. we return the quadruple $(s, d, n, top.h)$ with the extracted hypernym $top.h \subset s$ for the *definiendum* d (line 8);

To illustrate further how our hypernym extraction works, in Table 1 we display all steps for the input triple (s, d, n) where: s = *"A perennial plant or simply perennial (from Latin per, meaning through, and annus, meaning year) is a plant that lives for more than two years...."*;[6] d = *"perennial plant"*;[7] n = http:// dbpedia.org/resource/Perennial_plant.

---

[5] A standard set of english stop words is used to select only content words.

[6] Corresponding to sentence http://www.ontologydesignpatterns.org/data/oke-chall enge/task-2/sentence-117#char=0,381.

[7] Corresponding to *definiendum* http://www.ontologydesignpatterns.org/data/oke-challenge/task-2/sentence-117#char=2,17.

Table 1. Example of tuples extracted for the *definiendum* "perennial plant"

$t_1.d =$	"perennial plant"
$t_1.h =$	"shrub"
$t_1.Context =$	"Forests where the dominant woody elements are shrubs i.e. perennial plants, generally more than 0.5 m and less than 5 m high on maturity and without a definite crown. Changes in forest cover classes as interpreted from forest cover mapping. are shrubs i.e. woody perennial plants, generally of more than 0.5 m and less than 5 m in Shrub It refers to vegetation types where the dominant woody elements are shrubs i.e. woody perennial plants, generally more than 0.5 m and less than 5 m in height at maturity and without a definite crown. . . "
$t_1.f =$	5
$t_2.d =$	"perennial plant"
$t_2.h =$	"plant"
$t_2.Context =$	"A perennial plant is a plant that lives longer than 2 years. A perennial plant or simply perennial is a plant that lives for more than two years. A perennial plant is a plant that lives more than 2 years, or a winter hardy plant. A perennial plant is a plant that lives for more than two years or produces in successive years. Such raw- materials can be derived from plants, such as annular or perennial plants, including straw, willow, energy hay, Miscanthos. . . "
$t_2.f =$	9
$t_3.d =$	"perennial plant"
$t_3.h =$	"shrub about 13 tall"
$t_3.Context =$	"perennial plant is a small much-branched shrub about 13' tall."
$t_3.f =$	1

Since no matches were produced by applying our set of Hearst-like patterns directly on the sentence s, we collected in H a total of 152 different "isa" relations for the *definiendum* d. In Table 1 we show an excerpt of three tuples from the 152 extracted for the *definiendum* "perennial plant".

To score all the extracted tuples we first create s' as an extension of s including some additional textual information we derive from n. In our example $s' =$ "A perennial plant or simply perennial (from Latin per, meaning through, and annus, meaning year) is a plant that lives for more than two years. The term is often used to differentiate a plant from shorter-lived annuals and biennials. The term is also widely used to distinguish plants with little or no woody growth from trees and shrubs, which are also technically perennials." We then assign to each extracted tuple t_i a score which combines: (i) the normalized number of common words between the bag of words of s' and of t_i; the string support of $t_i.h$ in s, and (iii) the tuple frequency $t_i.f$. For the tuples enumerated in Table 1 we obtain: $rank(t_1) = 10.71$, $rank(t_2) = 65.62$ and $rank(t_3) = 0.0$. We finally identify the top ranking tuple t_2 and output $(s, d, n, \text{"plant"})$.

Algorithm 2. Entity Typing

Require: - (s, d, n, h): s the definitional sentence for d, n the corresponding DBpedia node for d and $h \subset s$ the hypernym of d; OWN: the OntoWordnet mapping.

Ensure: - (s, d, n, h, c): s the definitional sentence for d, n the corresponding DBpedia node for d, $h \subset s$ the hypernym of d and c the most suitable DOLCE class for d

1: $I = OWN.getAllInstances()$
2: **for** $wn = (synonyms, DolceClassAncestors) \in I$, h right substring of at least one $ws \in wn.synonyms$ **do**
3:      **for** $dc \in wnDolceClassAncestors$ **do**
4:          $rank(dc) = rank(dc) + 1$
5: **if** $rank.isEmpty()$ **then**
6:      $H' = extractWebIsaTuples(h)$
7:      $s' = s + additionalContextWords(n)$
8:      **for** $t_i = \{h, h', Context, f\} \in H'$ **do**
9:          **for** $wn = \{synonyms, DolceClassAncestors\} \in I$, h' right substring includes at least one word $ws \in wn.synonyms$ **do**
10:             **for** $dc \in wnDolceClassAncestors$ **do**
11:                $rank(dc) = \frac{|getBoW(s') \cup getBoW(t_i.Context)|}{|getBoW(t_i.Context)|}$
12:                $rank(dc) = rank(dc) * stringCoverage(h, s)$
13:                $rank(dc) = rank(t_i) * t_i.f$
14: $top = \text{argmax}_{dc}\, rank(dc)$
15: **return** (s, d, n, h, top)

3.2 Entity Typing

For the next subtask of identifying the appropriate DOLCE class, following the naming scheme from previous subsection, we take as input the quadruple (s, d, n, h) where s is the definitional sentence for the *definiendum* d, n is the corresponding DBpedia node for d and h is the most suitable hypernym for d in s; and output the same tuple with an additional component c which is the most suitable DOLCE class for d in the context of s. To identify the most suitable DOLCE class c for d in the context of the sentence s we proceed as follows:

1. we first create the set I of all the linked instances from OntoWordnet (line 1); Each element of I contains an "isa" pair $(synonyms, DolceClassAncestors)$, where $synonyms$ is a set of word expression for the same WordNet concept and $DolceClassAncestors$ is a set of DOLCE classes mapping to the ancestors of w in WordNet;
2. we search all the pairs $wn \in I$ where at least one word $\in wn.synonyms$ is right string included by the hypernym h, for all these pair wn we count all the corresponding classes $dc \in wn.DolceClassAncestors$ (lines 2–4). The right string inclusion is an effective heuristic also adopted in the taxonomy induction process to project hypernyms into core ontology concepts [27].
3. if no DOLCE classes were identified on the previous step (line 5) we proceed by creating the set H' containing all the hypernyms h' in the WebIsADb for h (line 6);

4. we follow the approach from hypernym ranking (Algorithm 1) and assign a score to all the DOLCE classes related to the instances $wn \in I$ where at least one word $\in wn.synonyms$ is right string included by the hypernym h'.

5. finally, we return the quintuple (s, d, n, h, top) where top is the top-ranked DOLCE class in our rank (lines 14 and 15);

Let us now extend the running example from Sect. 3.1 with the class induction step. Let (s, d, n, h) be the output tuple from our Hypernym Extraction algorithm, the Class Induction phase then works as follows:

– we create the set I, which is composed by all the pair extracted from OntoWordNet of the kind $(synonyms, DolceClassAncestors)$;
– we analyse all the pairs wn from I where at least one word in $wn.synonyms$ is right-string included in h, for example the pairs: $w_1 =$ "(herb, herbaceous plant,Organism)", $w_2 =$ "(plant, flora, plant life,Organism)" and $w_3 =$ "(plant, works, industrial plant,PhysicalObject)" and create the following rank: $rank(Organism) = 2$ and $rank(PhysicalObject) = 1$;
– since a non-empty rank is obtained, the most suitable DOLCE class for the $definiendum$ $d =$ "perennial plant" is the class $Organism$;

When h does not "project" its string into any WordNet synset, we search for all the hypernyms of h in the WebIsADb and apply for them the same right-string inclusion heuristic to candidate WordNet synsets and rank the corresponding DOLCE class as expressed in the OntoWordnet ontology.

4 Evaluation

We follow the standard evaluation protocol of the OKE challenge. We developed a Web-service which is consumed by the General Entity Annotator Benchmark (GERBIL)[8] [25]. By calling our Web-service the above system send a series of requests in the NLP Interchange Format (NIF) [8] and our Web-service replies to the request using the same format. An example excerpt of such request follows:

```
. . .
@prefix nif: <http://persistence.uni-leipzig.org/nlp2rdf/ontologies/nif-core#> .
@prefix d0: <http://ontologydesignpatterns.org/ont/wikipedia/d0.owl#>
@prefix dul: <http://www.ontologydesignpatterns.org/ont/dul/DUL.owl#>
@prefix dbpedia: <http://dbpedia.org/resource/> .
@prefix gerbil: <http://www.aksw.org/gerbil/NifWebService/>.
. . .
gerbil:request_1#char=0,150
    a                   nif:RFC5147String , nif:String , nif:Context ;
    nif:isString        "Brian Banner is a fictional villain from the Marvel Comics
                        Universe created by Bill Mantlo and Mike Mignola and first
                        appearing in print in late 1985." ;
    nif:beginIndex      "0"^^xsd:int ;
    nif:endIndex        "150"^^xsd:int .

gerbil:request_1#char=0,12
    a                   nif:RFC5147String , nif:String ;
    nif:anchorOf        "Brian Banner"@en ;
```

---

[8] http://aksw.org/Projects/GERBIL.html.

```
nif:referenceContext   gerbil:request_1#char=0,150 ;
nif:beginIndex         "0"^^xsd:int ;
nif:endIndex           "12"^^xsd:int ;
itsrdf:taIdentRef      dbpedia:Brian_Banner .
```

The expected answer for the above request is:

```
...
@prefix oke: <http://www.ontologydesignpatterns.org/data/oke-challenge/task-2/> .
...

oke:sentence-1#char=0,150
      a                nif:RFC5147String , nif:String , nif:Context ;
      nif:isString     "Brian Banner is a fictional villain from the Marvel Comics
                       Universe created by Bill Mantlo and Mike Mignola and first
                       appearing in print in late 1985." ;
      nif:beginIndex   "0"^^xsd:int ;
      nif:endIndex     "150"^^xsd:int .

oke:sentence-1#char=0,12
      a                nif:RFC5147String , nif:String ;
      nif:anchorOf     "Brian Banner"@en ;
      nif:referenceContext  oke:entence-1#char=0,150 ;
      nif:beginIndex   "0"^^xsd:int ;
      nif:endIndex     "12"^^xsd:int ;
      itsrdf:taIdentRef  dbpedia:Brian_Banner .

dbpedia:Brian_Banner
      a                oke:FictionalVillain;
      rdfs:label       "Brian Banner"@en .

oke:FictionalVillain
      a                owl:Class ;
      rdfs:label       "fictional villain"@en ;
      rdfs:subClassOf  dul:Personification .

oke:sentence-1#char=18,35
      a                nif:RFC5147String , nif:String ;
      nif:anchorOf     "fictional villain"@en ;
      nif:referenceContext  oke:entence-1#char=0,150 ;
      nif:beginIndex   "18"^^xsd:int ;
      nif:endIndex     "35"^^xsd:int ;
      itsrdf:taIdentRef  oke:FictionalVillain .
```

In Table 2 we show the performance of our approach on the 2015 challenge
gold standard sample dataset. For the Entity Recognition step we report two
kinds of experiment:

1. *weak*: which evaluates the extracted hypernyms by means of a partial string
   match;
2. *strong*: where, instead, extracted hypernyms are compared against gold standard hypernyms with a perfect string match.

GERBIL computes micro and macro precision, recall and F-measure (for
each of the approaches). In Table 3 we also show the final results of the 2016
challenge. The performance here are compared against:

- the winning team "WestLab" [10]: an annotator which combines both
  SPARQL patterns and DBpedia;
- the first edition winning and unbeaten annotator system "Cetus" [20]: Cetus
  includes a parser built on top of grammatical rules aimed at discover the entity
  type directly from the contextual sentences.

Table 2. Performances on Entity Recognition and Entity Typing on the gold-standard sample from the 2015 edition of the challenge.

Sub-task	Micro			Macro		
	F1	P	R	F1	P	R
Entity Recognition (weak)	57.8	80.0	45.3	47.8	48.5	47.5
Entity Recognition (strong)	30.3	42.4	23.6	24.9	25.2	24.7
Entity Typing	29.5	29.0	29.9	28.4	28.6	29.5

Table 3. Results of the 2016 edition of the challenge (https://github.com/anuzzolese/oke-challenge-2016).

Annotator	Sub-task	Micro			Macro		
		F1	P	R	F1	P	R
DWS	Entity Recognition	72.3	77.3	68.0	64.7	63.0	68.0
DWS	Entity Typing	0.0	0.0	0.0	0.0	0.0	0.0
DWS	Global	36.2	38.6	34.0	32.3	31.5	34.0
WestLab	Entity Recognition	86.0	86.0	86.0	86.0	86.0	86.0
WestLab	Entity Typing	7.2	8.0	6.7	7.0	8.0	6.7
WestLab	Global	46.6	47.0	46.3	46.5	47.0	46.3
Cetus	Entity Recognition	77.2	68.7	88.0	77.3	72.0	88.0
Cetus	Entity Typing	23.3	22.2	24.5	19.9	22.2	24.5
Cetus	Global	**50.2**	**45.5**	**56.2**	**48.6**	**47.1**	**56.2**

In the sub-task of "Enty Recognition" the team WestLab reached very high performances while for the sub-task of "Enty Typing" the "Cetus" annotator remain unbeaten. Our annotator performances on "Entity Recognition" are aligned to the general state of the art performances reachable by using Hearst-like patterns. Unexpectedly, in the "Entity Typing" subtask we did not match any of the types in the dataset.

In general the error analysis on the two dataset shows that most of the errors our system makes originate from one of the following issues:

- errors derived from the part-of-speech tagging system [24] we used during the application of lexico-syntactic patterns (see Algorithm 1, line 1), in particular errors come from Verb/Noun mismatches and proper nouns from different languages;
- presence of erroneous "isa" relations extracted by lexico-syntactic patterns, which, in turn, may introduce wrong taxonomic relations;
- complex definitional sentence structures that can not be matched using Hearst-like pattern, and challenging, highly polysemous definiendum like, for instance, "god", which is a concept hard to classify also for human;

– the low recall of Entity Typing may be improved by extending/replacing the OntoWordNet mapping between WordNet and the foundational DOLCE with a more rich sense inventory, i.e. DBpedia.

An additional analysis of our results on "Entity Typing" shows that our system (based on OntoWordNet) often tends to wrongly annotate with the "d0:CognitiveEntity".

5   Conclusion

In this paper we presented the annotator we developed for the task number 2 of the 2016 edition of the Open Knowledge Extraction (OKE) Challenge. In line with the general aims of the challenge and with the Open Knowledge Extraction paradigm, our system:

– performs hypernym extraction by applying Hearst-like patterns on the input definitional contexts and harvesting "isa" relations from a large collection of Hearst-like patterns matches extracted from the Web;
– in order to discover the most suitable class for the extracted hypernyms, it combines WordNet and OntoWordNet to target the DOLCE ontology model.

We discussed the results of the challenge and analysed the nature of the errors. We also performed an evaluation on the 2015 gold standard sample dataset.

Our contributions indicate that: (i) Hearst-like patterns are a versatile solution for the hypernym extraction; (ii) but potentially lead to "noisy/erroneous" relations during the class induction phase; (iii) richer mappings than OntoWordNet should be involved to improve the coverage of the Entity Typing sub-task. We also consider the Web scale of the proposed approach an initial step to investigate a multilingual settings of the Open Knowledge Extraction paradigm.

Further work will focus on the implementation of more reliable "isa" relation filtering techniques aimed at removing noisy taxonomical relations extracted from the Web. Performance may benefit from a mapping between common senses and foundational ontologies with a higher sense coverage.

Acknowledgement. This work was partially funded by the Junior-professor funding programme of the Ministry of Science, Research and the Arts of the state of Baden-Württemberg, Germany (project "Deep semantic models for high-end NLP applications").

References

1. Agirre, E., Ansa, O., Hovy, E.H., Martínez, D.: Enriching very large ontologies using the WWW. In: ECAI Workshop on Ontology Learning, Berlin, Germany, pp. 28–33 (2000)
2. Biemann, C.: Ontology learning from text: a survey of methods. LDV Forum **20**(2), 75–93 (2005)
3. Buitelaar, P., Cimiano, P., Frank, A., Hartung, M., Racioppa, S.: Ontology-based information extraction and integration from heterogeneous data sources. Int. J. Hum.-Comput. Stud. **66**(11), 759–788 (2008)
4. Fellbaum, C.: WordNet: An Electronic Lexical Database. Bradford Books, Cambridge (1998)
5. Gandon, F., Cabrio, E., Stankovic, M., Zimmermann, A. (eds.): Semantic Web Evaluation Challenges. CCIS, vol. 548. Springer, Berlin (2015)
6. Gangemi, A., Navigli, R., Velardi, P.: The OntoWordNet project: extension and axiomatization of conceptual relations in WordNet. In: Meersman, R., Schmidt, D.C. (eds.) CoopIS 2003, DOA 2003, and ODBASE 2003. LNCS, vol. 2888, pp. 820–838. Springer, Heidelberg (2003)
7. Hearst, M.: Automatic acquisition of hyponyms from large text corpora. In: Proceedings of 14th Conference on Computational Linguistics, pp. 539–545. Association for Computational Linguistics (1992)
8. Hellmann, S., Lehmann, J., Auer, S., Brümmer, M.: Integrating NLP using linked data. In: Alani, H., et al. (eds.) ISWC 2013, Part II. LNCS, vol. 8219, pp. 98–113. Springer, Heidelberg (2013)
9. Kozareva, Z., Hovy, E.: Learning arguments and supertypes of semantic relations using recursive patterns. In: Proceedings of ACL-2010, Uppsala, Sweden, pp. 1482–1491 (2010)
10. Lara, H.A., Ludovic, F., Amal, Z., Michel, G.: Entity typing and linking using SPARQL patterns and DBpedia. In: 2nd Open Knowledge Extraction Challenge @ 13th European Semantic Web Conference (ESWC 2016) (2016)
11. Maedche, A., Staab, S.: Ontology learning. Handbook on Ontologies, pp. 245–268. Springer, Berlin (2009)
12. Masolo, C., Borgo, S., Gangemi, A., Guarino, N., Oltramari, A.: WonderWeb deliverable D18 ontology library (final). Technical report, IST Project 2001–33052 WonderWeb: Ontology Infrastructure for the Semantic Web (2003)
13. Meusel, R., Petrovski, P., Bizer, C.: The WebDataCommons microdata, RDFa and microformat dataset series. In: Mika, P., et al. (eds.) ISWC 2014, Part I. LNCS, vol. 8796, pp. 277–292. Springer, Heidelberg (2014)
14. Navigli, R., Velardi, P.: Learning word-class lattices for definition and hypernym extraction. In: Proceedings of 48th Annual Meeting of the Association for Computational Linguistics, pp. 1318–1327. Association for Computational Linguistics, Uppsala (2010)
15. Nuzzolese, A.G., Gentile, A.L., Presutti, V., Gangemi, A., Garigliotti, D., Navigli, R.: Open knowledge extraction challenge. In: Gandon, F., Cabrio, E., Stankovic, M., Zimmermann, A. (eds.) SemWebEval 2015. CCIS, vol. 548, pp. 3–15. Springer, Heidelberg (2015). doi:10.1007/978-3-319-25518-7_1
16. Oakes, M.P.: Using Hearst's rules for the automatic acquisition of hyponyms for mining a pharmaceutical corpus. In: RANLP Text Mining Workshop 2005, Borovets, Bulgaria, pp. 63–67 (2005)

17. Perez, G.A., Mancho, M.D.: A survey of ontology learning methods and techniques. OntoWeb Deliverable 1.5 (2003)
18. Petasis, G., Karkaletsis, V., Paliouras, G., Krithara, A., Zavitsanos, E.: Ontology population and enrichment: state of the art. In: Paliouras, G., Spyropoulos, C.D., Tsatsaronis, G. (eds.) Knowledge-Driven Multimedia Information Extraction and Ontology Evolution. LNCS, vol. 6050, pp. 134–166. Springer, Heidelberg (2011)
19. Ritter, A., Soderland, S., Etzioni, O.: What is this, anyway: automatic hypernym discovery. In: Proceedings of 2009 AAAI Spring Symposium on Learning by Reading and Learning to Read, Palo Alto, California, pp. 88–93 (2009)
20. Röder, M., Usbeck, R., Speck, R., Ngomo, A.C.N.: CETUS – a baseline approach to type extraction. In: Gandon, F., Cabrio, E., Stankovic, M., Zimmermann, A. (eds.) ESWC 2015. CCIS, vol. 548, pp. 16–27. Springer, Berlin (2015)
21. Seitner, J., Bizer, C., Eckert, K., Faralli, S., Meusel, R., Paulheim, H., Ponzetto, S.: A large database of hypernymy relations extracted from the web. In: Proceedings of 10th edition of Language Resources and Evaluation Conference, Portorož Slovenia (2016)
22. Snow, R., Jurafsky, D., Ng, A.: Semantic taxonomy induction from heterogeneous evidence. In: Proceedings of ACL-COLING-2006, Sydney, Australia, pp. 801–808 (2006)
23. Snow, R., Jurafsky, D., Ng, A.Y.: Learning syntactic patterns for automatic hypernym discovery. In: Saul, L.K., Weiss, Y., Bottou, L. (eds.) Proceedings of NIPS-2004, pp. 1297–1304. MIT Press, Cambridge (2004)
24. Toutanova, K., Klein, D., Manning, C.D., Singer, Y.: Feature-rich part-of-speech tagging with a cyclic dependency network. In: Proceedings of Human Language Technology Conference of the North American Chapter of the Association for Computational Linguistics (HLT-NAACL 2003), pp. 173–180. ACL (2003)
25. Usbeck, R., Röder, M., Ngomo, A.N., Baron, C., Both, A., Brümmer, M., Ceccarelli, D., Cornolti, M., Cherix, D., Eickmann, B., Ferragina, P., Lemke, C., Moro, A., Navigli, R., Piccinno, F., Rizzo, G., Sack, H., Speck, R., Troncy, R., Waitelonis, J., Wesemann, L.: GERBIL: general entity annotator benchmarking framework. In: Proceedings of 24th International Conference on World Wide Web, WWW 2015, Florence, Italy, 18–22 May 2015, pp. 1133–1143 (2015)
26. Velardi, P., D'Antonio, F., Cucchiarelli, A.: Open domain knowledge extraction: inference on a web scale. In: Proceedings of 3rd International Conference on Web Intelligence, Mining and Semantics, WIMS 2013, pp. 35:1–35:8. ACM, New York (2013)
27. Velardi, P., Faralli, S., Navigli, R.: Ontolearn reloaded: a graph-based algorithm for taxonomy induction. Comput. Linguist. **39**(3), 665–707 (2013)
28. Yang, H., Callan, J.: A metric-based framework for automatic taxonomy induction. In: Proceedings of ACL-2009, Suntec, Singapore, pp. 271–279 (2009)

Entity Typing and Linking Using SPARQL Patterns and DBpedia

Lara Haidar-Ahmad[1], Ludovic Font[1], Amal Zouaq[1,2],
and Michel Gagnon[1(✉)]

[1] Department of Computer and Software Engineering,
Ecole Polytechnique de Montreal, Montreal, Canada
{lara.haidar-ahmad,ludovic.font,
michel.gagnon}@polymtl.ca, azouaq@uottawa.ca
[2] School of Electrical Engineering and Computer Science,
University of Ottawa, Ottawa, Canada

Abstract. The automatic extraction of entities and their types from text, coupled with entity linking to LOD datasets, are fundamental challenges for the evolution of the Semantic Web. In this paper, we describe an approach to automatically process natural language definitions to (a) extract entity types and (b) align those types to the DOLCE+DUL ontology. We propose SPARQL patterns based on recurring dependency representations between entities and their candidate types. For the alignment subtask, we essentially rely on a pipeline of strategies that exploit the DBpedia knowledge base and we discuss some limitations of DBpedia in this context.

1 Introduction

The growth of the Semantic Web depends on the ability to handle automatically the extraction of structured information from texts and the alignment of this information to linked datasets. The first OKE Challenge competition [1] targeted these two issues and is a welcome initiative to advance the state of the art of open information extraction for the Semantic Web. In this paper, we present our service for entity typing and linking using SPARQL patterns and DBpedia[1]. This service is the winner of the OKE challenge 2016 Task 2.

Besides a participation to the OKE challenge, one aim of this research is to provide a task-based evaluation of the DBpedia knowledge base. Hence our linking strategies exploit both the DBpedia ontology and the DBpedia knowledge base to extract *rdfs:subClassOf* relationships between natural language types and DBpedia types.

This paper is structured as follows: Sect. 2 presents some related work. Sections 3 and 4 describe the two subtasks of our service: type recognition and extraction from text, and type alignment using the ontology Dolce+DUL. In Sect. 5, we present the evaluation of our system. We discuss our results in Sect. 6.

---

[1] http://westlab.polymtl.ca/OkeTask2/rest/annotate/post.

© Springer International Publishing Switzerland 2016
H. Sack et al. (Eds.): SemWebEval 2016, CCIS 641, pp. 61–75, 2016.
DOI: 10.1007/978-3-319-46565-4_5

2   Related Work

Several tasks are related to the challenge of entity typing and alignment, among which we can cite named entity recognition [2], relation extraction [3–5], ontology learning [6] and entity linking [7–9]. Due to space constraints, this state of the art will be limited to the participants of the previous OKE challenge [1].

2.1   Type Extraction

The automatic extraction of *taxonomical* and *instance-of* relations from text has been a long-term challenge. Overall, state-of-the-art approaches that target the extraction of relations from text are mainly pattern-based approaches. In the first edition of the 2015 OKE challenge, there were three participating systems for the task of type extraction from natural language definitions: CETUS [10], OAK@Sheffield [11] and FRED [12]. CETUS relies on grammar rules based on parts of speech (POS) to extract an entity type from text. OAK uses machine learning to learn to recognize the sentences' portions that express the entity type, and then uses a POS pattern grammar for type annotation. FRED uses the system Boxer [13] and Discourse Representation Theory, and thus relies on a complex architecture for ontology extraction that is not limited to type extraction. Compared to previous pattern-based approaches in the OKE competition [10, 11], our system differs by the nature of the patterns, which exploit a dependency grammar representation. One particular novelty is the use of SPARQL to model and search for patterns occurrences. Overall, we believe that our approach represents a middle ground between patterns based on a superficial representation of sentences (usually parts of speech) and approaches such as FRED [12] which depend on complex first-order logic and frame semantics.

2.2   Type Alignment

In the context of the Semantic Web, the challenge of entity typing is coupled with the difficulty of finding an alignment with linked datasets. Among the three systems of the OKE challenge 2015 mentioned previously, the authors of CETUS [10] developed an alignment between Yago and Dolce + DnS Ultralite; FRED [12] uses an already existing API that exploits Dolce, WordNet and VerbNet; OAK [11] relies on the existence of *dul* types in DBpedia, using a method similar to our method 2 (see Sect. 4.1). In our approach, we chose to use the existing mappings DBpedia - Dolce + DnS Ultralite [14] and Yago wordnet - Dolce + DnS Ultralite [15].

Our main contribution in this subtask is the exploitation of several strategies that consider either the DBpedia ontology (T-box) or the DBpedia knowledge base (A-box) to find a DBpedia type. We exploit both the knowledge about the entity and the type given as input. When there is not any direct type information linked to the DBpedia ontology or Yago, we revert to type inference methods. Among the strategies described in Sect. 4.1, method 6 is based on our previous work [16] to infer types using predicates' domain and range, while method 2 is similar to the one used by OAK [11].

However, we also introduce a novel approach based on DBpedia categories and propose a pipeline of strategies that aggregates several methods.

3 Entity Type Extraction

Entity type extraction consists in finding the natural language type of an entity, given its textual definition. Our approach relies on pattern extraction using a dependency-based syntactic analysis. The extraction of an entity type is processed in two steps: sentence representation in RDF and pattern occurrence identification using SPARQL queries.

3.1 Sentence Graph Representation

First, we extract grammatical dependencies from the definitions using the Stanford parser [17] and build an RDF graph representing each sentence. Before the parsing step, we identify the input DBpedia entity in the sentence and aggregate multi-words entities with an underscore between the words. For instance, in the sentence *All's Well That Ends Well is a play by William Shakespeare*, we identify *All's Well That Ends Well* (the input DBpedia resource) as one single entity and simply modify the sentence to obtain *All's_Well_That_Ends_Well is a play by William Shakespeare*.

We then construct an RDF graph representing the dependency structure of the definition. Thus we specify the label and part of speech of each word in addition to its grammatical relations with the other words. This RDF graph allows us to look for pattern occurrences using SPARQL requests in the following step. Figure 1 presents the RDF graph of the definition *Skara Cathedral is a church in the Swedish city of Skara*.

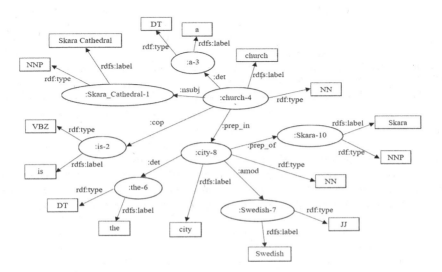

Fig. 1. The RDF Representation of the definition of *Skara Cathedral*.

3.2   Pattern Identification

As for the detection of patterns, based on the train dataset[2] distributed in the OKE challenge, we manually identified several recurring syntactic and grammatical structures between the entities and their respective types. Table 1 presents the most common patterns that we identified in the dataset.

Table 1. Most frequent patterns describing an entity/type relationship.

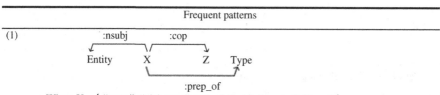

Frequent patterns

(1)
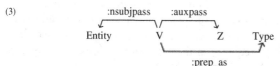

Where X = { "name", "nickname", "alias", "one", "species", "form" }
Sant'Elmo is the name of both a hill and a fortress in Naples, located near the Certosa di San Martino.
Entity = *Sant'Elmo ;* **Type** = *Hill*

(2)

:nsubj   :cop

Entity   Type   Z

El Oso, released in 1998, is an album by the New York City band Soul Coughing.
Entity = *El Oso ;* **Type** = *Album*

(3)

:nsubjpass   :auxpass

Entity   V   Z   Type

:prep_as

Bromius in ancient Greece was used as an epithet of Dionysus/Bacchus.
Entity = *Bromius ;* **Type** = *Epithet*

(4)

x

Entity   Type

Where x = { :amod, :appos, :nn}
The AES11 standard published by the Audio Engineering Society provides a systematic approach to the synchronization of digital audio signals.
Entity = *AES11 ;* **Type** = *Standard*

We created a pipeline of SPARQL requests with a specific processing order as shown in Table 1. In fact, Pattern 1 has a higher priority than Pattern 2. For example, in the sentence *Sant'Elmo is the name of both a hill and a fortress in Naples, located near the Certosa di San Martino,* if the second pattern was processed before the first one, we

---

[2] https://github.com/anuzzolese/oke-challenge-2016/blob/master/GoldStandard_sampleData/task2/dataset_task_2.ttl.

would wrongly extract the type *name*. Similarly, Pattern 4 is the pattern with the lowest priority, which is executed only after all the other patterns are tested.

Each pattern is modeled using a single SPARQL request. The following is an example of the SPARQL implementation of Pattern 2:

```
SELECT ?typeLabel
WHERE {
        ?type :nsubj ?entity.
        ?type :cop ?cop.
?entity rdfs:label ?entityLabel.
?type rdfs:label ?typeLabel.
FILTER(REGEX(?entityLabel,
'^THE_LABEL_OF_THE_DBPEDIA_ENTITY', 'i'))
}
```

As this SPARQL request shows, we search for the type of an entity, where the entity's label is a perfect match with the input DBpedia entity's label. We first execute all the SPARQL patterns in this "full match" mode. In case all requests fail to return a type, we then look for occurrences of the same patterns using a partial match of the entity's label.

Once we find a candidate type, we create an OWL class representing this type. We remove all the accents and special characters and extract the lemma of the types in plural form. Overall, we adopted the singular as a convention for our entity types. For instance, in *Alvorninha is one of the sixteen civil parishes that make up the municipality of Caldas da Rainha, Portugal*, we extract the type *oke:Parish* from the string *parishes*. Finally, we create a *rdf:type* relation between the entity and the returned type.

4   Type Alignment

In this paper, we refer to the namespaces http://dbpedia.org/ontology/ and http://dbpedia.org/page/ as *dbo* and *dbr* respectively. The ontologies http://www.ontologydesignpatterns.org/ont/dul/DUL.owl and http://ontologydesign-patterns.org/ont/wikipedia/d0.owl are represented by the prefixes *dul* and *d0* respectively. Besides, we use "Dolce" as a shortcut for "Dolce + DnS Ultralite".

Once the *natural language type* of a given DBPedia entity is identified, for instance [*dbr:Brian_Banner, oke:Villain*], where the first element represents the entity and the second element the natural language type, the second part of the OKE challenge task 2 is to align the identified type to a set of given types in the Dolce ontology[3]. The objective is to link the natural language type to a super-type in the ontology using an *rdfs:subClassOf* link. For instance, *dul:Person* would be a possible super-class for *oke:Villain*.

---

[3] https://github.com/anuzzolese/oke-challenge-2016#task-2.

Our alignment strategy relies only on the DBpedia knowledge base and its links to external knowledge bases, when applicable, and exploits available mappings between DBpedia and Dolce, and Yago and Dolce. In fact, besides the OKE challenge in itself, one objective of this research is to determine whether the DBpedia knowledge base, one of the main hubs on the Linked Open Data cloud, is a suitable resource for the entity linking task. Thus our goal is to find a link, either directly or indirectly, between the *oke* type (e.g. *oke:Village*) returned by the first subtask and the DBpedia ontology and/or Yago and/or Dolce.

4.1   DBpedia Type Identification

Our global alignment strategy first queries the DBpedia ontology (http://dbpedia.org/ontology/[*Input Type*]). If the type is not found as a DBpedia class, we query DBpedia resources (http://dbpedia.org/resources/[*Input Type*]) and either find direct types or infer candidate types using several strategies. Our queries result in three possible outputs:

1. There is a *dbo* resource for the input type. In our dataset, this case occurred in 83 out of 198 cases (42 %).
2. There is only a *dbr* resource for this type. In this case, we attempt to find a predicate *rdf:type* between the natural language type and some type that can be aligned with Dolce + DnS Ultralite, i.e. a type in the DBpedia ontology, Yago-Wordnet or DUL. In our dataset, this case occurred in 68 % of the cases.
3. There is neither a *dbo* nor a *dbr* resource for this type (e.g. *oke:Villain*). In this case, we cannot infer any type and we rely solely on the entity page (*dbr:Brian_Banner*) to identify a potential type when possible. In our training data set, this case never occurred.

Next, we assign a score to our candidate types based on the number of instances available for these types. Finally, we return the Dolce + DnS Ultralite type that is equivalent or is a super-type of the chosen DBpedia type. The following sections describe the various implemented strategies for type alignment.

Method 1: Alignment Based on the DBpedia Ontology: The first method checks if there is full match between the natural language type and a class in the DBpedia ontology (e.g. for the input "*oke:Villain*", we look for the URI *dbo:Villain*). If such a class exists, we simply align this type with Dolce.

Method 2: Alignment Based on the Type of Instances: In this step, the idea is to exploit the "informal" types available in the *dbr* namespace using the predicate *dbo:type*. For instance, even though *dbr:Village* is not defined as a class, we can find the triple *dbr:Bogoria,_Poland dbo:type dbr:Village*. Thus, given that *dbr:Bogoria, _Poland* is also of type *dbo:Place*, our general hypothesis is that we can consider *dbr: Village* to be a subclass of *dbo:Place*. To choose among all the candidates, we consider all the instances (using *dbo:type*) of *dbr:Village*, and assign a score to each of their types (available through *rdf:type*) depending on the number of times in which they appear in relation with the instances of *dbr:Village*.

Method 3: Alignment Based on the Entity Type: In this strategy, we exploit the information available in the DBpedia entity page itself. In fact, if the given natural language type does not have any DBpedia page, or if that page does not contain any information that could allow us to infer a valid type, we search for a direct *rdf:type* relation in the **entity** description. For instance, for the input [*dbr:Adalgis, oke:King*], our assumption is that the triple: *dbr:Adalgis rdf:type dul:NaturalPerson* implies *oke: King rdfs:subClassOf dul:NaturalPerson*. All the (rdf) types of the entity represent our candidate types with an initial score of 1.

Method 4: Alignment Based on Direct Types: Here we query the DBpedia resource corresponding to the natural language type (e.g. *dbr:Club*) and find the triples of the form *dbr:Club rdf:type [Type]* and return *[Type]*. Like in Method 3, all the candidates returned by this method have an initial score of 1.

Method 5: Alignment Based on Categories: This strategy exploits Wikipedia categories represented by the http://dbpedia.org/page/Category: namespace (*dbc*). Categories are indicated in most pages using the predicate *dct:subject*. The idea here is to look at all the categories in which a given type is included (for instance *dbc:Administrative_divisions*, *dbc:Villages*, etc. for *dbr:Village*), and then find the type(s) of all the elements in each of these categories. In this example, the category *dbc:Villages* contains several villages (such as *dbr:Mallekan*) of type *dbo:Place*. *dbo:Place* is therefore a candidate type for *dbr:Village*. Like in previous methods, this approach returns many candidates. Each type is given a score equal to the number of triples in which it appears.

Method 6: Alignment Based on Predicates' Domain and Range: This method infers a type for an entity by examining the *rdfs:domain* and *rdfs:range* of predicates that are used in the description of the DBpedia page associated with the natural language type. For instance, the two triples:

$$dbr:Marko\_Vovchok \quad dbo:birthplace \quad dbr:Village$$
$$dbo:birthPlace \quad rdfs:range \quad dbo:Place$$

allow us to infer *dbr:Village rdf:type dbo:Place* using the information available in the range of the predicate. In this approach, we only take into account the *dbo* predicates, as the *dbp* (http://dbpedia.org/property) predicates typically do not have any domain or range specified. Like in method 2, we give each inferred type a score equal to the number of triples in which the type is used.

4.2 Dolce+DUL Alignment

Following all our type identification methods, we obtain a set of candidate types with a score. Next, we rely on the alignment between the DBpedia ontology and Dolce + DnS Ultralite to replace each *dbo* type with their *dul/d0* counterpart. The same is done to replace *yago* types with *dul/d0* types. However, as the set of types used by the OKE challenge does not include all *dul* types, we modify this alignment in the following way: if a *dul* type is not included in the set of OKE challenge types, we replace it by its closest ancestor that is included in the set. For instance, *dul:SocialPerson* is not an element of the OKE challenge set, but its super class, *dul:Person*, is available. Therefore, if our alignment returns *dbo:Band rdfs:subClassOf dul:SocialPerson*, our final output is *dbo:Band rdfs:subClassOf dul:Person*. In our experiments on the OKE dataset, this strategy did not work well only with the *dul* types *dul:Concept* and *dul: Agent*, which do not have any parent in the OKE challenge set.

Next, the obtained set of *dul* candidates is very often a set of classes that have some taxonomical link among themselves. Given that our objective is to find the most precise candidate, the score of each candidate is modified by adding the score of its ancestors among this set, thus effectively favoring classes that are deeper in the taxonomy. Finally, the chosen candidate is the *dul* type with the highest score.

Here is a full example of our process for method 2 (instances) with the input (*dbr: Calvarrasa_de_Abajo*, "*oke:Village*"). First, we retrieve all the URIs that appear in a triple of the form *[subject] dbo:type dbr:Village*. Then, we retrieve all the types (*rdf:type*) of URIs of the form: *[subject] rdf:type [dbo_type]*. Each of these types' score increases by 1 every time it appears. In this example, the final list contains 26 types, with the best (score-wise) being: *dbo:Place* (480), *dbo:Location* (480), *dbo:PopulatedPlace* (480), *dbo:Settlement* (480), *dbo:Village* (460), *yago:location* (436) and *yago:object* (436). After the DOLCE alignment, this list becomes *d0:Location* (1905), *dul:PhysicalObject* (437), *dul:Object* (436) and *dul:Region* (436). During this step, if several types are aligned to the same *dul/d0* type, their scores are combined.

Finally, we check if our candidates include *dul* types that are not available in the OKE challenge set, and replace them by their equivalent (if available) or closest ancestor type that is available in the OKE set. Here, *dul:Region* is replaced by *d0: Characteristic*. We end up with *d0:Location* (1905), *dul:PhysicalObject* (873), *dul: Object* (436) and *d0:Characteristic* (436). The type with the highest score is *d0: Location* (1905), therefore we return *oke:Village rdfs:subClassOf d0:Location*.

5   Evaluation

5.1   Type Extraction Evaluation

Our first evaluation calculates the precision and recall of the natural language type identification subtask. We consider a type as a true positive only when its lemmatized oke type is a perfect match with at least one of the lemmatized oke types of the OKE gold standard (See footnote 2). Using this evaluation method on the 2016 train dataset, our precision and recall for the type extraction subtask is 87 % as shown in Table 2a, and 80 % on the evaluation dataset as shown in Table 2b. Table 4 presents the official evaluation on the 2016 train dataset using Gerbil[4] [18]. We can notice some decrease in performance using Gerbil (Tables 2a and 4), some of which can be explained by the existence of OKE types in plural form in the gold standard (e.g. *oke:Awards* versus

---

[4] http://gerbil.aksw.org/gerbil.

oke:Award). In fact, contrary to Table 4, the results in Table 2a take into account the lemmatization of both the natural language types and the gold standard types.

Table 2a. Statistics for the type extraction evaluation on the train dataset

Pattern	(1)	(2)	(3)	(4)	Total
Found types	10	156	2	4	172
Total occurrences	14	173	2	9	198
Precision/recall	71 %	90 %	100 %	44 %	87 %

Table 2b. Statistics for the type extraction evaluation on the evaluation dataset

Pattern	(1)	(2)	(3)	(4)	Total
Found types	1	39	0	0	39
Total occurrences	1	47	0	2	50
Precision/recall	100 %	82.98 %	–	0 %	80 %

We performed an analysis of the unsuccessful sentences and identified few potential sources of errors. A good proportion of errors arise from grammatical ambiguities and incorrect syntactic analyses of sentences. This is the case for sentences like *Brad Sihvon was a Canadian film and television actor*, for which we find the type *oke:Film* instead of *oke:Actor*, due to an error in the parsing process. Similar errors can also occur, in some rare cases, for long sentences like *Bradycardia, also known as bradyarrhythmia, is a slow heart rate, namely, a resting heart rate of under 60 beats per minute (BPM) in adults* for which we extract the type *oke:Heart* instead of *oke: Rate*. In some cases, the errors are debatable. We list as examples the sentence *Gimli Glider is the nickname of an Air Canada aircraft that was involved in an unusual aviation incident*, for which we extract the type *oke:Aircraft* instead of *oke:Nickname*, or *Caatinga is a type of desert vegetation, and an ecoregion characterized by this vegetation in interior northeastern Brazil*, for which we find the type *oke:Vegetation* instead of *oke:TypeOfDesertVegetation*.

We also evaluated the precision and recall of our patterns separately. Given that the precision and recall are the same, Tables 2a and b show the results for each pattern based on the 2016 OKE train and evaluation datasets.

5.2 Type Alignment Evaluation

To assess the efficiency of each type alignment method, we compared the obtained types to those present in the gold standard. Table 3 shows the number of returned types, the number of correct types, as well as the precision, recall and F-measure for each method on the OKE challenge train dataset. Taken individually, most methods achieve limited or poor performance. However, we also implemented a pipeline strategy to combine these methods, thus increasing the recall of our approach. The pipeline is based on the most successful to the least successful strategies (in terms of precision) based on the results of individual methods. In this pipeline, a strategy is executed only if the previous one was unsuccessful in returning a type.

Table 3. Comparison of each method for type alignment on the OKE challenge train dataset

Method	Returned types	Correct types	Precision	Recall	F-measure
1: ontology	83	59	71 %	30 %	42 %
2: instances	57	38	67 %	19 %	30 %
3: entity	140	67	48 %	34 %	40 %
4: direct type	17	6	35 %	3 %	6 %
5: category	130	22	15 %	10 %	12 %
6: predicates	89	11	12 %	6 %	8 %
Pipeline 1–6	172	96	56 %	48 %	52 %

5.3   Overall Results

The overall results of the task of entity typing and alignment on the OKE training dataset using the evaluation framework Gerbil are shown in Table 4. These results rely on the pipeline strategy for the type alignment subtask.

We can notice a slight decrease in performance compared to our local evaluation on the 2016 train dataset.

Table 4. Overall precision, recall and F-measure computed using Gerbil on the 2016 train dataset

Task	Micro precision	Micro recall	Micro F-measure	Macro precision	Macro recall	Macro F-measure
Type extraction	82.32 %	75.81 %	78.93 %	82.32 %	78.96 %	80.05 %
Type alignment	49.63 %	45.47 %	47.45 %	49.62 %	45.42 %	46.47 %
Total (average)	65.97 %	60.64 %	63.19 %	65.97 %	62.19 %	63.26 %

The system was also tested on the 2016 test dataset; the result for this dataset are presented in Table 5.

Table 5. Overall precision, recall and F-measure computed using Gerbil on the 2016 test dataset

Task	Micro precision	Micro recall	Micro F-measure	Macro precision	Macro recall	Macro F-measure
Type extraction	81.63 %	73.39 %	77.29 %	80.81 %	76.60 %	77.95 %
Type alignment	46.46 %	42.51 %	44.40 %	46.46 %	42.51 %	43.53 %
Total (average)	64.05 %	57.95 %	60.85 %	63.64 %	59.55 %	60.74 %

For comparison purposes, we also report the results of two competing systems on the evaluation dataset in Table 6. These systems are Mannheim [19], a participant to the OKE 2016 challenge and CETUS [10], the baseline system and the winner of the

OKE 2015 challenge. Mannheim uses taxonomical relation ("isa") extraction based on Hearst-like patterns in text to find the entity type, then chooses one of these isa relations and exploits a mapping between OntoWordnet, Wordnet and DOLCE to infer the super class in the OKE types set. CETUS uses pattern extraction to identify potential types, then creates a hierarchy between these types. Finally, it proposes two approaches to align the type with DOLCE: the first one is based on a mapping with Yago, and the second on an entity recognition tool (FOX).

Table 6. Overall precision, recall and F-measure for the two participating systems and the baseline CETUS on the test dataset

System	Task	Micro			Macro		
		Precision	Recall	F1	Precision	Recall	F1
CETUS	Extr.	68.75 %	88.00 %	77.19 %	72.00 %	88.00 %	77.33 %
	Align.	22.17 %	24.47 %	23.26 %	22.17 %	24.47 %	19.89 %
	Total	45.46 %	56.24 %	50.23 %	47.08 %	56.24 %	48.61 %
Mannheim	Extr.	77.27 %	68.00 %	72.34 %	63.00 %	68.00 %	64.67 %
	Align.	0.00 %	0.00 %	0.00 %	0.00 %	0.00 %	0.00 %
	Total	38.64 %	34.00 %	36.17 %	31.50 %	34.00 %	32.33 %
WestLab	Extr.	86.00 %	86.00 %	86.00 %	86.00 %	86.00 %	86.00 %
	Align.	8.00 %	6.67 %	7.27 %	8.00 %	6.67 %	7.00 %
	Total	47.00 %	46.33 %	46.64 %	47.00 %	46.33 %	46.5 %

Our system WestLab obtains satisfying results in terms of precision for the type extraction task; we obtain a Micro and Macro values of 86 %, whereas the baseline CETUS obtains 68.75 % (micro) and 72 % (macro), and Mannheim obtains 77.27 % (micro) and 63 % (macro).

As for the recall, we obtain 86 % (micro and macro), which is better than Mannheim's recall of 68 % (micro and macro), but lower than CETUS which obtains 88 % (micro and macro).

Our Micro and Macro F-Measures are both 86 %. These results are higher than CETUS', which are 77.19 % and 77.33 % respectively, and Mannheim's, that obtains 72.34 % and 64.67 % respectively, for the type extraction.

Thus, we can conclude that we outperform other systems when taking into account precision but we note that our recall is lower than the one obtained by the baseline CETUS. These results also show that our patterns do not always detect and extract the type of the entity, which is an indicator that the patterns set must be extended in our future work. However, our patterns rarely extract types that are false positives, which shows that they are well defined and accurate.

Concerning the type alignment, there have been some issues with the test dataset distributed by the OKE challenge organizers at the time of the evaluation, which have been corrected later. This explains the very low performance shown in Table 6 for the type alignment subtask. Given this modification, we are able to provide results only for our system and the CETUS baseline on the corrected evaluation dataset. At the time of this publication, we don't have the updated results for the Mannheim system. Table 7

Table 7. Overall precision, recall and F-measure for the two participating systems and the baseline CETUS on the corrected test dataset

System	Task	Micro			Macro		
		Precision	Recall	F1	Precision	Recall	F1
CETUS	Extr.	68.75 %	88.00 %	77.19 %	72.00 %	88.00 %	77.33 %
	Align.	22.17 %	24.47 %	23.26 %	22.17 %	24.47 %	19.89 %
	Total	45.46 %	56.24 %	50.23 %	47.08 %	56.24 %	48.61 %
WestLab	Extr.	81.63 %	73.39 %	77.29 %	80.81 %	76.60 %	77.95 %
	Align.	46.46 %	42.51 %	44.40 %	46.46 %	42.51 %	43.53 %
	Total	64.05 %	57.95 %	60.85 %	63.64 %	59.55 %	60.74 %

provides our results on the updated dataset, compared with the baseline. Overall, we can notice a huge improvement on the corrected dataset. In fact, the WestLab system obtains an F-Measure of 44.4 % (micro) and 43.53 % (macro) for the type alignment subtask, whereas CETUS obtains 23.26 % and 19.89 %. These results constitute a considerable improvement for the type alignment task, even though they are still under the threshold of 50 %.

For the overall results including both type extraction and alignment, we outperform all systems and obtain F-Measures of 60.85 % (micro) and 60.74 % (macro), whereas CETUS obtains F-Measures of 50.23 % (micro) and 48.61 % (macro). We cannot compare our system with Mannheim on the corrected test dataset, except on the type recognition subtask, for the reasons mentioned in the previous paragraph.

6 Discussion

Type Extraction. One limitation of our approach for natural language type identification is the small number of implemented patterns, which does not guarantee to find an entity type. However, our proposal of SPARQL patterns, coupled with an RDF representation of definitions, represents an elegant and simple solution which facilitates the addition of new patterns. Another limitation comes from the fact that our system relies on a syntactic analysis. Thus, errors that occur in the parsing process also affect our system. However, according to our preliminary results, this approach displays a satisfactory precision and recall values compared to previous approaches in the OKE competition.

DBpedia for Type Alignment. Task alignment requires the discovery of *rdfs:-subClassOf* links between natural language types and ontological classes. One of our research objectives was to assess how well a type alignment could be performed based on the structured knowledge available in the DBpedia ontology and resources. Some of our methods exploit the grey zone around the notion of subclass and instance in DBpedia. In fact, DBpedia resources (A-box) cannot be normally expected to use the *rdfs:subClassOf* predicate. However, some of the resources employ the predicate *dbo:type*. For example, *dbr:Bogoria,_Poland dbo:type dbr:Village*. Thus *dbr:Village* can be effectively considered as a *class* based on RDFS semantics. There were 57 (out of 198) similar cases in our train dataset. Based on this line of thought, if we found *dbr:Village rdf:type dbo:Place,* we inferred *dbr:Village rdfs:subClassOf dbo:Place.* These examples show that DBpedia resources (A-box) are also described using an informal or implicit schema. This further highlights the need of describing these resources in the ontology rather than in the knowledge base.

Due to the lack of directly exploitable type information in DBpedia, we relied on type inference methods (M2 - instances, M6 - predicates, M5 - categories) in few cases (27 out of 172 types are retrieved using these methods). More specifically, we employed these strategies when an input type does not have a *dbo* page or when its *dbr* page does not contain any *rdf:type* predicate. However, these methods often give poor results. Finally we did not process the disambiguation pages (e.g. *dbr:Motion*) that are sometimes returned by our methods. Altogether, our system failed to return any type in 14 % of the cases. In this case, it returns *owl:Thing.*

Examples of Problematic Cases. Most of our errors boil down to two error sources: (a) inaccurate, noisy, or plain false information and (b) unavailable information in DBpedia. In the following, we give a few examples of problematic cases in some of the alignment methods.

M2 – instances: According to the gold standard, *dbr:Court* should be a *dul:Organization.* However, in DBpedia, *dbr:Court* instances, as depicted by the *rdf:type* predicate, are inaccurate (e.g. *dbr:Mansion_in_Grabowo_Krolewskie*) or refer to broken links. Our type alignment based on these links wrongly concludes that *dbr:Court* is a *d0:Location.*

M6 – predicates: dbr:Season should be a *dul:Situation.* However, in DBpedia, there is a confusion between a season (time of the year) and seasonal music (such as Christmas songs) which does not have a *dbr* resource. Therefore, the resource *dbr:Season* is used erroneously, instead of the non-existing *dbr:Seasonal_Music* page. This leads to triples such as *dbr:Christmas _(Kenny_Rogers_album) dbo:genre dbr:Season.* Given that the predicate method exploits *dbo:genre rdfs:range dbo:Genre,* we erroneously conclude that a *dbr:Season* is a subclass of *dbo:Genre.*

M5 – categories: dbr:Tournament is part of only one category, *dbc:Tournament_systems,* containing pages such as *dbr:Round-robin_tournament* or *dbr:Double-elimination_tournament.* All of these resources have a type in Yago (*artifact*) that is aligned to *dul:PhysicalObject,* which makes us conclude that a *dbr:Tournament* is a *dul:PhysicalObject.* Here, the error is double: *dbr:Tournament* should not be in the category *dbc:Tournament_systems,* and the resources should not be typed as *yago:Artifact.*

In all the above examples, the correct answer is never present in our candidates list. This observation confirms that DBpedia resources are often poorly described [16]. Despite these limitations, our pipeline, which is based on a set of methods ordered from the most trustworthy to ·the least one, obtains a micro precision of 49.6 % on the training dataset and 46.5 % on the test dataset, and micro recall of 45.5 % on the training dataset and 42.5 % on the test dataset, which we consider as reasonable given the complexity of the task.

Gold Standard. We had some issues when comparing our results with the gold standard. Quite often, our results could be considered as correct, but are different from the ones in the gold standard as they are based on the DBpedia ontology. For instance, we infer that a *oke:Meeting* is a subclass of *dul:Event* (*dul:Event:* "Any physical, social, or mental process, event, or state"), but the gold standard states that a *oke: Meeting* is a subclass of *dul:Activity*. Both answers could be acceptable. In the OKE train dataset, we identified 20 "borderline" cases out of 198 in the alignment subtask. In the natural language type extraction subtask, we identified some potentially questionable types in the gold standard of the form "Set_Of_X" or "Type_Of_X". For instance, in the sentence *Caatinga is a type of desert vegetation...* our position is that the type could be *oke:DesertVegetation* rather than *oke:TypeOfDesertVegetation*.

Future Work. For the type alignment sub-task, our next step will consider the problem of the disambiguation pages. Such pages represent a non-negligible portion of the data set (26 %), and systematically constitute a source of errors. The objective is to choose the correct type among all the possible disambiguations. For instance, given the input [*dbr:Babylonia, oke:State*], the returned type *dbr:State* is a disambiguation page, linking to pages such as *dbr:Nation_state*, *dbr:State_(functional_analysis)* or *dbr: Chemical_state*.

7   Conclusion

This paper describes our approach for the extraction of entity types from text and the alignment of these types to the Dolce+DUL ontology. The patterns used to extract natural language types from textual definitions achieved high precision and recall values. As for the type alignment, the strength of our approach is based on the multiplicity of strategies which exploit both the DBpedia ontology and knowledge base and rely on DBpedia large coverage. Our experiments highlight the necessity of a better linkage between DBpedia resources and the DBpedia ontology and the need for restructuring some DBpedia resources as ontological classes.

Acknowledgement. This research has been funded by the NSERC Discovery Grant Program.

References

1. OKE 2015 Challenge Description. https://github.com/anuzzolese/oke-challenge/blob/master/participating%20systems/OKE2015_challengeDescription.pdf. Accessed 20 Mar 2016
2. Nadeau, D., Sekine, S.: A survey of named entity recognition and classification. Lingvist. Investig. **30**(1), 3–26 (2007)
3. Hearst, M.A.: Automatic acquisition of hyponyms from large text corpora (1992)
4. Min, B., Shi, S., Grishman, R., Lin, C.Y.: Ensemble semantics for large-scale unsupervised relation extraction. In: 2012 Joint Conference on Empirical Methods in Natural Language Processing and Computational Natural Language Learning, pp. 1027–1037. Association for Computational Linguistics (2012)

5. Mintz, M., Bills, S., Snow, R., Jurafsky, D.: Distant supervision for relation extraction without labeled data. In: Proceedings of the Joint Conference of the 47th Annual Meeting of the ACL, pp. 1003–1011. ACL, Association for Computational Linguistics (2009)
6. Otero-Cerdeira, L., Rodriguez-Martinez, F.J., Gomez-Rodriguez, A.: Ontology matching: a literature review. Expert Syst. Appl. **42**(2), 949–971 (2015)
7. Bizer, C., Heath, T., Ayers, D., Raimond, Y.: Interlinking open data on the web. Media **79**(1), 31–35 (2007)
8. Giunchiglia, F., Shvaiko, P., Yatskevich, M.: Discovering missing background knowledge in ontology matching. In: Proceedings of the 2006 Conference on ECAI 2006: 17th European Conference on Artificial Intelligence, pp. 382–386. IOS Press (2006)
9. Kachroudi, M., Moussa, E.B., Zghal, S., Ben, S.: Ldoa results for OAEI 2011. In: Ontology Matching, p. 148 (2011)
10. Röder, M., Usbeck, R., Ngonga Ngomo, A.: CETUS - a baseline approach to type extraction. In: OKE Challenge 2015 Co-located with the 12th Extended Semantic Web Conference (ESWC) (2015)
11. Gao, J., Mazumdar, S.: Exploiting linked open data to uncover entity types. In: OKE Challenge 2015 Co-located with the 12th Extended Semantic Web Conference (ESWC) (2015)
12. Consoli, S., Reforgiato, D.: Using FRED for named entity resolution, linking and typing for knowledge base population. In: OKE Challenge 2015 Co-located with the 12th Extended Semantic Web Conference (ESWC) (2015)
13. Bos, J.: Wide-coverage semantic analysis with boxer. In: Bos, J., Delmonte, R. (eds.) Semantics in Text Processing, pp. 277–286. College Publications (2008)
14. Ontology Design Patterns (2014). http://ontologydesignpatterns.org/ont/dbpedia_2014_imports.owl. Accessed 20 Mar 2016
15. CETUS. DolCE_YAGO_mapping (2015). http://github.com/AKSW/Cetus/blob/master/DOLCE_YAGO_links.nt. Accessed 20 Mar 2016
16. Font, L., Zouaq, A., Gagnon, M.: Assessing the quality of domain concepts description in DBpedia. In: 11th International Conference on Signal-Image Technology and Internet-Based Systems, pp. 254–261. IEEE (2015)
17. De Marneffe, M-C., Manning, C.D.: The stanford typed dependencies representation. In: Proceedings of the Workshop on Cross-Framework and Cross-Domain Parser Evaluation, ACL (2008)
18. Usbeck, R., Röder, M., Ngonga Ngomo, A.-C., Baron, C., Both, A., Brümmer, M., Ceccarelli, D., Cornolti, M., Cherix, D., Eickmann, B., Ferragina, P., Lemke, C., Moro, A., Navigli, R., Piccinno, F., Rizzo, G., Sack, H., Speck, R., Troncy, R., Waitelonis, Jö., Wesemann, L.: GERBIL – General Entity Annotation Benchmark Framework. In: 24th WWW Conference (2015)
19. Faralli, S., Ponzetto, S.: DWS at the 2016 open knowledge extraction challenge: a hearst-like pattern-based approach to hypernym extraction and class induction. In: OKE Challenge 2016 Co-located with the 13th Extended Semantic Web Conference (ESWC) (2016)

Challenge on Semantic Sentiment Analysis

Challenge on Fine-Grained Sentiment Analysis Within ESWC2016

Mauro Dragoni[1] and Diego Reforgiato Recupero[2]([⊠])

[1] Fondazione Bruno Kessler, Trento, Italy
dragoni@fbk.eu
[2] Universitá di Cagliari, Cagliari, Italy
diego.reforgiato@unica.it

Abstract. The wide spread of the social media has given users a means to express and share their opinions and thoughts on a large range of topics and events. The number of opinions, emotions, sentiments that are being expressed within social media grows at an exponential rate; all these data can be exploited in order to come up with useful insights, analytics, etc. Initial Sentiment Analysis systems used lexical and statistical resources to automatically assess polarities of opinions and sentiment. With the raise of the Semantic Web, it has been proved that Sentiment Analysis techniques can have higher performances if they use semantic features. This generated further opportunities for the research domain as well as the market domain where key stakeholders need to catch up with the latest technology if they want to be compelling. Therefore, deep understanding of natural language text and the related semantics are urgent matter to be familiar with. Following the first two editions, the third edition of the Fine-Grained Sentiment Analysis challenge aims at providing a stimulus toward this direction. On the one hand, it represents an event where researchers can learn and share their methods and how they employed Semantics for Sentiment Analysis. On the other hand, it offers an occasion for stakeholders to get an idea of what research is being developed and where the research is headed to plan future strategies within the domain of Sentiment Analysis.

1 Introduction

Social media evolution has given users one important opportunity for expressing, sharing and commenting their thoughts and opinions online. The information thus produced is related to many different areas such as commerce, tourism, education, health and causes the size of the Social Web to expand exponentially.

There is a great opportunity that arises from this amount of information which is the one to automatically detect and mine the opinions of the users [1,2]. This has raised further interest within the scientific community where open challenges still exist and the business world where social analysis brings substantial benefits. According to an IDC survey [3], the amount of unstructured text occupies 80 % of the digital space with respect to the 20 % of the structured text.

© Springer International Publishing Switzerland 2016
H. Sack et al. (Eds.): SemWebEval 2016, CCIS 641, pp. 79–94, 2016.
DOI: 10.1007/978-3-319-46565-4_6

Besides, there are not so many solutions that can accurately analyse the text and present insights in an understandable manner as this task is still extremely difficult. In fact, mining opinions and sentiments from natural language involves a deep understanding of most of the explicit and implicit, regular and irregular, syntactical and semantic rules proper of a language.

Existing approaches are mainly focused on the identification of parts of the text where opinions and sentiments can be explicitly expressed such as polarity terms, expressions, affect words. They usually adopt purely syntactical approaches and are heavily dependant on the source language of the input text [4]. It follows that they miss many language patterns where opinions can be expressed because this would involve a deep analysis of the semantics of a sentence.

One example is constituted by the presence of implicit opinions deriving from particular use of verbs that are difficult to catch using a classical sentiment analysis tool.

As an example, a sentence such as *Players of that soccer team are happy that the president has fired the coach* includes the verb *are* that carries a positive sentiment tone *happy* of the expressed opinion. However, this is not enough for a complete sentiment analysis of this sentence. With classical sentiment analysis approaches we can only state that the sentence expresses a positive opinion on the event *has fired* and nothing else. With fine-grained sentiment analysis [5–8] we can go one step deeper as it focuses on a semantic analysis of text through the use of web ontologies, semantic resources, or semantic networks, allowing the identification of opinion data which with only natural language techniques would be very difficult. Fine-Grained sentiment analysis allows, in the example above, detecting a negative opinions of the holder *Players* towards the subject *coach*.

Understanding the semantics of a sentence offers an exciting research opportunity and challenge to the Semantic Web community as well. In fact, the Fine-Grained Sentiment Analysis Challenge aims to go beyond a mere word-level analysis of text and provides novel methods to opinion mining and sentiment analysis that can transform more efficiently unstructured textual information to structured machine-processable data, in potentially any domain.

By relying on large semantic knowledge bases and Semantic Web best practices and techniques, fine-grained sentiment analysis steps away from blind use of keywords, simple statistical analysis based on syntactical rules, but rather relies on the implicit features associated with natural language concepts. Unlike purely syntactical techniques, semantic sentiment analysis approaches are able to detect also sentiments that are implicitly expressed within the text.

The third edition of the Fine-Grained Sentiment Analysis Challenge[1] followed the success, experience and best practices of the first two. It provided further stimulus and motivations for research within the Semantic Sentiment Analysis area. The Aspect Based Sentiment Analysis task, introduced in the 2014 edition,

---

[1] https://github.com/diegoref/SSA2016.

and proposed in the SEMEVAL 2015 workshop for the first time[2], was once again confirmed.

The third edition of the challenge focused on further development of novel approaches for semantic sentiment analysis. Participants had to design a concept-level opinion-mining engine that exploited Linked Data and Semantic Web ontologies, such as DBPedia[3].

The authors of the competing systems showed how they employed semantics to obtain valuable information that would not be caught with traditional senti-ment analysis methods. Accepted systems were based on natural language and statistical approaches with an embedded semantics module, in the core approach. As happened within the first edition [9] and second edition [10] of the challenge, a few systems merely based on syntax/word-count were excluded.

The third challenge benefited from a Google Group that we created and named Semantic Sentiment Analysis Intiative[4] and that we opened before the Challenge proposal. Currently, the group consists of more than 150 participants and we leverage that to disseminate and promote our initiatives related to the Sentiment Analysis domain. Moreover, the third edition of the challenge could also benefit from a Workshop[5] we chaired at ESWC 2016 related to the same topics. Challenge had therefore an additional strength provided by the mutual support between the two events. Challenge systems were in fact invited to a Workshop dedicated session for discussing open issues and research directions showing the last technological advancements. This dual action stimulated and encouraged participants to present their work at the two events.

The rest of the chapter is organized as follows. Related work and background in semantic sentiment analysis is included in Sect. 2. The tasks we have proposed in the third edition of the challenge are detailed in Sect. 3 where we have also described the annotation process of the datasets we have used for the evaluation phase. The systems submitted by the challengers are explained in Sect. 4 and their results are showed in Sect. 5. Finally, conclusions and considerations gained from this challenge are depicted in Sect. 6 where we also sketch our plans for the next edition of the challenge.

2   Related Work

After the successes of the 2014 and 2015 editions, the ESWC conference[6] included again a challenge call with a dedicated session. The challenge on Fine-Grained Sentiment Analysis has been proposed and accepted for the third time on a row in the 2016 ESWC program.

---

[2] http://alt.qcri.org/semeval2015/task12/.

[3] http://dbpedia.org.

[4] Publicly accessible at https://groups.google.com/forum/#!forum/semantic-sentime nt-analysis.

[5] http://www.maurodragoni.com/research/opinionmining/events/.

[6] http://2016.eswc-conferences.org/.

The 2014 and 2015 editions of the ESWC challenges have been published in books [11,12] where each challenge, its tasks, evaluation process have been introduced and each system participating to each challenge has been described, detailed and results and comparisons have been shown. The Fine-Grained Sentiment Analysis challenge has been included in the two volumes above [9,10]. The 2014 edition of the challenge was also the first edition in parallel with a workshop at ESWC of the same domain that hosted around 20 participants [13]. The 2016 edition of the challenge repeated the success of the dual events of the 2014 edition and run in parallel with the Semantic Sentiment Analysis workshop whose proceedings are in the process of publication.

Besides the Semantic Sentiment Analysis challenge described in this chapter and its previous editions, there are a few number of relevant events and challenges that is worth to mention.

SemEval (Semantic Evaluation)[7] consists of a series of evaluations workshops of computational semantic analysis systems. It is now in its tenth edition[8] and it has been collocated with the 15th Annual Conference of the North America Chapter of the Association for Computational Linguistics: Human Language Technologies[9]. Since 2007 the workshop has covered the sentiment analysis topic. During the last edition, SemEval2016 included four tasks for the sentiment analysis track:

- Sentiment Analysis in Twitter. It was subdivided in five subtasks related to classification and quantification of the polarity of sentiment according to a two or five point scale;
- Aspect-Based Sentiment Analysis. Divided in three subtasks including opinionated document and customer reviews about a target entity and one subtask supporting the French language;
- Detecting Stance in Tweets. It consisted of two tasks, depending on whether the training set was provided or not;
- Determining Sentiment Intensity of English and Arabic Phrases. Divided in three subtasks including English modifiers, English mixed polarity and Arabic tweets to process.

We are proud to mention that the aspect-based sentiment analysis task was introduced in SemEval after it was proposed in the first edition of the Concept-Level sentiment analysis challenge at ESWC2014.

Sentiment analysis initiatives and events are usually listed within the SenticNet page[10]. SenticNet [14] is a concept-level knowledge base providing a set of semantics sentics, and polarity associated with 30k natural language concepts. The SENTIRE[11] and WISDOM[12] workshops series listed within the SenticNet

---

[7] https://en.wikipedia.org/wiki/SemEval.

[8] http://alt.qcri.org/semeval2016/.

[9] http://naacl.org/naacl-hlt-2016/index.html.

[10] http://sentic.net/.

[11] http://sentic.net/sentire/.

[12] http://sentic.net/wisdom/.

page are worth to mention and focus on analysing the effect of the crowds on opinionated web documents.

The Sentiment Analysis Symposium[13] has been one of the first conference to address the business value of sentiment, opinion, emotion within the social media. Every year developers, key stakeholders and researchers participate to the event presenting their ideas and recent advancements. The symposium includes both a description of state of art technology, solution and business presentations and a tutorial for in-depth technical content.

The Semantic Web Challenge[14] aims at extending the human-readable web using semantic web best practices and resources.

It is run within the International Semantic Web Conference (which is currently in its 15th edition[15] and it started in 2003). The goal of the challenge goal is to build systems using Semantic Web technologies in order to come up with information needed to assist users in performing certain tasks. There are not specified tasks defined nor training set, data set, or domains where the proposed systems must be focused. There are a small number of requirements and additional desirable features that the systems need to meet. The challenge is divided into two tracks, the open track and the big data track. The latter has the goal to demonstrate the capabilities of semantic web technologies to process large amount of web information whereas the former is open to any domain. So far, only within the 2014 edition a system related Semantic Sentiment Analysis was proposed. It was *SHELDON: Semantic Holistic framEwork for LinkeD ONtology data*, a semantic framework that can be used and called over REST API for several purposes. One of the features of SHELDON performed semantic sentiment analysis as it included a Sentic Computing approach called Sentilo, [5], to detect holders and topic of opinion sentences.

Recently, solutions based on the use of information retrieval strategies for building sentiment analysis systems have been proposed [15]. The authors presented also a system using fuzzy logic for representing uncertainty associated with each word and its different polarity, related to different domains [16].

3   Tasks, Datasets and Evaluation Measures

The third edition of the Fine-Grained Sentiment Analysis challenge included five tasks: Polarity Detection, Aspect-Based Sentiment Analysis, Semantic Sentiment Retrieval, Frame Entities Identification, Implicit Opinions related to Verbnet verbs and roles. One more task was represented by the Most Innovative Approach. Participants had to submit an abstract of no more than 200 words and a 4 pages paper including the details of their systems, why it is innovative, which features it provides, which design choices were made, what lessons were learnt, which tasks it addressed and how the semantics was employed. After a round of

---

[13] http://sentimentsymposium.com.

[14] http://challenge.semanticweb.org/.

[15] http://iswc2016.semanticweb.org/.

reviews, only sentiment analysis systems including a semantic flavor (e.g. semantic web resources such as ontologies) were accepted and they had to provide a full description of their system, web access or a link where the system could be downloaded together with a short set of instructions.

Following we will describe each task and, in particular, will detail datasets and evaluation methodologies we have provided for tasks 1 and 2, those targeted by the submitted systems.

3.1   Task 1: Polarity Detection

The basic task of the challenge was binary polarity detection. The proposed semantic opinion-mining engines were assessed according to precision, recall and F-measure of the detected polarity values (positive OR negative) for each review of the evaluation dataset. As an example, for the sentence *These sunglasses are all right. They were a little crooked, but still cool..*, the correct answer that a sentiment analysis system needed to give was *positive* and therefore it had to write *positive* between the $<polarity>$, $</polarity>$ tags of the output. Figure 1 shows an example of the output schema for task1.

This task was pretty straightforward to evaluate. A precision/recall analysis was implemented to compute the accuracy of the output for this task. A true positive (tp) was defined when a sentence was correctly classified as positive. On the other hand, a false positive (fp) is a positive sentence which was classified as negative. Then, a true negative (tn) is detected when a negative sentence was correctly identified as such. Finally, a false negative (fn) happens when a negative sentence was erroneously classified as positive. With the above definitions, we defined the precision as

$$precision = \frac{tp}{tp + fp}$$

```xml
<?xml version="1.0" encoding="UTF-8" standalone="yes"?>
<Sentences>
    <sentence id="apparel_0">
        <text>
            GOOD LOOKING KICKS IF YOUR KICKIN IT OLD SCHOOL LIKE ME. AND COMFORTABLE.
            AND RELATIVELY CHEAP. I'LL ALWAYS KEEP A PAIR OF STAN SMITH'S
            AROUND FOR WEEKENDS
        </text>
        <polarity>
        positive
        </polarity>
    </sentence>
    <sentence id="apparel_1">
        <text>
            These sunglasses are all right. They were a little crooked, but still cool..
        </text>
        <polarity>
        positive
        </polarity>
    </sentence>
```

Fig. 1. Task 1 output example. Input is the same without the polarity tag.

the recall as

$$recall = \frac{tp}{tp + fn}$$

the F1 measure as

$$F1 = \frac{2 \times precision \times recall}{precision + recall}$$

and the accuracy as

$$accuracy = \frac{tp + tn}{tp + fp + fn + tn}$$

As training set has been used the *DRANZIERA* [17] dataset consisting in one million reviews crawled from the Amazon website. All reviews are split in 20 subsets representing as many domains. For each domain, we had a polarity balanced scenario where 25,000 reviews were positive and 25,000 were negative.

While, for the test set, 33,361 reviews (12,337 positive and 21,024 negative) have been submitted to participants that were asked to infer their polarity.

3.2 Task 2: Aspect-Based Sentiment Analysis

The output of this task was a set of aspects of the reviewed product and a binary polarity value associated to each of such aspects. So, for example, while for the task 1 an overall polarity (positive or negative) was expected for a review about a mobile phone, this task required a set of aspects (such as *speaker, touchscreen, camera*, etc.) and a polarity value (positive or negative) associated with each of such aspects. Engines were assessed according to both aspect extraction and aspect polarity detection using precision, recall, f-measure, and accuracy similarly as performed during the first edition of the Concept-Level Sentiment Analysis challenge held during ESWC2014 and re-proposed at SemEval 2015 Task 12[16]. Please refer to SemEval 2016 Task 5[17] for details on the precision-recall analysis. Figure 2 shows an example of the output schema for task2.

The training set was composed by 5,058 sentences coming from two different domains: "Laptop" (3,048 sentences) and "Restaurant" (2,000 sentences). While, the test set was composed by 891 sentences coming from the "Laptop" (728 sentences) and "Hotels" (163 sentence). The reason for which we decided to use the "Hotels" domain in the test set with respect to the "Restaurant" one was to observe the capability of the participant systems to be general purpose with respect to the training set.

3.3 Task 3: Semantic Sentiment Retrieval

The output of this task was a list of entities ordered by strength of positive judgements of any of their features. As an example, given an input list of reviews on smartphones, create a structured output of each review where smartphones

---

[16] http://www.alt.qcri.org/semeval2015/task12/.
[17] http://alt.qcri.org/semeval2016/task5/.

```xml
<?xml version="1.0" encoding="UTF-8" standalone="yes"?>
<Review rid="1">
    <sentences>
        <sentence id="348:0">
            <text>Most everything is fine with this machine: speed, capacity, build.</text>
                <Opinions>
                    <Opinion aspect="MACHINE" polarity="positive"/>
                </Opinions>
        </sentence>
        <sentence id="348:1">
            <text>The only thing I don't understand is that the resolution of the
            screen isn't high enough for some pages, such as Yahoo!Mail.
            </text>
            <Opinions>
                <Opinion aspect="SCREEN" polarity="negative"/>
            </Opinions>
        </sentence>
        <sentence id="277:2">
            <text>The screen takes some getting use to, because it is smaller
            than the laptop.</text>
            <Opinions>
                <Opinion aspect="SCREEN" polarity="negative"/>
            </Opinions>
        </sentence>
    </sentences>
</Review>
```

Fig. 2. Task 2 output example. Input is the same without the opinion tag and its descendant nodes.

are listed together with their features and opinions on each of them. This task included Information Retrieval (detect features of given entities), Named Entity Recognition (detect smartphone models within the review possibly using some sort of knowledge base), Sentiment Analysis (aggregate features opinions for the entity sentiment for either overall or feature based retrieval). An example input is given in Fig. 3 and the related output is shown in Fig. 4.

```xml
<?xml version="1.0" encoding="UTF-8" standalone="yes"?>
<Sentences>
    <sentence id="0">
        <text>So far so good. My wife just loves the new Samsung S5: the display is awesome
        and the colors are very brilliant. However, further memory is necessary for storing
        everything.</text>
    </sentence>
    <sentence id="1">
        <text>All the LG G3 have problems with videos: they often are not able to connect
        with tv and when they can, the quality of the image is poor. The only strong point
        is the amount of memory coming from the factory.</text>
    </sentence>
</Sentences>
```

Fig. 3. Task 3 input example.

```
<?xml version="1.0" encoding="UTF-8" standalone="yes"?>
<Ranks>
    <rank quality="display">
        <position value="1" name="Samsung Galaxy S5"/>
        <position value="2" name="LG G3"/>
    </rank>
    <rank quality="memory">
        <position value="1" name="LG G3"/>
        <position value="2" name="Samsung Galaxy S5"/>
    </rank>
    <rank quality="GENERAL">
        <position value="1" name="Samsung Galaxy S5"/>
        <position value="2" name="LG G3"/>
    </rank>
</Ranks>
```

Fig. 4. Task 3 output example.

3.4   Task 4: Frame Entities Identification

The challenge focused on semantic fine-grained sentiment analysis. The proposed engines needed to work beyond word/syntax level, and addressing a concepts/semantics perspective. This task evaluated the capability of the proposed systems to identify the objects involved in a typical opinion frame according to their role: holders, topics, opinion concepts (i.e. terms referring to highly polarized concepts). For example, in a sentence such as *The mayor is loved by the people in the city, but he has been criticised by the state government* (taken from [18]), a system should be able to identify that *the people* and *state government* are opinion holders, that *is loved* and *has been criticized* are opinion concepts, and that *The mayor* is a topic (or subject) of the opinion. This task is proposed for the second time in the Semantic Sentiment Analysis challenge. Details about the winner of last year can be found in [19].

Figure 5 shows an example of annotation for the sentence: *"The mayor is loved by the people in the city, but he has been criticized by the state government."*, including the two opinion frames that a system should be able to identify.

3.5   Task 5: Implicit Opinions Related to Verbnet Verbs and Roles

A human would easily understand that the people referred to by the sentence *People hope that the President will resign* have a rather negative opinion on *the President* because they envision his/her resignation. This simple sentence however lacks of terms explicitly indicating a positive or negative opinion, e.g. about the President, making it hard for a NLP-based tool to catch it. However, the term *hope* evokes a positive attitude towards what is referred to by the subordinate proposition *the President will resign*. This means that *people* refers to the holder of a positive opinion about a possible resign event (i.e., main topic) whose agent is *the President* (i.e. a subtopic). Intuitively, a subtopic is an entity that is indirectly targeted by an opinion sentence. In this case the opinion holder indirectly expresses an opinion on *the President*, while it directly expresses an

```
<?xml version="1.0" encoding="UTF-8" standalone="yes"?>
<Sentences>
    <sentence id="348:0">
        <text>The mayor is loved by the people in the city,
        but he has been criticized by the state government.
        </text>
        <Frames>
            <Frame>
                <holder start="22" end="32" value="the people"/>
                <topic start="0" end="9" value="The mayor"/>
                <opinion start="10" end="18" value="is loved"/>
                <polarity>positive</polarity>
            </Frame>
            <Frame>
                <holder start="76" end="96" value="the state government"/>
                <topic start="0" end="9" value="The mayor"/>
                <opinion start="53" end="72" value="has been criticized"/>
                <polarity>negative</polarity>
            </Frame>
        </Frames>
    </sentence>
</Sentences>
```

Fig. 5. Task 4 annotated sentence example.

opinion on a resign event. Being a resignation a generally negative event for its agent, a positive judgement of it implies a negative one on its agent. In this task a list of VerbNet verbs roles were annotated and the proposed systems had to take into account the annotated resource in order to answers to sentences as the ones mentioned at the beginning of the section. Basically each verb's role should have an annotation (positive, negative or neutral) indicating whether that role can be affected by an opinion on that verb. Challengers might have a look at a similar research paper [5] where a resources called *Sentilonet* and a tool Sentilo have been developed for the same purpose. The expected output should be a polarity value (positive, negative) on detected VerbNet roles of identified verbs in the sentences included in the list of selected verbs. An example output format for such a task is shown in Fig. 6.

```
<?xml version="1.0" encoding="UTF-8" standalone="yes"?>
<Sentences>
    <sentence id="348:0">
        <text>Tom is happy that the President and the VicePresident were condemned.</text>
        <Sentilonet>
            <opinion targetverb="were condemned" verbrole="Theme"
                     roleobject="President" polarity="negative"/>
            <opinion targetverb="were condemned" verbrole="Theme"
                 roleobject="VicePresident" polarity="negative"/>
        </Sentilonet>
    </sentence>
</Sentences>
```

Fig. 6. Task 5 annotated sentence example.

3.6   The Most Innovative Approach Task

This task aimed at awarding the most innovative system that in this context was identified based on a number of criteria: the use of common-sense knowledge, how the semantics was applied, the computational time, the number of features that was possible to query, the usability of the system, the appealing of the user interface, and the innovative nature of the approach, including multi-language capabilities.

4   Submitted Systems

At the time of the call for participation, we received 7 expressions of interest to submission to the third edition of the Fine-Grained Sentiment Analysis challenge. The challenge chairs used the Google group mentioned above as a forum where interested people asked questions and followed discussions before the submission deadline about the requirements that needed to be satisfied and to check whether the semantics that they were employing in their systems was fine. One submitted system used as core engine a classical statistical method for sentiment analysis without any employment of semantic web capabilities and, therefore, it was discouraged to apply. The timing problem we faced during the first two editions of the challenge has been finally solved. In fact, we used the Google group to post, much ahead of time, all the details related to the challenge thus letting potential authors know about the challenge, its tasks and the different deadlines. Therefore, authors had the time to finalise their systems and submit their proposal according to the deadlines of the challenge. We did not experience any delays during the submission phase. Table 1 shows the details (title, authors, tasks participating into) of the submitted systems.

During the ESWC conference all the participants had the opportunity to present a poster and a demo of their systems within a dedicated session, which was aimed at fostering brainstorming, research and network activities.

5   Results

A week before the ESWC conference, the two evaluation datasets (including only the sentences), one for Task 1 and the other for Task 2, were published. Participants had to run their systems and send to the challenge chairs their results by the next two days. Computing the accuracy was pretty straightforward as accuracy scripts were already prepared and available to download within the website of the past editions of the challenge. In the following, we will show the results of the participants' systems.

5.1   Task 1

In Table 2 we show the precision-recall analysis of the four systems competing for Task 1. The system of *Efstratios Sygkounas, Xianglei Li, Giuseppe Rizzo and Raphaël Troncy* had the best f-measure and, therefore, was awarded with a Springer voucher of the value of 150 euros, as the winner of the task.

Table 1. The systems participating at the third edition of the Fine-Grained Sentiment Analysis challenge and the tasks they addressed.

System	Task 1	Task 2	Most Inn. Approach
Emanuele Di Rosa and Alberto Durante **App2Check extension for Sentiment Analysis of Amazon Products Reviews** [20]	X		X
Marco Federici and Mauro Dragoni **A Knowledge-based Approach For Aspect-Based Opinion Mining** [21]		X	X
Efstratios Sygkounas, Xianglei Li, Giuseppe Rizzo and Raphael Troncy **The SentiME System at the SSA Challenge Task 1** [22]	X		X
Soufian Jebbara and Philipp Cimiano **Aspect-Based Sentiment Analysis Using a Two-Step Neural Network Architectures** [23]		X	X
Andi Rexha, Mark Kröll, Mauro Dragoni and Roman Kern **Exploiting Propositions for Opinion Mining** [24]	X	X	X
Giulio Petrucci and Mauro Dragoni **An Information Retrieval-based System For Multi-Domain Sentiment Analysis** [25]	X		X

5.2   Task 2

Table 3 shows the precision-recall analysis for the system competing for Task 2. The system presented by *Soufian Jebbara and Philipp Cimiano* obtained the best performance in both the extraction of aspects and in computing the associated polarity. Hence, it was awarded with a Springer voucher of the value of 150 euros.

5.3   The Most Innovative Approach Task

The Innovation Prize, consisting of a Springer voucher of 150 euros, was awarded to *Soufian Jebbara and Philipp Cimiano* with their presented contribution "Aspect-Based Sentiment Analysis Using a Two-Step Neural Network Architecture". The system combines, for the first time, the deep learning paradigm and semantic resources (SenticNet and WordNet) for extracting aspects from sentences and for inferring the polarity associated with each aspect. The polarity has been inferred by aggregating the values of opinion words associated with each aspect.

Table 2. Precision-recall analysis and winners for Task 1.

System	Precision	Recall	F-measure
Efstratios Sygkounas, Xianglei Li *Giuseppe Rizzo and Raphaël Troncy* **The SentiME System at the SSA Challenge Task 1** [22]	0.85686	0.90541	0.88046
Emanuele Di Rosa and Alberto Durante **App2Check extension for Sentiment Analysis of Amazon Products Reviews** [20]	0.82777	0.90789	0.87142
Giulio Petrucci and Mauro Dragoni **An Information Retrieval-based System For Multi-Domain Sentiment Analysis** [25]	0.81837	0.89198	0.85359
Andi Rexha, Mark Kröll *Mauro Dragoni and Roman Kern* **Exploiting Propositions for Opinion Mining** [24]	0.50494	0.81665	0.62403

6 Conclusions

Following the success of the first two editions, the third edition of the Fine-Grained Sentiment Analysis challenge attracted researchers and industrials from both the semantic web community and the traditional sentiment analysis research area. Researchers belonging to the former are usually new to the Sentiment Analysis. Researchers belonging to the latter have investigated new opportunities provided by the Semantic Web world and have adapted their systems with Semantic Web best practices and technologies.

Following the positive example of the first edition, we coupled the challenge with a dedicated Workshop in order for the two events to benefit from each other. The workshop was a full day event attracting around 20 people and included four presentations and an invited talk in the morning session and participants from the challenge were invited to describe their systems during the afternoon session.

We noticed that in the current edition, only the first two tasks were targeted by the participants. We believed that the other three tasks were too peculiar and therefore did not attract submissions. One idea we will apply in the future editions will be to reduce the number of tasks and define a number of subtasks (similar to the main task with minor differences involving the language, continuous polarity scoring vs. binary polarity scoring, etc.).

We will keep the precision-recall script together with the annotated test set in the future proposals so that participants can test their results and the evaluation script we will be using to select the winners. A further evolution concerning evaluation will be the introduction of the confusion matrix for validating the participant systems instead of a single precision-recall measure.

Table 3. Precision-recall analysis and winners for Task 2.

System	Precision	Recall	F-mesure	Accuracy
Soufian Jebbara and Philipp Cimiano **Aspect-Based Sentiment Analysis Using** **a Two-Step Neural Network Architectures** [23]	0.41471	0.45196	0.43253	0.87356
Marco Federici and Mauro Dragoni **A Knowledge-based Approach** **For Aspect-Based Opinion Mining** [21]	0.34820	0.35745	0.35276	0.84925
Andi Rexha, Mark Kröll *Mauro Dragoni and Roman Kern* **Exploiting Propositions for Opinion Mining** [24]	N/A	N/A	N/A	N/A[a]

[a] Disqualified for sending output results after the deadline.

One more winning action has been the exploitation of the Google group that reached more than 150 participants and it is still growing. This filled the need of dissemination and promotion actions we had faced during the last editions of the challenge.

It is intention of the challenge chairs to propose again the challenge and the workshop as a dual event and to keep stimulating the development, testing, and competitions of systems for semantic sentiment analysis.

Acknowledgement. Challenge Organizers want to thank Springer for supporting the provided awards also for this year edition. Moreover, the research leading to these results has received funding from the European Union Horizons 2020 the Framework Programme for Research and Innovation (2014–2020) under grant agreement 643808 Project MARIO Managing active and healthy aging with use of caring service robots.

References

1. Subrahmanian, V.S., Reforgiato, D.: AVA: adjective-verb-adverb combinations for sentiment analysis. IEEE Intell. Syst. **23**, 43–50 (2008)
2. Benamara, F., Cesarano, C., Picariello, A., Reforgiato, D., Subrahmanian, V.S.: Sentiment analysis: adjectives and adverbs are better than adjectives alone. In: Proceedings of the International Conference on Weblogs and Social Media (ICWSM), Short paper (2007)
3. Gan tzandetal, J.: The expanding digital universe: a forecast of world wide information growth through, 2007 (2010)
4. Petrucci, G., Dragoni, M.: An information retrieval-based system for multi-domain sentiment analysis. In: Gandon, F., et al. (eds.) SemWebEval 2015. CCIS, vol. 548, pp. 234–243. Springer, Heidelberg (2015). doi:10.1007/978-3-319-25518-7_20. Revised Selected Papers

5. Recupero, D.R., Presutti, V., Consoli, S., Gangemi, A., Nuzzolese, A.: Sentilo: frame-based sentiment analysis. Cogn. Comput. **7**(2), 211–225 (2014)
6. Consoli, S., Gangemi, A., Nuzzolese, A.G., Reforgiato Recupero, D., Spampinato, D.: Extraction of topics-events semantic relationships for opinion propagation in sentiment analysis. In: Proceedings of Extended Semantic Web Conference (ESWC), Crete, GR (2014)
7. Gangemi, A., Presutti, V., Reforgiato Recupero, D.: Frame-based detection of opinion holders, topics: a model and a tool. IEEE Comput. Intell. Mag. **9**(1), 20–30 (2014)
8. Dragoni, M., Tettamanzi, A.G.B., Pereira, C.: A fuzzy system for concept-level sentiment analysis. In: Presutti, V., et al. (eds.) Semantic Web Evaluation Challenge. CCIS, vol. 475, pp. 21–27. Springer, Heidelberg (2014)
9. Recupero, D.R., Cambria, E.: ESWC 2014 challenge: concept-level sentiment analysis. SemWebEval@ESWC 2014, pp. 3–20, May 2014. http://challenges.2014. eswc-conferences.org/index.php/SemSA
10. Recupero, D.R., Dragoni, M., Presutti, V.: ESWC15 challenge on concept-level sentiment analysis. SemWebEval@ESWC (2011) Observation of Strains, pp. 211–222, May 2015
11. Presutti, V., et al. (eds.): Semantic Web Evaluation Challenge. CCIS, vol. 475. Springer, Heidelberg (2014)
12. Gandon, F., Cabrio, E., Stankovic, M., Zimmermann, A.: Semantic Web Evaluation Challenges. Second SemWebEval Challenge at ESWC, Portoroz, Slovenia, May 31-June 4, Revised Selected Papers. Springer (2015)
13. Gangemi, A., Alani, H., Nissim, M., Cambria, E., Recupero, D.R., Lanfranchi, V., Kauppinen, T.: Joint Proceedings of the 1th Workshop on Semantic Sentiment Analysis (SSA2014), and the Workshop on Social Media and Linked Data for Emergency Response (SMILE 2014), Co-located with 11th European Semantic Web Conference (ESWC 2014), 25 May 2014, Crete, Greece (2014). http://ceur-ws. org/Vol-1329/
14. Cambria, E., Olsher, D., Rajagopal, D.: Senticnet 3: a common and common-sense knowledge base for cognition-driven sentiment analysis. In: Brodley, C.E., Stone, P. (eds.) Twenty-Eight AAAI Conference on Artificial Intelligence, pp. 1515–1521. AAAI Press, Palo Alto, July 2014
15. Dragoni, M.: SHELLFBK: an information retrieval-based system for multi-domain sentiment analysis. In: Proceedings of the 9th International Workshop on Semantic Evaluation, SemEval 2015, pp. 502–509. Association for Computational Linguistics, Denver, June 2015
16. Dragoni, M., Tettamanzi, A.G.B., da Costa Pereira, C.: Propagating and aggregating fuzzy polarities for concept-level sentiment analysis. Cogn. Comput. **7**(2), 186–197 (2015)
17. Dragoni, M., Tettamanzi, A., Pereira, C.D.C.: DRANZIERA: an evaluation protocol for multi-domain opinion mining. In: Calzolari, N. (Conference Chair), Choukri, K., Declerck, T., Goggi, S., Grobelnik, M., Maegaard, B., Mariani, J., Mazo, H., Moreno, A., Odijk, J., Piperidis, S. (eds) Proceedings of the Tenth International Conference on Language Resources and Evaluation (LREC 2016), Paris, France. European Language Resources Association (ELRA), May 2016
18. Liu, B.: Sentiment Analysis and Opinion Mining. Synthesis Lectures on Human Language Technologies. Morgan & Claypool Publishers, San Rafael (2012)

19. Aprosio, A.P., Corcoglioniti, F., Dragoni, M., Rospocher, M.: Supervised opinion frames detection with RAID. In: Gandon, F., et al. (eds.) SemWebEval 2015. CCIS, vol. 548, pp. 251–263. Springer, Heidelberg (2015). doi:10.1007/978-3-319-25518-7_22. Revised Selected Papers

20. Rosa, E.D., Durante, A.: App2check extension for sentiment analysis of amazon products reviews. In: Sack et al. [26], pp. 95–107

21. Federici, M., Dragoni, M.: A knowledge-based approach for aspect-based opinion mining. In: Sack et al. [26], pp. 141–152

22. Sygkounas, E., Li, X., Rizzo, G., Troncy, R.: Sentiment polarity detection from amazon reviews: an experimental study. In: Sack et al. [26], pp. 108–120

23. Jebbara, S., Cimiano, P.: Aspect-based sentiment analysis using a two-step neural network architectures. In: Sack et al. [26], pp. 153–167

24. Rexha, A., Kröll, M., Dragoni, M., Kerns, R.: Exploiting propositions for opinion mining. In: Sack et al. [26], pp. 121–125

25. Petrucci, G., Dragoni, M.: The IRMUDOSA system at ESWC challenge on semantic sentiment analysis. In: Sack et al. [26], pp. 126–140

26. Sack, H., Dietze, S., Tordai, A., Lange, C. (eds.): SemWebEval 2016. CCIS, vol. 641. Springer, Heidelberg (2016)

App2Check Extension for Sentiment Analysis of Amazon Products Reviews

Emanuele Di Rosa$^{(\boxtimes)}$ and Alberto Durante

Machine Learning and Semantic Analysis Department,
Finsa s.p.a., Genova, Italy
{emanuele.dirosa,alberto.durante}@finsa.it

Abstract. App2Check is a web application and an engine for opinion mining applied to user comments evaluating apps published in app stores. It includes features ranging from topic extraction, sentiment analysis of user reviews and topics, sentiment vs rating chronological trend, sentiment trend comparison between competitors, and many others. App2Check goal is to help app owners and makers to evaluate in real time their own apps, compare them with the apps available in the market, and extract from this analysis useful insights to perform a continuous improvement during both design and maintenance process. In this paper we describe App2Check features, by focusing in particular on the ones applying semantic and sentiment analysis to apps reviews, and we present an experimental comparison respect to 19 research tools. Then we show App2Check performance when applied to Amazon products reviews. In this experimental evaluation, we show App2Check performance with and without a specific training on Amazon products reviews, and we compare our results with two state-of-the-art research tools.

Keywords: Sentiment analysis · Machine learning · App2Check · Tweet2Check · Amazon products reviews

1 Introduction

Sentiment Analysis has nowadays a crucial role in social media analysis and, more generally, in analyzing user opinions about general topics or user reviews about product/services, enabling a huge number of applications. For instance, sentiment analysis can be applied to monitoring the reputation or opinion of a company or a brand with the analysis of reviews of consumer products or services [8]. App stores can be seen as another, not yet well explored, field of application of sentiment analysis. Indeed, they are another social media where users can freely express their own opinion through app reviews about a product, i.e. the specific app under evaluation, or a service, to which the considered app is connecting the user (e.g., a mobile banking app connects users to mobile banking services). In addition, reading user reviews on app stores shows that people frequently talk about and evaluate also the brand associated to the app under review: thus, it is possible to extract people opinion about a brand or

© Springer International Publishing Switzerland 2016
H. Sack et al. (Eds.): SemWebEval 2016, CCIS 641, pp. 95–107, 2016.
DOI: 10.1007/978-3-319-46565-4_7

the sentiment about a company or the provided service quality. App2Check [5] is an innovative web application implementing machine learning and semantic/sentiment analysis algorithms in its core engine in order to analyze unstructured information available within user comments related to apps, with the aim to enable a real time monitoring/evaluation of such apps and make possible a comparison with apps available in the market and, finally, performing a continuous improvement of the app under evaluation. Opinion mining is thus the main problem targeted by the underlying engine, involving sentiment evaluation of sentences, topic extraction and sentiment quantification of the extracted topics. In this paper, we describe App2Check web application main features allowing to evaluate in a quantitative way apps published in the main app stores, and show a brief experimental evaluation of its sentiment engine with respect to 19 state-of-the-art research tools. The innovative aspects of our tool are both related to the evaluation of apps through the extracted main topics and their sentiment evaluation, and to the internal sentiment engine showing that its performance is always the best for Accuracy and Macro-F1 compared to 19 state-of-the-art research tools.

App2Check provides the following main features:

1. App search in the main app stores (Apple App Store, Google Play Store and Microsoft Marketplace) and their adding to the system to keep them monitored.
2. App Topics Extraction and their overall sentiment evaluation, constituting general app pros and cons.
3. Chronological Sentiment vs Rating plot
4. All comments related to a specific topic
5. Comment-level Semantic Analysis: topics extraction (and highlighting), comment and sub-comments sentiment quantification
6. Overall most recurrent app topics
7. Overall most interesting app topics from sentiment vs rating perspective
8. Main topics per month and their sentiment evaluation
9. Sentiment/Rating comparison with competitors
10. Continuous learning and improvement
11. App Statistics Comparison

The ones shown from 2 to 10 are features related to semantic/sentiment analysis and will be described in a deeper way in the following section, while features 1 and 11 are more related to structured data and will not be shown here.

In this paper we also show App2Check performance when applied to a different kind of users reviews: Amazon products reviews. In this specific experimental evaluation, we show App2Check performance with and without a specific training on Amazon products reviews, and we compare its results with two state-of-the-art research tools.

This paper is structured as follows: in Sect. 2 we describe App2Check semantic-related features (mainly sentiment analysis and topic extraction), in Sect. 3 we present an experimental evaluation on apps reviews on both Italian and English and an evaluation on Amazon products reviews; then we provide paper conclusions in Sect. 4.

2   Sentiment Analysis and Topic Extraction

App2Check implements supervised learning techniques that allowed us to create a predictive model for sentiment quantification specialized on apps reviews. Training of predictive models is performed by considering a huge variety of language domains and different kinds of user reviews. This model provides, as answer to a sentence in Italian or English language, a quantification of the sentiment polarity scored from 1 to 5, according to the most recent trend shown in the last sentiment evaluation SemEval [10], where tracks considering quantification have been introduced. Thus, we consider the following quantification: as positive, sentences with score 4 (positive) or 5 (very positive); as negative, sentences with score 1 (very negative) or 2 (negative); as neutral, sentences with score 3.

Semantic Resources. App2Check also takes into account some semantic resources and, in particular, some WordNet [9] features like parts of speech tagging are used. They are applied mostly as a support in the creation of custom lexical resources that we build and use before the training process, together with other natural language processing techniques. We choose to dynamically generate adaptable lexicons instead of using state-of-the-art lexical resources because we believe that, at least in the case of predicting sentiment and extracting topics from text, a good lexical resource should be adapted to the specific domain to reduce the need for disambiguation: e.g. looking for synsets for the verb "execute" with WordNet produces seven different results, including "murder in a planned fashion" and "run a program", which contain different sentiment. By using a domain-specific resource which has been dynamically adapted to the context, it is possible to find and apply only one of these meanings. Similar examples can be found on topic extraction. It is not possible to give more details on the engine due to non-disclosure restrictions.

2.1   General App Pros and Cons

In Fig. 1 we show the main app pros and cons which are reported as the best and worst app topics. In particular, these are recurrent topics with positive and negative average sentiment quantification referred, respectively, to pros and cons. More positive and negative topics are available clicking the "Read all" button, under the tables. Such general pros and cons are calculated by averaging the sentiment score of each topic occurrence in all the sub-sentences (i.e. portions of sentences as presented in Sect. 2.4 and shown in Fig. 3 at the green underlined sub-sentence - fourth row in the table).

Fig. 1. Home page for Google+: Pros, Cons and main statistics.

2.2   Chronological Sentiment vs Rating Plot

In Fig. 2 we show a plot reporting two time-series, showing for each month the average App Rating and app-level Sentiment score.

The plot is divided in three areas of the same width, so that it is easier to understand whether a point in the graph is positive, neutral or negative.

This graph is a visual aid to understand how customers reacted to changes in the app development through time. Knowing the dates and the content of releases could help understand which features the customers liked most or less. Moreover, clicking on a point of the series it is possible to focus on the topics on the selected month (see Sect. 2.7 for further details).

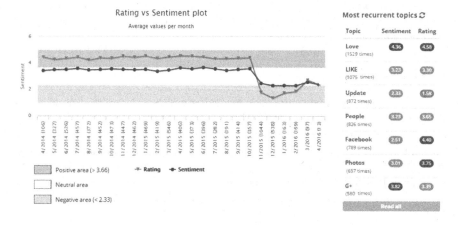

Fig. 2. Sentiment vs Rating plot *(left)* and Most recurrent topics *(right)* for Google+.

2.3 All Comments Related to a Specific Topic

In Fig. 3 we show an example of the list of comments associated to keyword "works". This is a useful feature to quickly investigate the reason of the user sentiment for a specified keyword/topic.

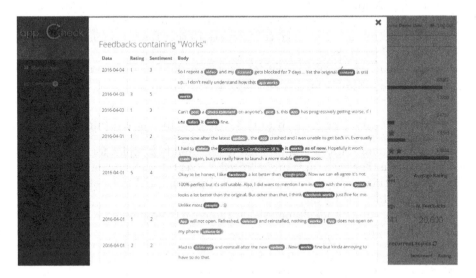

Fig. 3. Detailed sentiment analysis and topic extraction on the reviews containing the topic "Works" for Facebook. (Color figure online)

2.4 Comment-Level Semantic Analysis

In Fig. 3 it is shown a more detailed analysis of user reviews. In particular, for each document both user rating and document-level sentiment quantification are reported. The topics contained in each review are highlighted and colored depending on their relative sentiment. Reviews have been split in sub-sentences/comments, and for each sub-sentence the sentiment score is obtained. It is possible to see the sentiment score and the confidence of each phrase moving the mouse over it, as it will be underlined and a dialogue box will appear. We can see that the sentiment score referred to a sub-sentence can in general differ from the whole document score.

2.5 Overall Most Recurrent App Topics

The table in Fig. 3 reports the topics mentioned most times by the users.

For each topic, both the overall sentiment score and rating average are present. As it can be seen in Fig. 3, a topic can have a positive sentiment score and a neutral rating average, or a negative sentiment and a positive rating, and so on. These differences are the main reason because it is important to evaluate sentiment on a topic instead of only considering the score a user assign. More topics can be seen clicking on the "Read all" button under the table.

2.6 Overall Most Interesting App Topics from Sentiment vs Rating Perspective

An additional table, that it is not present in the pictures, contains the most interesting topic from the point of view of the differences between app rating and sentiment score. The main reason because we choose to add this table is to highlight the topics that users still find positive when the overall evaluation of the app is negative, or, vice-versa, the weaknesses that the users found in an app otherwise considered good.

In particular, we highlight the topics whose difference between sentiment score and average app rating is more than 2.

2.7 Main Topics Per Month and Their Sentiment Evaluation

Clicking on a point on the graph, a list of the trending topics for the selected month is shown. For each topic are reported the sentiment score on that month, the average rating of the reviews in which it is present, and the number of times it is mentioned by the users.

As usual, a more detailed analysis on the reviews containing a topic can be seen clicking on it.

2.8 Continuous Learning and Improvement

Since sentiment quantification on short sentences can be sometimes difficult to correctly perform, App2Check offers a feature through which is possible to click on a sub-sentence to update its sentiment score. In such a way App2Check predictive model is automatically updated and can continuously improve along time.

This feature also allows us to extend our model on a specific domain, in order to reach a higher accuracy, to fulfill customer's demands. In this cases, our tool – and in general – machine learning-based approaches – allow to meet this important goal by performing learning on a subset of sentences of the required domain, while different approaches with no learning feature cannot reach it or anyway reach it with more effort.

For an example of how much our model can improve, see Subsect. 3.2.

2.9 Sentiment/Rating Comparison with Competitors

App2Check allows also to compare multiple selected apps in a plot of monthly averaged sentiment and rating. In Fig. 4 we show a plot comparing Facebook and Google+ apps, allowing to evaluate their relative performance along time. It is easy to see that Google+ app was better than Facebook app in 2015/10 and with rating higher than sentiment, but at the end of the plot they are evaluated almost in the same – quite negative – way by the users.

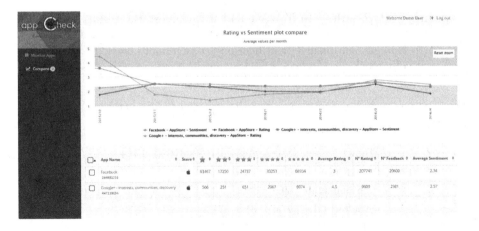

Fig. 4. Comparing Facebook and Google+ apps respect to Sentiment and Rating.

3   Experimental Evaluation

In order to give an brief insight about App2Check sentiment analysis performance (a deeper analysis has been shown in [5] and more papers under review process), in Subsect. 3.1 we compared it with 19 state-of-the-art research tools using iFeel [2,3], a research web platform allowing to run many sentiment analysis tools simultaneously. As per testset we considered 4000 user reviews related to the Candy Crush Saga and Gmail apps for both their Italian and English (US) version: we used 1000 (most recent) user comments per app per language. Our benchmark testset is released with the paper to make the results repeatable. In Subsect. 3.2 we compare App2Check performance on a subset of the dataset made available for this challenge, composed by reviews from Amazon in English language, with the performance of two of the most famous research tools for sentiment analysis: SentiStrength [13] and Stanford Deep Learning [12].

The following tables show columns macro F1 (MF1), accuracy (Acc), and F1 on each classes (resp. F1($-$) for negative, and F1($+$) for positive), and are sorted by macro F1, considered as main evaluation index for the tools.

3.1   Evaluation on App Reviews

iFeel is a research web platform [2] allowing to run 19 state-of-the art research tools for sentiment analysis on the specified list of sentences. It allows to natively run tools supporting English and to first translate sentences from other languages into English and then run the underlying tools on the English translated sentences. It has been experimentally shown in [3] that well known language specific methods do not have a significant advantage over a simple machine translation approach. The tools included in iFeel are the ones shown in the following tables.

Experimental Results. In Tables 1 and 2 we report the results related to Candy Crush Saga app for both English and Italian, respectively, evaluated with respect to Human Sentiment Quantification (i.e. manually labeled sentences by a trained volunteer).

Table 1. Comparison on 1000 app reviews in English from Candy Crush Saga with respect to Human Sentiment Quantification.

Tool	M-F1	Acc	F1(−)	F1(x)	F1(+)
App2Check	**62.0**	**74.6**	**82.1**	23.3	**80.6**
Sentiment140	59.9	70.1	77.2	29.7	73.0
Stanford DL	57.6	61.9	71.2	33.3	68.3
NRC Hashtag	56.8	64.9	75.6	31.7	63.2
Umigon	56.4	59.9	59.1	33.3	76.8
Op. Lexicon	53.0	55.6	54.7	**36.5**	67.7
AFINN	52.8	57.6	54.1	33.6	70.9
SentiWordNet	52.5	58.9	63.4	31.5	62.5
SentiStrength	46.3	49.9	44.0	29.1	65.9
CogitoAPI	45.9	46.9	53.0	28.7	55.9
SO-CAL	43.1	44.5	50.4	27.2	51.6
Emolex	41.3	43.1	38.0	32.6	53.4
Vader	40.4	43.0	24.8	31.8	64.7
SASA	40.0	44.4	40.0	19.2	60.7
Senticnet	38.1	43.0	32.9	29.0	52.4
H. Index	33.9	40.1	19.7	26.4	55.7
Op. Finder	27.9	27.7	29.5	26.3	28.0
SANN	19.2	22.8	0.9	29.9	26.8
Panas-t	11.1	17.7	3.6	29.1	0.5
Emoticon DS	0.0	39.9	0.0	15.8	56.2
Emoticons	0.0	16.8	0.0	28.8	0.0

Since rating and Human Sentiment Quantification (HSQ) agreed in about 80 % of cases, in Tables 3 and 4 we compared tools with respect to app rating, which can be seen as an approximation of HSQ; in this case we changed the app domain by taking into account app with user reviews about an utility app (Gmail) instead of a game. App2Check is the best tool for both Italian and English, even if it shows a better performance for Italian where the other tools receive a translated version of the comments as in [3], since they officially do not manage Italian.

Table 2. Comparison on 1000 app reviews in Italian from Candy Crush Saga with respect to Human Sentiment Quantification.

Tool	M-F1	Acc	F1(−)	F1(x)	F1(+)
App2Check	**65.8**	**81.8**	**85.9**	25.2	**86.4**
Sentiment140	58.1	71.6	80.4	21.4	72.5
SentiWordNet	57.2	67.3	73.5	**26.6**	71.4
Stanford DL	53.7	60.5	70.0	21.2	69.9
NRC Hashtag	52.9	65.5	76.4	16.4	66.0
Umigon	50.9	56.4	54.0	20.5	78.1
SentiStrength	50.4	57.8	47.5	25.6	78.0
Op. Lexicon	49.8	54.2	53.1	23.7	72.6
AFINN	49.7	57.4	52.4	21.8	74.9
Vader	40.9	42.8	31.9	22.8	68.1
Senticnet	40.2	50.6	37.4	19.7	63.6
Emolex	40.0	43.3	43.1	16.1	60.9
SASA	39.3	44.3	40.8	15.8	61.3
SO-CAL	39.0	40.8	45.2	17.0	54.8
H. Index	32.3	39.8	16.9	18.6	61.4
Op. Finder	23.6	22.9	24.9	18.1	27.9
Emoticon DS	21.5	41.6	1.8	4.1	58.7
SANN	16.5	17.8	1.8	19.9	27.7
Panas-t	7.4	11.0	1.3	18.7	2.2
Emoticons	-	10.3	-	18.6	-

3.2 Evaluation on Amazon Reviews

In order to show the power of the continuous learning of our models, we tested our model on the 20 % of the ESWC 2016 Semantic Sentiment Analysis Challenge dataset [1,4,6,7,11] before and after updating it with Amazon reviews, and making in this way also an App2Check extended version that is specific for such kind of reviews. We report the results in Table 5, together with the reference results obtained by Stanford Deep Learning, the best machine learning-based tool in Table 1, and SentiStrength, which uses a different approach described in the following, on the same testset.

SentiStrength [13] was produced as part of the CyberEmotions project, supported by EU FP7. It estimates the strength of positive and negative sentiment in short texts, even for informal language. According to the authors, it has human-level accuracy for short social web texts in English, except political texts. SentiStrength authors make available many different versions of the tools, supporting other languages or domain-specific.

Table 3. Comparison on 1000 app reviews in English from Gmail with respect to App Rating.

Tool	M-F1	Acc	F1(−)	F1(x)	F1(+)
App2Check	**55.2**	**61.0**	**65.6**	27.7	**72.1**
Umigon	49.0	51.7	52.1	29.1	65.9
Stanford DL	46.1	51.1	62.4	22.2	53.6
Op. Lexicon	45.9	49.2	45.8	27.8	64.1
AFINN	45.3	49.4	43.2	30.2	62.4
SO-CAL	44.2	47.0	51.6	24.2	56.7
SentiStrength	43.5	46.0	40.0	30.2	60.4
NRC Hashtag	42.2	54.2	65.1	7.7	53.8
Sentiment140	40.9	53.3	65.1	8.8	48.7
SentiWordNet	40.0	51.5	53.8	5.9	60.2
Vader	38.2	40.1	21.8	**33.7**	59.2
SANN	36.9	38.5	32.6	26.4	51.7
SASA	36.9	38.5	32.6	26.4	51.7
Emolex	35.6	36.8	34.5	25.0	47.1
H. Index	34.7	37.1	25.4	29.3	49.4
Op. Finder	34.0	33.9	32.7	27.3	42.1
senticnet	31.4	43.4	29.4	7.6	57.2
Emoticon DS	19.2	39.0	0.5	0.9	56.1
Panas-t	12.0	20.2	1.5	33.0	1.5
Emoticons	-	20.0	-	33.3	-

Stanford Deep Learning model [12] builds up a representation of whole sentences based on a type of Recursive Neural Network that builds on top of grammatical structures, so that it takes into account the order of the words in a sentence. It computes the sentiment based on how words compose the meaning of longer phrases.

Results in Table 5 show that App2Check version without any specific training for Amazon products reviews is anyway better than SentiStrength and Stanford Deep Learning, reaching a macro F1 almost 9 % higher. Its extended version, instead, which has been updated with reviews from Amazon, outperforms all of the other tools considered in this analysis, reaching 86.5 % of macro-F1 and 86.4 % of accuracy. This shows that App2Check can easily improve its performance when a new domain is introduced.

Table 4. Comparison on 1000 app reviews in Italian from Gmail with respect to App Rating.

Tool	M-F1	Acc	F1(−)	F1(x)	F1(+)
App2Check	**72.7**	**85.1**	**76.1**	**50.3**	**91.8**
SentiWordNet	43.7	63.8	50.9	2.9	77.1
SentiStrength	43.3	57.0	35.0	21.6	73.2
Stanford DL	42.3	50.3	48.6	14.3	64.1
Op. Lexicon	42.1	55.4	35.3	18.2	73.0
SO-CAL	41.2	50.5	42.7	14.0	66.9
AFINN	41.1	57.9	36.2	12.2	74.9
Sentiment140	41.1	55.3	47.0	8.0	68.2
NRC Hashtag	39.1	52.2	44.5	7.4	65.5
senticnet	38.5	66.7	30.9	3.9	80.6
SASA	37.6	50.7	24.5	20.1	68.1
Umigon	37.0	43.9	37.5	13.5	60.0
Emolex	35.5	44.0	33.5	12.5	60.7
Vader	35.4	45.7	21.4	20.3	64.5
Emoticon DS	27.8	67.9	1.1	1.3	81.0
Op. Finder	26.8	27.5	26.4	17.0	37.1
H. Index	25.1	31.5	18.6	8.7	48.0
SANN	13.8	18.5	1.1	20.2	20.2
Emoticons	-	10.6	-	19.2	-
Panas-t	-	10.6	-	19.2	-

Table 5. Comparison on Amazon reviews.

Tool	M-F1	Acc	F1(−)	F1(+)
App2check extended	86.5	86.4	86.9	86.0
App2Check	71.9	71.0	71.6	72.2
SentiStrength	63.0	55.2	56.8	69.2
Stanford	60.2	60.4	70.5	49.8

4   Conclusion

In this paper we described App2Check features, by focusing in particular on the ones applying Semantic and Sentiment Analysis to apps reviews, and we present an experimental evaluation on apps reviews and a comparison with 19 state-of-the-art research tools, showing that its performance is always the best for Accuracy and Macro-F1 for both Italian and English. Then we showed App2Check performance when applied to Amazon products reviews. In this experimental evaluation, we show App2Check performance with and without a specific training on Amazon products reviews, and we compare our results with two state-of-the-art research tools.

References

1. Palmero Aprosio, A., Corcoglioniti, F., Dragoni, M., Rospocher, M.: Supervised opinion frames detection with RAID. In: Gandon, F., et al. (eds.) SemWebEval 2015. CCIS, vol. 548, pp. 251–263. Springer, Heidelberg (2015). doi:10.1007/978-3-319-25518-7_22

2. Araújo, M., Gonçalves, P., Cha, M., Benevenuto, F.: iFeel: a system that compares and combines sentiment analysis methods. In: Proceedings of the 23rd International Conference on World Wide Web, WWW 2014 Companion, pp. 75–78. ACM, New York (2014). http://doi.acm.org/10.1145/2567948.2577013

3. Araújo, M., dos Reis, J.C., Pereira, A.M., Benevenuto, F.: An evaluation of machine translation for multilingual sentence-level sentiment analysis. In: Proceedings of the 31st Annual ACM Symposium on Applied Computing, Pisa, Italy, 4–8 April 2016, pp. 1140–1145 (2016). http://doi.acm.org/10.1145/2851613.2851817

4. Chung, J.K.-C., Wu, C.-E., Tsai, R.T.-H.: Polarity detection of online reviews using sentiment concepts: NCU IISR Team at ESWC-14 challenge on concept-level sentiment analysis. In: Presutti, V., et al. (eds.) SemWebEval 2014. CCIS, vol. 475, pp. 53–58. Springer, Heidelberg (2014)

5. Di Rosa, E., Durante, A.: App2check: a machine learning-based system for sentiment analysis of app reviews in Italian language. In: Proceedings of the International Workshop on Social Media World Sensors (Sideways)- Held in conjunction with LREC 2016, pp. 8–11 (2016). http://www.lrec-conf.org/proceedings/lrec2016/workshops/LREC2016Workshop-Sideways_Proceedings.pdf

6. Dragoni, M., Tettamanzi, A., da Costa Pereira, C.: Dranziera: an evaluation protocol for multi-domain opinion mining. In: Chair, N.C.C., Choukri, K., Declerck, T., Goggi, S., Grobelnik, M., Maegaard, B., Mariani, J., Mazo, H., Moreno, A., Odijk, J., Piperidis, S. (eds.) Proceedings of the Tenth International Conference on Language Resources and Evaluation (LREC 2016), European Language Resources Association (ELRA), Paris, France, May 2016

7. Dragoni, M., Tettamanzi, A.G.B., da Costa Pereira, C.: A fuzzy system for concept-level sentiment analysis. In: Presutti, V., et al. (eds.) SemWebEval 2014. CCIS, vol. 475, pp. 21–27. Springer, Heidelberg (2014)

8. Hu, M., Liu, B.: Mining and summarizing customer reviews. In: KDD, pp. 168–177. ACM (2004)

9. Miller, G.A.: Wordnet: a lexical database for English. Commun. ACM $38(11)$, 39–41 (1995). http://doi.acm.org/10.1145/219717.219748

10. Nakov, P., Ritter, A., Sara, R., Sebastiani, F., Stoyanov, V.: Semeval-2016 task 4: sentiment analysis in Twitter. In: Proceedings of the 10th International Workshop on Semantic Evaluation, Association for Computational Linguistics (2016). http://alt.qcri.org/semeval2016/task4/

11. Schouten, K., Frasincar, F.: The benefit of concept-based features for sentiment analysis. In: Gandon, F., et al. (eds.) SemWebEval 2015. CCIS, vol. 548, pp. 223–233. Springer, Heidelberg (2015). doi:10.1007/978-3-319-25518-7_19

12. Socher, R., Perelygin, A., Wu, J., Chuang, J., Manning, C.D., Ng, A.Y., Potts, C.: Recursive deep models for semantic compositionality over a sentiment treebank. In: Proceedings of the 2013 Conference on Empirical Methods in Natural Language Processing, pp. 1631–1642. Association for Computational Linguistics, Stroudsburg, October 2013
13. Thelwall, M., Buckley, K., Paltoglou, G., Cai, D., Kappas, A.: Sentiment strength detection in short informal text. JASIST **61**(12), 2544–2558 (2010). http://dx.doi.org/10.1002/asi.21416

Sentiment Polarity Detection from Amazon Reviews: An Experimental Study

Efstratios Sygkounas[1], Giuseppe Rizzo[2], and Raphaël Troncy[1(✉)]

[1] EURECOM, Sophia Antipolis, France
{efstratios.sygkounas,raphael.troncy}@eurecom.fr
[2] ISMB, Turin, Italy
giuseppe.rizzo@ismb.it

Abstract. With the ever increasing number of electronic commerce portals and the selling of goods on the Web, customers' reviews are usually used as means to grasp the goodness of products. Mining and understanding the polarity of reviews is therefore crucially important for future customers that seek opinions and sentiments to support their decision buying process. This paper proposes an experimental study of SentiME, our approach for extracting the sentiment polarity of a message, on the Amazon review-based corpus provided by the ESWC SSA challenge. We use an Ensemble Learning algorithm implementing five state-of-the-art classifiers that are known to well perform in the domains of tweets and movie reviews. The Ensemble Learning is trained with a Bootstrapping Aggregating process using a set of linguistic (such as ngrams), and semantics (such as dictionary-based of polarity values for emojis) features. The approach presented in this paper has been first successfully tested on the SemEval Twitter-based corpora. It has then been tested in the ESWC Semantic Sentiment Analysis 2016 challenge, where properly trained, it reaches a F-measure of 88.05 % over the test set for the detection of positive and negative polarity, which ranks our approach as the first system among the ones competing in this challenge.

1 Introduction

Nowadays, people frequently purchase goods on the Web. One of the most popular web site where goods are bought is Amazon. After a purchase, the buyer is generally invited to publish a review about the product. These reviews are useful for future customers that seek opinions and sentiments to support them in their decision buying process.

Sentiment analysis has already been widely successfully applied on tweets. For instance, the SemEval Task 10 [10] is a competition where participants must classify the message polarity of tweets among three classes (positive, neutral, negative). Numerous systems have been proposed over the series of the SemEval Sentiment Analysis challenges. We have decided to further study and improve the performance of the best performing system named Webis [5], which is based on an ensemble of 4 classifiers and that won the 2015 SemEval Sentiment Analysis Task

© Springer International Publishing Switzerland 2016
H. Sack et al. (Eds.): SemWebEval 2016, CCIS 641, pp. 108–120, 2016.
DOI: 10.1007/978-3-319-46565-4_8

on polarity detection. Our improvements include the usage of a different training method relying on the Bootstrap Aggregating algorithm and the integration in the model of an additional state-of-the-art classifier. As outcome, we observed an improvement in F-measure score of 1 % for the general tweet corpus and of 6.5 % for the sarcasm corpus of SemEval 2015 [13].

Inspired by these results, we apply a similar approach to tackle the first task of the ESWC Semantic Sentiment Analysis (SSA) 2016 challenge, which is about binary polarity detection of customer reviews coming from Amazon. A participant system should take as input an XML document containing textual comments within the text tag, and should generate another XML document that adds, for each sentence, the result of the classification enclosed in a polarity tag. Precision, recall and F-measure of the detected polarity values (positive or negative) are the metrics used for evaluating the participant system on each review of the evaluation dataset. Our experimental study showed the robustness of our approach in this domain and we observed the importance of the ensemble learning and how all classifiers of our model complement each other with the broad variety of the features they implement. We also learned that the bagging algorithm introduced in our system improves the training process, avoiding over-fitting due to the random creation when sampling the training dataset. Concerning the limitations of our system, we encounter scalability problems which occurred when reviews are long (in terms of characters).

The remainder of this paper is organized as follows. In Sect. 2, we present some related work in the field of sentiment polarity and opinion mining across different domains such as Twitter and movie reviews. We then detail our approach implemented in the SentiME system (Sect. 3). We describe the corpus proposed by the SSA 2016 challenge providing some useful statistics (Sect. 4). We discuss the experimental study we carried out and the results that our approach has achieved (Sect. 5). We present some limitations and lessons learned in Sect. 6 and we conclude with future plans in Sect. 7.

2   Related Work

Numerous research efforts have been proposed for performing sentiment analysis on tweets. The SemEval series of challenges on Sentiment Analysis from tweets is the major venue where practitioners can compare their systems on a common benchmark. In this section, we further describe the top three systems of the 2015 SemEval challenge [10]: Webis [5], Unitn [11] and Lsislif [6].

Webis is the winning system of the 2015 SemEval challenge Task 10. The system implements an ensemble system composed of four classifiers, each of them having participated in previous editions of the SemEval challenges. The four classifiers are NRC-Canada [8], which ranked 1st in SemEval 2013, GU-MLT-LT [4] ranking 2nd in SemEval 2013, KLUE [9] raking 5th in SemEval 2013 and TeamX [7] ranking 1st in SemEval 2014. Webis trains those four classifiers separately. When evaluating the ensemble system, Webis uses a linear function which averages the classification distributions provided by the four sub-classifiers and produce the final classification according to the maximum value of the labels in the average classification distribution.

Unitn is a deep learning system that implements a three-step process to train model that is used for the classification. First, a neural language model trained on a large unsupervised tweet dataset is used for initializing the word embeddings. Second, a convolutional neural network is used to further refine the embeddings on a large distant supervised corpus. In the end, word embeddings and other parameters from the previous steps are used to initialize the neural network that is trained with the supervised training dataset [11].

Lsislif uses a logistic regression classifier that is trained with different weighting schema for each domain for positive and negative labels and several groups of features are extracted including lexical, syntactic, semantic, lexicon and Z score features. Z score can be considered as a standardization of the term frequency using multinomial distribution and it can distinguish the importance of each term in each class [6].

Beyond the studies of how to extract sentiment analysis from tweets, numerous research efforts have also studied the same task on reviews. For instance, [3] explains how to use NLP techniques to categorize Amazon reviews according to their sentiment. Similarly, the Stanford Sentiment System [12] has been proposed recently in the domain of movie reviews. It contains the Stanford Tree Parser, a machine-learning model that parses the input text into the Stanford Tree format and uses some existing models, some of them being trained especially for parsing tweets. The Stanford Sentiment Classifier is at the heart of the system. This classifier takes as input Stanford Trees and outputs their classification results. The Stanford Sentiment Classifier provides also useful detailed results such as classification label and classification distribution on all the nodes in the Stanford Tree. The Stanford Sentiment System is a Recursive Neural Tensor Network trained on the Stanford Sentiment TreeBank that is the first corpus with fully labeled parse trees which makes possible training a model with large and labeled dataset. This model stores the information for compositional vector representations, its size of parameters is not very large and the computation cost is empirically tested as feasible in the movie review domain. In addition, the Stanford Sentiment System captures the meaning of longer phrases and shows a great strength in classifying negative sentences. It beats the bag of word approaches when predicting fine-grained sentiment labels.

Despite the vast and mature research results for detecting the polarity of tweets, in terms of precision and scalability, we encountered scalability issues when parsing Amazon reviews due to the length of the review text. The use of conventional machine learning approaches based on linguistic and semantic features showed therefore some limitations. The Stanford Sentiment System, as we mentioned above, uses the Stanford Sentiment TreeBank which creates a large tree containing the classification for each word. This becomes troublesome when sentences become large (more than 300 characters), the computation of the final classification of the sentence taking then more time. In our approach, we have addressed this scalability issues by performing statistical sampling of the reviews making sure to respect the distributions of features and the relevancy of the final sample.

3   SentiME: A Sentiment Analysis System for Tweets

We develop the SentiME system which implements an ensemble learning of 5 classifiers [13]. It is inspired by and built upon the Webis [5] system that implements 4 state-of-the-art classifiers. We extend this system using the Bootstrap Aggregating Algorithm [1] (referred as bagging) for training, and adding a fifth classifier, namely the Stanford Sentiment System [12], that is used as an off-the-shelf classifier[1] trained with a corpus of movie reviews and best performing, from our observation, in sarcasm detection.

The Stanford Sentiment System implements a recursive neural tensor network parsed by the Stanford Tree Bank. It is significantly different from all the other classifiers used on tweets polarity prediction and it shows great performance on negative classification. Hence, the negative recall of the Stanford Sentiment System is over 90 % on average that makes it trustworthy to detect negation (Table 1). We want to investigate whether the addition of this new classifier in the ensemble actually improves the SentiME system.

Table 1. Negative recall of the sole Stanford Sentiment System on SemEval datasets.

Corpus	Negative recall
SemEval2014-test-gold-B	91.09
SemEval2015-gold-B	89.81

The classification distribution provided by the Stanford Sentiment Classifier consists of five labels: very positive, positive, neutral, negative, and very negative. Consequently, we map these five labels into the two classes expected by the SemEval Sentiment Analysis challenge for a consistent integration with our system. We only extract the root classification distribution because it represents the classification distribution of the entire tweet text. We have tested different configurations for mapping the Stanford Sentiment System classification to the conventional Positive, Negative, and Neutral classes. According to the results of these tests, we have decided to use the following mapping algorithm: very positive and positive are mapped to Positive, neutral are mapped to Neutral and negative and very negative are mapped to Negative.

There are multiple ways to do an ensemble of different systems. In the case of SentiME, we propose to use the Stanford Sentiment System as an off-the-shelf classifier that will not be re-trained. We also propose to use a bagging algorithm for boosting the training of the four other classifiers. To perform bagging,

---

[1] The Stanford Sentiment System is used with the default model provided by the Stanford Sentiment System.

we generate new training dataset for the classifiers using uniformly sampling with replacement. In the whole procedure of selecting the documents used as training, the random selection process generates $(1 - \frac{1}{e})$ unique sentences, while the remaining ones are duplicated documents. Due to the fact that bagging introduces some randomness into the training process, and that the size of the bootstrap samples are not fixed, we performed multiple experiments with different sizes ranging from 33 % to 175 %. We perform this process three times and get three models to test for each size. We observed that doing bagging with 150 % of the initial dataset size leads to the best performance in terms of F1 score (Table 2).

Table 2. Experiments performed with different bagging sizes on SemEval2013-train+dev-B (11338 tweets initial size) training dataset. As model, we define the features created at the end of the training process. The percentages represent the percentage of the sentences chosen from the initial dataset during the bagging process.

Model	19842 (175 %)	17007 (150 %)	14173 (125 %)	11338 (100 %)
Model 1	64,76	65,45	65,26	65,05
Model 2	64,65	65,81	64,40	64,15
Model 3	64,50	65,71	64,29	65,25
	9000 (80 %)	7525 (66 %)	5644 (50 %)	3780 (33 %)
Model 1	63,37	63,80	64,64	62,81
Model 2	64,54	64,92	62,93	62,85
Model 3	63,54	64,67	63,85	61,65

To combine the results of the classifiers, we use the linear regression, averaging the sum of the classifier values by the total number of them. In detail, the linear aggregating function averages the classification distributions of each classifiers and choose the polarity which holds the maximum value among the average classification distributions as the final classification.

We performed four different experiments to evaluate the performance of SentiME compared to a baseline system. The corpora used for this experiment are SemEval2014-test, SemEval2014-sarcasm, SemEval2015-test and SemEval2015-sarcasm (Table 3).

1. Baseline system: this is the Webis system using re-trained models as explained in Sect. 2;
2. SentiME system: this is the ensemble system composed of the four sub-classifiers used by Webis plus the Stanford Sentiment System. The ensemble uses a bagging approach for the training phase;
3. Baseline system without TeamX;
4. SentiME system without TeamX;

The third and fourth experiments are variations of the first two experiments respectively where we simply remove the TeamX classifier, based on the observation that this particular classifier plays a similar role than the Stanford Sentiment system.

Table 3 reports the F-measure scores for the four different setups we described above on four different datasets: two datasets contain regular tweets and two datasets contain sarcasm tweets. We evaluated the four sub-classifiers on SemEval2014 test data set, SemEval2014 sarcasm data set, SemEval2015 test data set and SemEval2015 sarcasm data set in order to figure out whether the Stanford Sentiment System has significant impacts on the performance of our ensemble system. The last row of the Table 3 presents the F-measure scores of the baseline system as reported in the authors' paper [5].

Table 3. F1 scores for the four systems on four different datasets. Highest scores are in bold

System	SemEval2014-test	SemEval2014-sarcasm	SemEval2015-test	SemEval2015-sarcasm
Baseline	**69,31**	60,00	66,57	54,19
SentiME system	68,27	**62,57**	**67,39**	**60,92**
Baseline without TeamX	68,56	62,04	66,19	56,86
SentiME system without TeamX	69,27	62,04	66,38	58,92

We observe that the SentiME system outperforms the baseline system on all datasets except on the SemEval2014-test, in which the SentiME system without the TeamX sub-classifier has almost the same performance than the Webis Replicate system. Concerning the performance on both sarcasm datasets, it is clear that the SentiME system improves the F1 score by respectively 2.5 % and 6.5 % on SemEval2014-sarcasm and SemEval2015-sarcasm datasets.

We notice that some features used in TeamX come from the Stanford NLP Core package and we assume that TeamX shares some common characteristics with the Stanford Sentiment System. Since the idea of our experiments was to figure out what benefits the Stanford Sentiment System can bring to our replicate system, we consider it is reasonable to exclude the TeamX sub-classifier from our replicate system.

The complete workflow of the SentiME system is depicted in Fig. 1. The SentiME system trains the four sub-classifiers independently and the results are stored into four different attribute files. Then, concerning the test process, each classifier reads the attribute files as well as the test dataset. The Stanford Sentiment System just reads the test dataset. The final step is to average the five different classification results of the classifiers in order to derive the final classification result.

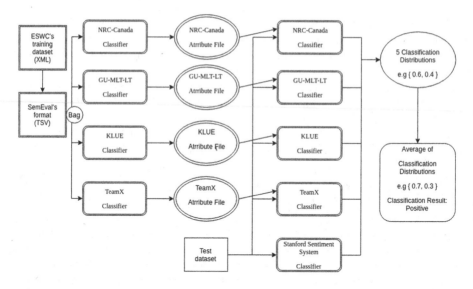

Fig. 1. Workflow of training and test for SentiME

4   The Semantic Sentiment Analysis Challenge Corpus

The training corpus of the SSA challenge 2016 [2] consists of one million Amazon reviews that are split in twenty different categories: Amazon Instant Video, Automotive, Baby, Beauty, Books, Clothing Accessories, Electronics, Health, Home Kitchen, Movies TV, Music, Office Products, Patio, Pet Supplies, Shoes, Software, Sports Outdoors, Tools Home Improvement, Toys Games, and Video Games. In each one of these twenty categories, there are ten xml files. There is a perfect balance between positive and negative reviews because each category contains five xml files of 5000 positive reviews each and another five xml files of 5000 negative reviews each. Consequently, each of the 20 category has 50000 reviews, 25000 positive and 25000 negative.

The training dataset consists of XML files. The structure of these files is as follows:

```
<?xml version="1.0" encoding="UTF-8" standalone="yes"?>
<Sentences>
    <sentence id="B000PVAFT0-A394FOR62LMMXO-1325203200-1">
        <domain>amazon_instant_video</domain>
        <polarity>positive</polarity>
        <summary>
        IMAX
        </summary>
        <text>
        This product works great and would definitely buy more of. I would
            definitely do business with this company again. Very Happy.
        </text>
```

</sentence>
...
...
</Sentences>

For each sentence, the file contains a sentence id, a polarity result, and the textual content of the review. We parse those XML files using a DOM parser and we create one file in the TSV format (following the structure of the SemEval format) that contains the one million reviews (Table 4). Then, we use this TSV file to perform the experiment. In the end, there is a conversion of the TSV file to the XML format so that the classification of the SentiME system produces can be evaluated against a gold standard.

Table 4. Dataset structure in the TSV format.

Sentence ID	Polarity	Sentence (review)
B000PVAFT0-A394FOR62LMMXO-1325203200-1	positive	This product works great ...
B00004OCL2-A3T7V207KRDE2O-1227398400-841	positive	We find it perfect for ...
B0002ZQB4M-A5IJ8JA8TMU57-1287705600-1701	negative	It's a fine pen, but ...

In the initial corpus, we encountered some problems while parsing the sentences due to Unicode encoding issues. We contacted the organisers letting them know about the problem and they fixed the corpus. We observed 73 duplicated sentences, i.e. sentences repeated more than once in the dataset. There are also two sentence ids with a blank review. We performed some computations on the corpus concerning the average number of words per sentence and the average number of characters per sentence. We perform these computations for both the positive and the negative sentences (Table 5).

Table 5. Statistics for the training dataset of 1 Million Amazon Reviews.

Corpus	Number of sentences	Average number of words per sentence	Average number of characters per sentence
Full corpus	1 M	90.39	489
Positive sentences	500 K	86.04	466
Negative sentences	500 K	94.75	513

5 Experimental Study with the SSA Corpus

In this section, we present our experimental setup and we discuss the different experiments we performed to evaluate the performance of SentiME with the SSA dataset. All the dataset configurations used in the following experiments keep a perfect balance of positive and negative sentences (50 %), since we applied a stratified sampling according to the polarity. We keep this balance since each category of the training dataset has equally positive and negative sentences.

We started our experiments by finding the best mapping for the Stanford Sentiment System. The original classification distribution provided by the Stanford Sentiment Classifier consists of five labels (very positive, positive, neutral, negative and very negative). For the SemEval experiments, we have mapped very positive and positive as Positive, neutral as Neutral and negative and very negative as Negative. According to the SSA's rules, the classification distribution should be binary (Positive or Negative). We performed a 10-fold cross validation with smaller random samples of the dataset by selecting sentences with the average length of 250 words. We preserved the 50 % balance, in a stratified setting. The result of this experiment mapped very positive and positive as Positive, and negative and very negative as Negative with a hybrid mapping for neutral which had the best performance regarding the final F-measure scores. We name hybrid mapping the case when neutral confident score is the greatest among the five confident scores of Stanford Sentiment System. In this case neutral is classified as Positive when the confident score of very positive and positive is greater than the confident score of negative and very negative. If it is not greater, the sentence is classified as Negative.

Figure 2 shows how we implemented the bagging process. Initially, we have the full corpus. Then, we translated the Amazon file which contains the percentage of sentences of the initial corpus which has been selected. Finally, we parse this "new" dataset to each sub-classifier separately.

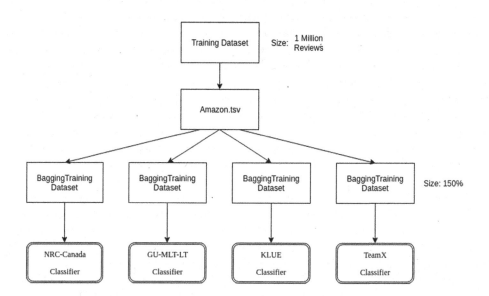

Fig. 2. Workflow for training SentiME with bagging

We experimented with different sizes of the training set for the bagging algorithm and we empirically assessed that the best performing setting of SentiME was with the 150 % of the corpus size as shown in Fig. 2. In detail, we used the SSA corpus in a 4-fold cross validation setting. We trained different models doing bagging for 150 %, 120 %, 100 %, 90 % and without bagging. We obtain a similar result when experimenting on the SemEval datasets [13]. Doing bagging for 150 % of the initial training dataset yields the best performance for the same test dataset regarding the final F1 score (Table 6).

Table 6. Performance of Stanford Sentiment System with different mapping in a 10-fold cross validation setup.

	VP + P + Neutral = positive & VN + N = negative	VP + P = positive & VN + N + Neutral = negative	VP + P = positive & VN + N = negative (Hybrid)
Average F1 score of 10-fold	0,870839208	0,871577973	**0,872691237**

The final experiment is about the F-measure scoring of SentiME in three different training datasets and two test datasets. We divided the training dataset in four parts in order to do a 4-fold cross-validation. We preferred 4-fold cross validation over a 10-fold cross validation because of the scalability issues we encountered. In each fold, we created three training datasets of ten thousand, fifty thousand and one hundred thousand sentences randomly selected among the 750,000 sentences we had available for training each time. Concerning the test datasets, we created two in each fold. The first was of one thousand and the second was of ten thousand sentences randomly selected from the 250,000 sentences we had available for testing in each fold. The experimental results are presented in Table 7.

Table 7. Results of experiment in a 4-fold cross validation setup.

Training–Test	Fold 1	Fold 2	Fold 3	Fold 4
10 K–1 K	0.8824	0.8964	0.8738	0.8784
50 K–10 K	0.8954	0.8969	0.8992	0.9011
100 K–10 K	0.9042	0.9065	0.9095	**0.9113**

Finally, we report the performance of SentiME over the test set in Table 8.

6   Lessons Learned from the Experimental Study

The SentiME system has demonstrated consistent results in the experimental setup. It seems to be agnostic to textual variations as it is the case in the sentences collected from a broad variety of categories. The addition of the Stanford

Table 8. Performance of teams competing on the SSA 2016 corpus.

Team	Precision	Recall	F1 score
1st SentiME	0.85686	0.90541	**0.88046**
2nd App2Check	0.82777	0.90789	0.87142
3rd IRS-MDSA	0.81837	0.89198	0.85359
4th EPOM	0.50494	0.81665	0.62403

Sentiment System in our ensemble improves the performance on sarcasm sentences, that are quite common when conveying sentiments. This happens because the native Stanford Sentiment System has a great strength to classify sentences whose golden standard is negative. This means that we can use Stanford Sentiment System to help our system to classify the sarcasm sentences. On the other hand, Stanford Sentiment System is heavily skewed towards negative. We address this problem in using the Bootstrap Aggregation Algorithm (bagging) with the size of 150 % of the original training set.

Amazon reviews are much longer in terms of number of characters (max 5000 characters) than tweets (max 140 characters). This transition made our system slower to compute the final classification due to the longer length of sentences. The limitation of our system is that the computing time for a big number of long sentences is huge. In the SSA training set, reviews consist of more than 3,000 characters while the average number of characters per sentence is 489 (Table 5). This number is almost four times bigger than the maximum tweet length. This has significantly challenged the computing time of the SentiME system as whole. In particular, the longer length of the content to analyze has negatively impacted the Stanford Sentiment System which uses the Stanford Tree Bank making the parsing of larger sentences slower. We need more processing power and to rely on parallel computing in order to make this work faster to scale (which was out of the scope of this experimental study).

We experiment with three different Stanford Sentiment System's mappings to the two classes only expected by the SSA challenge. While we did not observe significant different in the F-measure score, we chose the one with the slightly better performance. The best mapping is the one we called "hybrid" which classifies a sentence with the bigger confidence score neutral in Standford Sentiment System, as positive or negative according to the dominant confident scores of very positive + positive and very negative + negative.

We tried to train models on the complete training dataset, but it results to be unfeasible. We decided to build training model based on smaller dataset, built by sampling reviews from the entire set in a statistical manner and preserving the distributions of the classes.

7   Conclusions and Future Work

We have presented an experimental study aiming to evaluate the performance of SentiME, a sentiment analysis system we developed, on both tweets and Amazon reviews. The results show the robustness of our system that implements an ensemble learning approach of 5 state-of-the-art classifiers, working on different feature sets and trained with different training set (by using the bagging algorithm). After doing multiple experiments in different datasets and setups, we observe that implementing a bagging technique in the initial dataset with the percentage of 150 % improves the training process. The integration of the Stanford Sentiment System in SentiME improves the results in sarcasm corpora by 6.5 %. This improvement indicates that the Stanford Sentiment System deals better with sarcasm and this is due to the very good classification of negative sentences.

As future work, we will perform a thorough error analysis in the test dataset and we will investigate why some sentences have been wrongly classified. We aim to play with the features used in each of the sub-classifiers in order to better understand their respective contribution in the classification result. Moreover, we will experiment with different aggregation functions such as a weighting schema that would give priority to a particular classifier. Last but not least, we will investigate further how to improve the scalability of the Stanford Sentiment System.

Acknowledgments. The authors would like to thank Xianglei Li for his earlier work on the SentiME system. This work was partially supported by the innovation activity 3cixty (14523) of EIT Digital and by the European Union's H2020 Framework Programme via the FREME Project (644771).

References

1. Breiman, L.: Bagging predictors. Mach. Learn. **24**(2), 123–140 (1996)
2. Dragoni, M., Tettamanzi, A., Pereira, C.: DRANZIERA: an evaluation protocol for multi-domain opinion mining. In: 10th International Conference on Language Resources and Evaluation (LREC) (2016)
3. Fang, X., Zhan, J.: Sentiment analysis using product review data. J. Big Data **2**, 5 (2015)
4. Günther, T., Furrer, L.: GU-MLT-LT: sentiment analysis of short messages using linguistic features and stochastic gradient descent. In: 7th International Workshop on Semantic Evaluation (SemEval) (2013)
5. Hagen, M., Potthast, M., Büchner, M., Stein, B.: Webis: an ensemble for Twitter sentiment detection. In: 9th International Workshop on Semantic Evaluation (SemEval) (2015)
6. Hamdan, H., Bellot, P., Bechet, F.: Lsislif: feature extraction and label weighting for sentiment analysis in Twitter. In: 9th International Workshop on Semantic Evaluation (SemEval) (2015)
7. Miura, Y., Sakaki, S., Hattori, K., Ohkuma, T.: TeamX: a sentiment analyzer with enhanced lexicon mapping and weighting scheme for unbalanced data. In: 8th International Workshop on Semantic Evaluation (SemEval) (2014)
8. Mohammad, S., Kiritchenko, S., Zhu, X.: NRC-Canada: building the state-of-the-art in sentiment analysis of tweets. In: 7th International Workshop on Semantic Evaluation (SemEval) (2013)

9. Proisl, T., Greiner, P., Evert, S., Kabashi, B.: KLUE: simple and robust methods for polarity classification. In: 7th International Workshop on Semantic Evaluation (SemEval) (2013)
10. Rosenthal, S., Nakov, P., Kiritchenko, S., Mohammad, S., Ritter, A., Stoyanovm, V.: SemEval-2015 task 10: sentiment analysis in Twitter. In: 9th International Workshop on Semantic Evaluation (SemEval) (2015)
11. Severyn, A., Moschitti, A.: UNITN: training deep convolutional neural network for Twitter sentiment classification. In: 9th International Workshop on Semantic Evaluation (SemEval) (2015)
12. Socher, R., Perelygin, A., Wu, J.Y., Chuang, J., Manning, C., Ng, A., Potts, C.: Recursive deep models for semantic compositionality over a sentiment treebank. In: Conference on Empirical Methods in Natural Language Processing (EMNLP) (2013)
13. Sygkounas, E., Rizzo, G., Troncy, R.: A replication study of the top performing systems in SemEval Twitter sentiment analysis. In: 15th International Semantic Web Conference (ISWC) (2016)

Exploiting Propositions for Opinion Mining

Andi Rexha[1(✉)], Mark Kröll[1], Mauro Dragoni[2], and Roman Kern[1]

[1] Know-Center GmbH, Graz, Austria
{arexha,mkroell,rkern}@know-center.at
[2] FBK-IRST, Trento, Italy
dragoni@fbk.eu

Abstract. With different social media and commercial platforms, users express their opinion about products in a textual form. Automatically extracting the polarity (i.e. whether the opinion is positive or negative) of a user can be useful for both actors: the online platform incorporating the feedback to improve their product as well as the client who might get recommendations according to his or her preferences. Different approaches for tackling the problem, have been suggested mainly using syntactic features. The "Challenge on Semantic Sentiment Analysis" aims to go beyond the word-level analysis by using semantic information. In this paper we propose a novel approach by employing the semantic information of grammatical unit called preposition. We try to drive the *target* of the review from the *summary information*, which serves as an input to identify the proposition in it. Our implementation relies on the hypothesis that the proposition expressing the target of the summary, usually containing the main polarity information.

1 Introduction

User's opinions can be found in various social media platforms and online stores in textual form. The length and style of the text can vary substantially, ranging from short Twitter messages to longer book reviews. They also refer to different aspects, from politics, to pictures and comercial products. The nature of these opinions change the way to analyze the text from automatic polarity detection systems. Twitter messages aren't expressed using syntactically correct text and require a different preprocessing than, for example, book reviews. For the "Challenge on Semantic Sentiment Analysis" the task (with the winners of recent years [1,2,5,8]) is to detect the polarity of user's opinions of products on Amazon.com reviews. The dataset [4] consists of a set of summaries about different topics. The review is compound by an *id*, a *summary* and a *textual description* and is represented in a XML format as shown in the Example 1.

Since the summary text is expressed in well formed text, Natural Language Processing (NLP) tools can be used to preprocess and analyze those reviews. For this challenge we use a two step approach. In the first step we isolate the syntactic information (proposition) in which the summary is expressed. In the second step we use a supervised approach in order to classify the reviews in positive or negative polarity.

© Springer International Publishing Switzerland 2016
H. Sack et al. (Eds.): SemWebEval 2016, CCIS 641, pp. 121–125, 2016.
DOI: 10.1007/978-3-319-46565-4_9

```
<summary>Transformers</summary>
<text>By most accounts, the Michael Bay-directed Transformers
    films to date films to date are not very good, but that
    hasnt stopped them from making gobs and gobs of cash.
</text>
<polarity>positive</polarity>
```

Example 1: Example of a single entry in the dataset provided by the challange

The paper is organized in three sections. In Sect. 2 we detail the approach used for the two steps and the features used for the supervised task. Finally, Sect. 3 describes the partial results and the discussion the advantages and drawback of the approach.

2   Approach and Features

Each review is composed of a *summary* and a *textual information*. One or more sentences form the textual information of the summary contain the detailed specification of the user experience. We base our approach in the hypothesis that the summary is extended in the textual information and its "isolated" content contains the main polarity information. After preprocessing the textual summary with NLP tools, we annotate the words of the summary in each sentence. Later we extract the most "prominent" sentence (to be defined in Subsect. 2.2) which contains the main target of the summary. From the "prominent" sentence we select the "best fitting" proposition (we define it in Subsect. 2.3) which contains the summary. From the proposition, we extract the polarity of each word and encode the distribution of the polarity in the proposition as features. As a final step we train a classifier in order to predict the polarity of the whole tweet. Recapping, the approach can be split in the following steps:

- Preprocessing
- Extract the prominent sentence
- Extract the prominent proposition
- Polarity extraction and feature encoding

Below, we describe each of these steps in more detail. For illustration purposes we get the following example:

Summary: Typical movie of Al Pacino

Text: This was a very good movie from Al Pacino but the music wasn't that nice. Just think about how bad other movies are! The music doesn't play any role for my review!

Example 2: Example of a summary and review

2.1   Preprocessing

For preprocessing the reviews, we select the Stanford Core NLP tool [7]. For each review we annotate all sentences, words and parse the syntactic dependency graph. As a final step we annotate the text in the review with the tokens from the summary.

In the example 2 this would be: This was a very good movie from Al Pacino but the music wasn't that nice.

2.2   Extract the Prominent Sentence

From the annotated text of the review we need to select the sentence best matching with the summary. We define the most "prominent" one as the sentence which contains more terms in a TermFrequency-InverseSentenceFrequency of the term. So, for each term of the summary we calculate it's frequency (i.e. the number of times it occurs in each sentence) and it's inverse sentence frequency (i.e. the inverse fraction of documents containing the word). This formula reflects the tf-idf(term frequency-inverse documetn frequency) score, but applied to the sentences, and we consider it a tf-isf. In the former example, the first sentence would be selected due to it containing two annotated words.

2.3   Extract the Prominent Proposition

For the "best fit" sentence we try to extract the *proposition* which capture best the main information. As a first step, we extract the propositions composing the sentence. We use the well known Open Information Extraction tool, ClausIE [3]. It extracts relation of the form (subject, predicate, object) called propositions.

Returning to the example of the prominent sentence "This was a very good movie from Al Pacino but the music wasn't that nice.", it can be splitted in the following propositions:

– This was a very good movie
– This was a very good movie from Al Pacino
– the music wasn't that nice

We define the "best matching" proposition as the shortest one (in terms of words) containing most of the terms in the summary. This mean that the selected one in our example would be: This was a very good movie from Al Pacino .

2.4   Features

In this challenge we use polarit features extracted from SentiWordNet [6]. SentiWordNet is a thesaurus which contains polarity information about words. To each word it is assigned a score between −1 and 1, which indicates whether the word has a negative or positive polarity. We model the proposition as a function of the sentiment expressed in the words. More precisely we identify the polarity of each word in the "best fit" proposition. We express the features of the "best fit" proposition as maximum, minimum, arithmetic mean, and standard deviation of the polarities of the words. As a additional feature we use the number of negation words expressed in the "best matching" sentence.

3 Results and Discussion

We try to learn our model from the features we have extracted and predict new unseen reviews. We use a Logistic Regression to learn from the results. In the Table 1 we present the precision, recall and F1-measure of the 10 fold cross-validation.

Table 1. Results from a 10 fold cross validation in the training dataset

	Precision	Recall	F1-measure
Positive	0.629	0.823	0.713
Negative	0.744	0.515	0.608
Average	0.686	0.669	0.661

As we can see from the tables, the results from the cross validation are not as good as expected. After an analyses of the dataset we believe that this discrepancy of the results from the expectation might be caused by the false assumption that the summary of the review is also expressed in the text.

Acknowledgment. This work is funded by the KIRAS program of the Austrian Research Promotion Agency (FFG) (project number 840824). The Know-Center is funded within the Austrian COMET Program under the auspices of the Austrian Ministry of Transport, Innovation and Technology, the Austrian Ministry of Economics and Labour and by the State of Styria. COMET is managed by the Austrian Research Promotion Agency FFG.

References

1. Palmero Aprosio, A., Corcoglioniti, F., Dragoni, M., Rospocher, M.: Supervised opinion frames detection with RAID. In: Gandon, F., et al. (eds.) SemWebEval 2015. CCIS, vol. 548, pp. 251–263. Springer, Heidelberg (2015). doi:10.1007/978-3-319-25518-7_22
2. Chung, J.K.-C., Wu, C.-E., Tsai, R.T.-H.: Polarity detection of online reviews using sentiment concepts: NCU IISR Team at ESWC-14 challenge on concept-level sentiment analysis. In: Presutti, V., et al. (eds.) SemWebEval 2014. CCIS, vol. 475, pp. 53–58. Springer, Heidelberg (2014). doi:10.1007/978-3-319-12024-9_7
3. Del Corro, L., Gemulla, R.: Clausie: clause-based open information extraction. In: Proceedings of the 22nd International Conference on World Wide Web, WWW 2013, pp. 355–366. ACM, New York (2013). http://doi.acm.org/10.1145/2488388.2488420
4. Dragoni, M., Tettamanzi, A., da Costa Pereira, C.: Dranziera: an evaluation protocol for multi-domain opinion mining. In: Chair, N.C.C., Choukri, K., Declerck, T., Goggi, S., Grobelnik, M., Maegaard, B., Mariani, J., Mazo, H., Moreno, A., Odijk, J., Piperidis, S. (eds.) Proceedings of the Tenth International Conference on Language Resources and Evaluation (LREC 2016), European Language Resources Association (ELRA), Paris, France, May 2016

5. Dragoni, M., Tettamanzi, A.G.B., da Costa Pereira, C.: A fuzzy system for concept-level sentiment analysis. In: Presutti, V., et al. (eds.) SemWebEval 2014. CCIS, vol. 475, pp. 21–27. Springer, Heidelberg (2014). doi:10.1007/978-3-319-12024-9_2

6. Esuli, A., Sebastiani, F.: Sentiwordnet: a publicly available lexical resource for opinion mining. In: Proceedings of the 5th Conference on Language Resources and Evaluation, LREC06, pp. 417–422 (2006)

7. Manning, C.D., Surdeanu, M., Bauer, J., Finkel, J., Inc, P., Bethard, S.J., Mcclosky, D.: The Stanford CoreNLP natural language processing toolkit. In: Proceedings of the 52nd Annual Meeting of the Association for Computational Linguistics: System Demonstrations, pp. 55–60 (2014)

8. Schouten, K., Frasincar, F.: The Benefit of Concept-Based Features for Sentiment Analysis. In: Gandon, F., et al. (eds.) SemWebEval 2015. CCIS, vol. 548, pp. 223–233. Springer, Heidelberg (2015). doi:10.1007/978-3-319-25518-7_19

The IRMUDOSA System at ESWC-2016 Challenge on Semantic Sentiment Analysis

Giulio Petrucci[1] and Mauro Dragoni[2(✉)]

[1] Universitá di Trento, Trento, Italy
petrucci@fbk.eu
[2] Fondazione Bruno Kessler, Trento, Italy
dragoni@fbk.eu

Abstract. Multi-domain opinion mining consists in estimating the polarity of a document by exploiting domain-specific information. One of the main issue of the approaches discussed in literature is their poor capability of being applied on domains that have not been used for building the opinion model. In this paper, we present an approach exploiting the linguistic overlap between domains for building models enabling the estimation of polarities for documents belonging to any other domain. The system implementing such an approach has been presented at the third edition of the Semantic Sentiment Analysis Challenge co-located with ESWC 2016. Fuzzy representation of features polarity supports the modeling of information uncertainty learned from training set and integrated with knowledge extracted from two well-known resources used in the opinion mining field, namely Sentic.Net and the General Inquirer. The proposed technique has been validated on a multi-domain dataset and the results demonstrated the effectiveness of the proposed approach by setting a plausible starting point for future work.

1 Introduction

Opinion mining is a natural language processing task aiming to classify documents according to the opinion (polarity) they express on a given subject [1]. Generally speaking, opinion mining aims at determining the attitude of a speaker or a writer with respect to a topic or the overall tonality of a document. This task has created a considerable interest due to its wide applications. In recent years, the exponential increase of the Web for exchanging public opinions about events, facts, products, etc., has led to an extensive usage of opinion mining approaches, especially for marketing purposes.

Most of the work available in the literature address the opinion mining problem without distinguishing the domains which documents, used for building models, come from. The necessity of investigating this problem from a multi-domain perspective is led by the different influence that a term might have in different contexts. Let us consider the following examples. In the first example, we have an "emotion-based" context where the adjective "cold" is used differently based on the feeling, or mood, of the opinion holder:

© Springer International Publishing Switzerland 2016
H. Sack et al. (Eds.): SemWebEval 2016, CCIS 641, pp. 126–140, 2016.
DOI: 10.1007/978-3-319-46565-4_10

1. That person always behaves in a very **cold** way with her colleagues.
2. A **cold** drink is the best thing we can drink when the temperature is very hot.

while in the second one, we have a "subjective-based" context where the adjective "small" is used differently based on the product category reviewed by a user:

1. The sideboard is **small** and it is not able to contain a lot of stuff.
2. The **small** dimensions of this decoder allow to move it easily.

In the first context, we considered two different "emotional" situations: in the first one a person is commenting about the behavior of his colleague by using the adjective "cold" with a "negative" polarity. Instead, in the second one, a person is referring to the adjective "cold" in a "positive" way as a good solution for a situation.

Instead, in the second context, we considered the interpretation of texts referring to two different domain: "Furnishings" and "Electronics". In the first one, the polarity of the adjective "small" is, for sure, negative because it highlights an issue of the described item. On the other hand, in the second domain, the polarity of such an adjective may be considered positive.

The multiple facets with which textual information can be analyzed in the context of opinion mining led to the design of approaches creating models able to address this scenario. The idea of adapting terms polarity to different domains emerged only recently [2]. In general, multi-domain opining mining approaches discussed in the literature (surveyed in Sect. 2) focus on building models for transferring information between pairs of domains [3]. While on one hand such approaches allow to propagate specific domain information to other, their drawback is the necessity of building new transfer models any time a new domain has to be addressed. This way, approaches include a poor generalization capability of analyzing text, because transfer models are limited to the N domains used for building the models.

This paper describes our approach exploiting the linguistic overlap between domains for building models enabling the estimation of polarities for documents. Due to this peculiarity, the proposed approach is innovative, to the best of our knowledge, with respect to the state of the art of multi-domain opinion mining.

The rest of the article is structured as follows. Section 2 presents a survey on works about opinion mining either in the single o multi domain environment. Section 3 provides the references to the knowledge resources used in the implementation of the proposed approach described in detail in Sect. 4. Section 5 reports the system evaluation and, finally, Sect. 6 concludes the article.

2   Related Work

The topic of opinion mining has been studied extensively in the literature [4,5], where several techniques have been proposed and validated.

The use of domain adaptation demonstrated that opinion classification is highly sensitive to the domain from which the training data is extracted.

A classifier trained using opinionated documents from one domain often performs poorly when applied or tested on opinionated documents from another domain, as we showed through the examples presented in Sect. 1. The reason is that using the same words and even the same language constructs can carry different opinions, depending on the domain.

The classic scenario is when the same word in one domain may have positive connotations, but in another domain may have negative connotations; therefore, domain adaptation is needed. In the literature, different approaches related to the Multi-domain sentiment analysis have been proposed. Briefly, two main categories may be identified: (i) the transfer of learned classifiers across different domains [2,3,6,7], and (ii) the use of propagation of labels through graph structures [8–10].

While on one side such approaches demonstrated their effectiveness in working in a multi-domain environment, on the other one, they suffer by the limitation in abstracting their usage within any domain different from the ones used for building the model. With respect to this issue, our approach, starting from a limited number of domain-based labeled data, can build a model capable to estimate polarities of texts belonging to other domains. This important aspect allows the use of the produced model in any environments including ones where document content is totally unknown.

Besides work about multi-domain, opinion mining has been investigated from several points of view.

Machine learning techniques are the most common approaches used for addressing this problem, given that any existing supervised method can be applied to opinion classification. For instance, in [1,11], the authors compared the performance of Naive-Bayes, Maximum Entropy, and Support Vector Machines in opinion mining on different features like considering only unigrams, bigrams, combination of both, incorporating parts of speech and position information or by taking only adjectives.

An obstacle to research in this direction is the need of labeled training data, whose preparation is a time-consuming activity. Therefore, in order to reduce the labeling effort, opinion words have been used for training procedures. In [12], the authors used opinion words to label portions of informative examples for training the classifiers. Opinion words have been exploited also for improving the accuracy of opinion classification, as presented in [13], where a framework incorporating lexical knowledge in supervised learning to enhance accuracy has been proposed.

All the approaches presented so far operate at the document-level [14,15]; while, for improving the accuracy of the opinion classification, a more fine-grained analysis of the text, i.e., the opinion classification of every single sentence has to be performed [16,17]. In the literature, we may find approaches ranging from the use of fuzzy logic [10,18] or computational intelligence [19] methods, to the use of aggregation techniques [20] for computing the score aggregation of opinion words. Systems implementing the capabilities of identifying opinion's holder, target, and polarity have been presented [21] too.

A particular attention should be given also to the application of opinion mining in social networks. More and more often, people use social networks for expressing their moods concerning their last purchase or, in general, about new products. Such a social network environment opened up new challenges due to the different ways people express their opinions, as described by [22,23], who mention "noisy data" as one of the biggest hurdles in analyzing social network texts.

One of the first studies on opinion mining on micro-blogging websites has been discussed in [24], where the authors present a distant supervision-based approach for opinion classification.

At the same time, the social dimension of the Web opens up the opportunity to combine computer science and social sciences to better recognize, interpret, and process opinions and sentiments expressed over it. Such multi-disciplinary approach has been called *sentic computing* [25].

3   Knowledge Resources

The proposed approach exploits the use of background knowledge for supporting the creation of the multi-domain model used for computing text polarities. Such a background knowledge is composed by two linguistic resources freely available to the research community. Below, we briefly describe them, while in Sect. 4, we present how they have been used for supporting the implementation of the proposed approach.

SenticNet. *SenticNet*[1] [26] is a publicly available resource for opinion mining that exploits both artificial intelligence and semantic Web techniques to infer the polarities associated with common-sense concepts and to represent them in a semantic-aware format. In particular, SenticNet uses dimensionality reduction to calculate the affective valence of a set of Open Mind[2] concepts and it represents them in a machine accessible and processable format. SenticNet contains more than 5,700 polarity concepts (nearly 40 % of the Open Mind corpus) and it may be connected with any kind of opinion mining application. For example, after the de-construction of the text into concepts through a semantic parser, SenticNet can be used to associate polarity values to these and, hence, infer the overall polarity of a clause, sentence, paragraph, or document by averaging such values.

General Inquirer. *General Inquirer dictionary*[3] [27] is an English-language dictionary containing almost 12,000 elements associated with their polarity in different contexts. Such dictionary is the result of the integration between the "Harvard" and the "Lasswell" general-purpose dictionaries as well as a dictionary of categories define by the dictionary creators. When necessary, for ambiguous words, specific polarity for each sense is specified. For every words, a set of

---

[1] http://sentic.net/.

[2] http://commons.media.mit.edu/en/.

[3] http://www.wjh.harvard.edu/~inquirer/spreadsheet_guide.htm.

tags is provided in the dictionary. Among them, only a subset is relevant to the opinion mining topic and have been exploited in this work: "valence categories", "semantic dimensions", "words of pleasure", and "words reflecting presence or lack of emotional expressiveness". Other categories indicating ascriptive social categories rather than references to places have been considered out of the scope of the opinion mining topic and have not been considered in the implementation of the approach.

4 Method

The main goal of the presented approach is to exploit domain overlap for compensating the lack of knowledge caused by building models using only a snapshot of the reality. When a domain-based model is built, part of the knowledge belonging to such a domain is not included in the model due to its missing in the adopted training set. Besides, when a system has to classify a text belonging to a domain that has not been used for building the model, it is possible that such a model does not contain enough information for estimating text opinion. For this reason, it is necessary to compensate this lack of knowledge by partially exploiting information coming from other domains.

In this section, we describe the steps adopted for building the models and the strategy we implemented for computing the polarity of a text by exploiting domain overlapping.

The model construction process is composed by three steps: (i) features are extracted from each text contained in the training set (Sect. 4.1); then, (ii) a preliminary fuzzy membership functions [28], modeled by using a triangular shape, is computed by analyzing only the explicit information contained in the dataset (Sect. 4.2); and, (iii) this shape is transformed into a trapezoid after a refinement operation performed by compensating the uncertainty inherited by the adoption of a training set, with information coming from external resources: the SenticNet knowledge base and the General Inquirer dictionary (Sect. 4.3).

Finally, when the inference of a text polarity is requested, the usage of the model for estimating text polarities is performed by aggregating the polarities associated with each feature detected from the text to polarize and by computing the final judgment (Sect. 4.4).

4.1 Feature Extraction

During the Feature Extraction (FE) phase, documents are analyzed and significant elements are extracted and used as features for building the model. As feature, we mean every text chunk that may have a meaning in the context of opinion mining and/or in domain detection. The first step that we performed for extracting the features is to parse the content of each document by using the Stanford NLP Parser [29]. The parser has been used for annotating terms with part of speech (POS) tags and for extracting the dependencies tree of each sentence.

Let's consider the following text marked with a positive polarity: "This smartphone is great. The display is awesome and the touch system works very well."

By parser the text, we obtained the analysis of the two detected sentences. Their POS-tagged versions are represented in the lines 1 and 2; while, the dependencies of the first sentence are shown in the lines from 3 to 6 and, finally, the dependencies of the second sentence are shown in the lines from 7 to 17.

```
1.  This/DT smartphone/NN is/VBZ great/JJ ./.
2.  The/DT display/NN is/VBZ awesome/JJ and/CC the/DT
    touch/NN system/NN works/VBZ very/RB well/RB ./.

3.  root ( ROOT-0 , great-4 )
4.  det ( smartphone-2 , This-1 )
5.  nsubj ( great-4 , smartphone-2 )
6.  cop ( great-4 , is-3 )

7.  root ( ROOT-0 , awesome-4 )
8.  det ( display-2 , The-1 )
9.  nsubj ( awesome-4 , display-2 )
10. cop ( awesome-4 , is-3 )
11. cc ( awesome-4 , and-5 )
12. det ( system-8 , the-6 )
13. compound ( system-8 , touch-7 )
14. nsubj ( works-9 , system-8 )
15. conj:and ( awesome-4 , works-9 )
16. advmod ( well-11 , very-10 )
17. advmod ( works-9 , well-11 )
```

From the parser output, we distinguished two type of features for building our model:

- Single concepts feature: nouns, adjectives, and verbs are stored in the model as single features. Nouns are used for building the domain detection component of our model; while, adjectives and verbs are used for building the opinion mining component. By considering as example the dependencies extracted from the first sentence, the term "smartphone" is inserted in the model with the role of supporting the domain detection, while the term "great" is inserted in the model with the role of supporting the definition about how positive polarity is modeled.
- Terms dependency feature: a selection of the dependencies extracted by the parser is stored within the model with the aim of incorporating domain specific contextual knowledge describing (i) how concepts are connected in a particular domain and (ii) how such connections are related to a particular polarity. The kind of dependencies took into account are "noun-adjective", "adjective-verb", "noun-verb", and "adjective-adverb". In the example above, lines containing significant terms dependency features are 5, 9, 10, 14, 16, and 17. From each "term dependency" feature we actually extract further features that are inserted in the model as well. Let's consider as example the dependency at line 16. From the dependency, we extract "well-very", "very-well", "well", and "very".

4.2 Preliminary Learning Phase

The Preliminary Learning (PL) phase aims at estimating the starting polarity and the domain belonging degree (DBD) of each feature. The estimation of these values is done by analyzing only the explicit information provided by the training set.

Concerning the estimation of the feature polarity, this phase allows to define the preliminary fuzzy membership functions representing the polarity of each feature extracted from the training set with respect to the domain containing such polarity. The feature polarity is estimated as:

$$p_i^E(F) = \frac{k_F^i}{T_F^i} \in [-1, 1] \quad \forall i = 1, \ldots, n, \tag{1}$$

where F is the feature taken into account, index i refers to domain D_i which the feature belongs to, n is the number of domains available in the training set, k_F^i is the arithmetic sum of the polarities observed for feature F in the training set restricted to domain D_i, T_F^i is the number of instances of the training set, restricted to domain D_i, in which feature F occurs, and E stays for "estimated". The shape of the fuzzy membership function generated during this phase is a triangle with the top vertex in the coordinates $(x, 1)$, where $x = p_i^{(E)}(F)$ and with the two bottom vertexes in the coordinates $(-1, 0)$ and $(1, 0)$ respectively. The rationale is that while we have one point (x) in which we have full confidence, our uncertainty covers the entire space because we do not have any information concerning the remaining polarity values. At this stage, the types of feature took into account are the terms dependency features and, as single concept features, adjectives and verbs.

Figure 1 shows a picture of the generated fuzzy triangle.

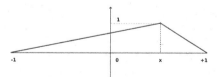

Fig. 1. The fuzzy triangle generated after the Preliminary Learning Phase.

After the polarity estimation, we computed the DBD of each feature. Such a value is exploited during the Polarity Aggregation and Decision Phase (described in Sect. 4.4) for computing the final polarity of a document.

The computation of the DBD is inspired by the well-known TF-IDF model [30] used in information retrieval, where the importance of a term is given by either the frequency of a term in a document contained in an index and the inverse of the number of documents in which such a term occurs.

In our case, the DBD of a feature is computed by summing two factors: the feature frequency associated with the domain in which the feature occurs and the "uniqueness" of the feature with respect to all domains. The domain-frequency is computed as:

$$\text{freq}_i(F) = \frac{k_i(F)}{N_i} \tag{2}$$

where F is the feature taken into account, i is the domain that is analyzed, k_i is the number of times that the feature F occurs in the domain i, and N_i is the total number of features contained in the domain i.

While, the feature uniqueness is computed as:

$$\mathrm{uniq}_i(F) = \frac{\mathrm{freq}_i(F)}{\sum\limits_{i=0}^{n} \mathrm{freq}_i(F)} \tag{3}$$

where F is the feature taken into account and n is the number of domains.

Finally the DBD of each feature is given by:

$$\mathrm{DBD}_i(F) = \mathrm{freq}_i(F) + \mathrm{uniq}_i(F) \tag{4}$$

4.3 Information Refinement Phase

Polarities estimated during the PL phase are refined by exploiting, for each feature, polarities extracted from the resources described in Sect. 3. The rationale behind this choice is to balance the polarity estimated from the training set (that represents only a snapshot of the world) with polarity information that are contained in supervised knowledge bases. When we estimate the polarity value of each feature, two scenarios may happen:

1. the estimated polarity "agrees" (i.e. it has the same orientation) with the one extracted from the knowledge bases;
2. the estimated polarity "disagrees" with the one extracted from the knowledge bases.

In the first case, the estimated polarity confirms, in terms of opinion orientation, what it is represented in the knowledge bases. The representation of this kind of uncertainty will be a tight shape.

On the contrary, in the second case, the estimated polarity is the opposite of what has been extracted from the knowledge bases. In this case, the uncertainty associated with the feature will produce a larger shape. Such a shape will model the contrast between what has been estimated from the training set and what has been defined by experts in the construction of the knowledge bases.

For this reason an Information Refinement (IR) phase is necessary in order to convert this uncertainty in a numerical representation that can be managed by the system.

Assume to have the following values associated to the feature F belonging to the domain i:

- p_s^F, represents the polarity of the feature F extracted from SenticNet;
- p_g^F, represents the polarity of the feature F extracted from the General Inquirer;
- avg_p is the average polarity computed among $p_i^E(F)$, p_s^F, and p_g^F;
- var_p is the variance computed between avg_p and the three polarities values $p_i^E(F)$, p_s^F, and p_g^F.

For "terms dependency" features, that are composed by two terms T_1 and T_2, the values p_s^F and p_g^F are the average of the single polarities computed on T_1 and T_2, respectively.

By starting from these values, the final shape of the inferred fuzzy membership functions, at the end of the IR phase, is a trapezoid whose core consists of the interval between the polarity value learned during the PL phase, $p_i^E(F)$, and avg_p. While, the support of the fuzzy shape is given, on both sides, by the variance var_p.

To sum up, for each domain D_i, $\mu_{F,i}$ is a trapezoid with parameters (a, b, c, d), where

$$a = \min\{p_i^E(F), avg_p\},$$
$$b = \max\{p_i^E(F), avg_p\},$$
$$c = \max\{-1, a - var_p\},$$
$$d = \min\{1, b + var_p\}.$$

The idea here is that the most likely values for the polarity of F for domain D_i are those comprised between the estimated value and average between the estimation of the training algorithm and the polarity values retrieved from supervised knowledge resources. The uncertainty modeled by the fuzzy shape is proportional to the level of "agreement" between the estimated polarity value, and the polarities retrieved from the supervised knowledge bases.

Figure 2 shows a picture of the generated fuzzy trapezoid.

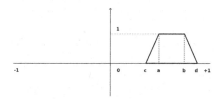

Fig. 2. The fuzzy trapezoid generated after the Information Refinement Phase.

4.4   Polarity Aggregation and Decision Phase

The fuzzy polarities of different features, resulting from the IR phase, are finally aggregated by a fuzzy averaging operator obtained by applying the extension principle [31] in order to compute fuzzy polarities for complex entities, like texts, which consist of a number of features and thus derive, so to speak, their polarity from them. When a crisp polarity value is needed, it may be computed from a fuzzy polarity by applying a defuzzification [32] method.

The operation of computing the entire polarity of a text is done not only on the model describing the domain which the document belongs, but also on the other domains. Indeed, one of the assumption of the proposed approach is to exploit possible linguistic overlaps for compensating missing knowledge of the training set. Therefore, given $p_i(D)$ as the polarity computed on document D with respect to the model of the domain i, and $DBD_i(D)$ as the domain belonging degree of document D to the domain i, we compute the following two vectors:

$$\langle \text{polarity}_D \rangle = [p_0(D), p_1(D), \ldots, p_n(D)] \tag{5}$$

and

$$\langle \text{domain}_D \rangle = [DBD_0(D), DBD_1(D), \ldots, DBD_n(D)] \tag{6}$$

where n is the number of domains contained in the model. The final polarity of a text is then computed by multiplying the two vectors as follow:

$$p_T = \frac{\langle \text{polarity}_D \rangle \times \langle \text{domain}_D \rangle}{N} \tag{7}$$

where N is the number of domains used for building the model.

5 System Evaluation

Here, we present the evaluation procedure adopted for validating the proposed approach.

5.1 The Dataset

The system has been trained by using the DRANZIERA dataset [33]. The dataset is composed by one million reviews crawled from product pages on the Amazon web site[4]. Such reviews belong to twenty different domains, we called in-model domains (IMD): Amazon Instant Video, Automotive, Baby, Beauty, Books, Clothing Accessories, Electronics, Health, Home Kitchen, Movies TV, Music, Office Products, Patio, Pet Supplies, Shoes, Software, Sports Outdoors, Tools Home Improvement, Toys Games, and Video Games.

For each domain, we extracted twenty-five thousands positive and twenty-five thousands negative reviews that have been split in five folds containing five thousand positive and five thousand negative reviews each. This way, the dataset is balanced with respect to either the polarities of the reviews and to the domain which they belong to. The choice between positive and negative documents has been inspired by the strategy used in [2] where reviews with 4 or 5 stars have been marked as positive, while the ones having 1 or 2 stars have been marked as negative.

Besides the twenty domains mentioned above, we used further 7 test sets for measuring the effectiveness of the approach in estimating polarities of texts belonging to domains different from the ones used to build the model, we called out-model domains (OMD). Such domains are: Cell Phones Accessories, Gourmet Foods, Industrial Scientific, Jewelry, Kindle Store, Musical Instruments, and Watches.

---

[4] All the material used for the evaluation and the built models are available at http://goo.gl/pj0nWS.

5.2   Evaluation Procedure

The approach has been evaluated through a 5-cross-fold evaluation procedure. For each execution we measured the precision and the recall and, at the end, we report their averages together with the standard deviation measured over the five executions.

The approach has been compared with three baselines:

- Most Frequent Polarity (MFP): results obtained by guessing always the same polarity for all instances contained in the test set.
- Domain Belonging Polarity (DBP): results obtained by computing the text polarity by using only the information of the domain the text belongs to. This means that the linguistic overlap between domains has not been considered.
- Domain Detection Polarity (DDP): results obtained by computing the text polarity by using only the information of the domain guessed as the most appropriate one for the text to evaluate. This means that the similarity between text content and domain is preferred with respect to the domain used for tagging the text.

The same baselines have been used for evaluating the OMD test sets. In this case, the DBP baseline has not been applied due to the mismatch between the domains used for building the model and the ones contained in the test sets. Each OMD test set has been applied to all five models built, and the scores averaged.

5.3   In-Vitro Results

Here, we show the results of the evaluation campaign conducted for validating the presented approach. Tables 1 and 2 present a summary of the performance obtained by our system and by the three baselines on IMD and OMD, respectively. First column contains the name of the approach, second, third and fourth ones contain the average precision, recall and F1 score computed over all domains; while, the fifth column contains the average standard deviation computed on the F1 score during the cross-fold validation. Finally, the sixth and seventh columns contain the minimum and the maximum F1 score measured during the evaluation.

Table 1 shows the results obtained on in-domain models; while in Table 2, we present the results obtained by testing our approach on out-model domains.

In Table 2, the results about the DBP baseline are not reported because the used test set contains texts belonging to domains that are not included in the model.

By considering the overall results obtained on the IMD (Table 1), we may observe how the proposed approach outperforms the provided baselines. The measured F1 scores (average, minimum, and maximum) are higher of about 4 % with respect to the baselines. The same happens for the Precision value, while for the Recall, all systems are very close to 100 % and now significant differences have been detected. Test instances for which polarity has not been estimated have been judged as "neutral". Concerning the stability of the approach, we can

Table 1. Comparison between the results obtained by the three baselines and the ones obtained by the proposed system on in-model domains.

Approach	Avg. precision	Avg. recall	Avg. F1
MFP	0.5000	1.0000	0.6667
DBP	0.7218	0.9931	0.8352
DDP	0.7115	0.9946	0.8290
DAP	**0.7686**	**0.9984**	**0.8679**
Approach	Avg. deviation	Avg. min. F1	Avg. max. F1
MFP	0.0000	0.6667	0.6667
DBP	0.5028	0.8153	0.8543
DDP	0.5584	0.8121	0.8456
DAP	**0.5954**	**0.8469**	**0.8881**

Table 2. Comparison between the results obtained by the two baselines and the ones obtained by the proposed system on out-model domains.

Approach	Avg. precision	Avg. recall	Avg. F1
MFP	0.5000	1.0000	0.6667
DBP	-	-	-
DDP	0.6766	0.9931	0.8045
DAP	**0.7508**	**0.9985**	**0.8564**
Approach	Avg. deviation	Avg. min. F1	Avg. max. F1
MFP	0.0000	0.6667	0.6667
DDP	0.5796	0.7906	0.8198
DAP	**0.6060**	**0.8389**	**0.8755**

notice that the exploitation of the domain information leads to a lower deviation over the five folds.

The second overall evaluation concerns the analysis of the results obtained on the set of OMD. The interesting aspect of this evaluation is to measure how the system is able to address the task of detecting polarities of documents coming from a different set of domains with respect to the ones used to build the model. Results are shown in Table 2. The first thing that we may observe is how the effectiveness obtained by the proposed system is very close to the one obtained in the IMD evaluation. Indeed, the difference between the two F1 averages is only around 1 %. This aspect remarked the capability of the proposed approach to work in a cross-domain environment and to exploit the linguistic overlaps between domains for estimating text polarities. In this second evaluation, it is also possible to notice how the DDP baseline obtained lower results with respect to the evaluation on the IMD. This result confirms that a solution based on exploiting information coming from a domain resulting the "most similar" to the text to analyze are inadequate for computing text polarities.

In light of these results, we may state that the exploitation of linguistic overlaps between domains is a suitable solution for compensating the possible lack of knowledge had by building opinion models on limited training sets.

5.4   The IRMUDOSA System at ESWC-2016 SSA Challenge Task #1

The system participated to the Task #1 of the Semantic Sentiment Analysis Challenge co-located with ESWC 2016. Table 3 shows the results of Task #1. We may observe that the system ranked third, not far from the best performer. This result confirm the viability of the implemented approach for future implementation after a study of the main error scenarios in which the fuzzy-based algorithm performed poorly.

Table 3. Precision-recall analysis and winners for Task 1.

System	Precision	Recall	F-Mesure
Efstratios Sygkounas, Xianglei Li, Giuseppe Rizzo and Raphaël Troncy	0.85686	0.90541	0.88046
Emanuele Di Rosa and Alberto Durante	0.82777	0.90789	0.87142
Giulio Petrucci and Mauro Dragoni	**0.81837**	**0.89198**	**0.85359**
Andi Rexha, Mark Kröll, Mauro Dragoni and Roman Kern	0.50494	0.81665	0.62403

6   Conclusion and Future Work

In this article, we presented an approach to multi-domain opinion mining exploiting linguistic overlaps between domains for estimating the polarity of texts. The approach is supported by the implementation of a fuzzy model used for representing either the polarity of each feature with respect to a particular domain and its associated uncertainty.

Models are built by combining information extracted from a training set with the knowledge contained in two supervised linguistic resources, Sentic.net and the General Inquirer. The estimation of polarities is performed by combining the degree which a text belongs to each domain with each domain-specific polarity information extracted from the model.

Results shown the effectiveness improvement of the proposed approach with respect to the baselines demonstrating its viability and the close gap between the proposed system and the best performer participated to the Task #1 of Semantic Sentiment Analysis Challenge proved the potential of the fuzzy-based solution. Moreover, the protocol used for the evaluation enables an easy reproducibility of the experiments and the comparison of obtained results with other systems.

Future work will focus either on the enrichment of the knowledge used for building the models and on the use of fuzzy membership functions. Finally, we foresee the integration of a concept extraction approach in order to equip the system with further semantic capabilities of extracting finer-grained information (i.e., single aspects and semantic information associated with them) which can be used during the model construction.

References

1. Pang, B., Lee, L., Vaithyanathan, S.: Thumbs up? Sentiment classification using machine learning techniques. In: Proceedings of EMNLP, pp. 79–86. Association for Computational Linguistics, Philadelphia, July 2002
2. Blitzer, J., Dredze, M., Pereira, F.: Biographies, bollywood, boom-boxes and blenders: domain adaptation for sentiment classification. In: ACL, pp. 187–205 (2007)
3. Pan, S.J., Ni, X., Sun, J.T., Yang, Q., Chen, Z.: Cross-domain sentiment classification via spectral feature alignment. In: WWW, pp. 751–760 (2010)
4. Pang, B., Lee, L.: Opinion mining and sentiment analysis. Found. Trends Inf. Retr. 2(1–2), 1–135 (2008)
5. Liu, B., Zhang, L.: A survey of opinion mining and sentiment analysis. In: Aggarwal, C.C., Zhai, C.X. (eds.) Mining Text Data, pp. 415–463. Springer, New York (2012)
6. Bollegala, D., Weir, D.J., Carroll, J.A.: Cross-domain sentiment classification using a sentiment sensitive thesaurus. IEEE Trans. Knowl. Data Eng. 25(8), 1719–1731 (2013)
7. Yoshida, Y., Hirao, T., Iwata, T., Nagata, M., Matsumoto, Y.: Transfer learning for multiple — domain sentiment analysis — identifying domain dependent/independent word polarity. In: AAAI, pp. 1286–1291 (2011)
8. Ponomareva, N., Thelwall, M.: Semi-supervised vs. cross-domain graphs for sentiment analysis. In: RANLP, pp. 571–578 (2013)
9. Huang, S., Niu, Z., Shi, C.: Automatic construction of domain-specific sentiment lexicon based on constrained label propagation. Knowl.-Based Syst. 56, 191–200 (2014)
10. Dragoni, M., Tettamanzi, A.G., da Costa Pereira, C.: Propagating and aggregating fuzzy polarities for concept-level sentiment analysis. Cogn. Comput. 7(2), 186–197 (2015)
11. Pang, B., Lee, L.: A sentimental education: sentiment analysis using subjectivity summarization based on minimum cuts. In: ACL, pp. 271–278 (2004)
12. Qiu, L., Zhang, W., Hu, C., Zhao, K.: SELC: a self-supervised model for sentiment classification. In: CIKM, pp. 929–936 (2009)
13. Melville, P., Gryc, W., Lawrence, R.D.: Sentiment analysis of blogs by combining lexical knowledge with text classification. In: KDD, pp. 1275–1284 (2009)
14. Dragoni, M.: SHELLFBK: an information retrieval-based system for multi-domain sentiment analysis. In: Proceedings of the 9th International Workshop on Semantic Evaluation, SemEval 2015, Denver, Colorado, pp. 502–509. Association for Computational Linguistics (2015)
15. Petrucci, G., Dragoni, M.: An information retrieval-based system for multi-domain sentiment analysis. In: Gandon, F., Cabrio, E., Stankovic, M., Zimmermann, A. (eds.) SemWebEval 2015. CCIS, vol. 548, pp. 234–243. Springer, Heidelberg (2015). doi:10.1007/978-3-319-25518-7_20

16. Riloff, E., Patwardhan, S., Wiebe, J.: Feature subsumption for opinion analysis. In: EMNLP, pp. 440–448 (2006)
17. Wilson, T., Wiebe, J., Hwa, R.: Recognizing strong and weak opinion clauses. Comput. Intell. **22**(2), 73–99 (2006)
18. Dragoni, M., Tettamanzi, A.G.B., da Costa Pereira, C.: A fuzzy system for concept-level sentiment analysis. In: Presutti, V., et al. (eds.) SemWebEval 2014. CCIS, vol. 475, pp. 21–27. Springer, Heidelberg (2014)
19. Dragoni, M., Azzini, A., Tettamanzi, A.G.B.: A novel similarity-based crossover for artificial neural network evolution. In: Schaefer, R., Cotta, C., Kołodziej, J., Rudolph, G. (eds.) PPSN XI. LNCS, vol. 6238, pp. 344–353. Springer, Heidelberg (2010)
20. da Costa Pereira, C., Dragoni, M., Pasi, G.: A prioritized "and" aggregation operator for multidimensional relevance assessment. In: Serra, R., Cucchiara, R. (eds.) AI*IA 2009. LNCS, vol. 5883, pp. 72–81. Springer, Heidelberg (2009)
21. Palmero Aprosio, A., Corcoglioniti, F., Dragoni, M., Rospocher, M.: Supervised opinion frames detection with RAID. In: Gandon, F., Cabrio, E., Stankovic, M., Zimmermann, A. (eds.) SemWebEval 2015. CCIS, vol. 548, pp. 251–263. Springer, Heidelberg (2015). doi:10.1007/978-3-319-25518-7_22
22. Barbosa, L., Feng, J.: Robust sentiment detection on Twitter from biased and noisy data. In: COLING (Posters), pp. 36–44 (2010)
23. Bermingham, A., Smeaton, A.F.: Classifying sentiment in microblogs: is brevity an advantage? In: CIKM, pp. 1833–1836 (2010)
24. Go, A., Bhayani, R., Huang, L.: Twitter sentiment classification using distant supervision. CS224N Project report, Standford University (2009)
25. Cambria, E., Hussain, A.: Sentic Computing: Techniques, Tools, and Applications. SpringerBriefs in Cognitive Computation, vol. 2. Springer, Netherlands (2012)
26. Cambria, E., Olsher, D., Rajagopal, D.: SenticNet 3: a common and common-sense knowledge base for cognition-driven sentiment analysis. In: AAAI, pp. 1515–1521 (2014)
27. Stone, P.J., Dunphy, D., Smith, M.S.: The General Inquirer: A Computer Approach to Content Analysis. M.I.T. Press, Oxford (1966)
28. Zadeh, L.A.: Fuzzy sets. Inf. Control **8**, 338–353 (1965)
29. Manning, C.D., Surdeanu, M., Bauer, J., Finkel, J., Bethard, S.J., McClosky, D.: The Stanford CoreNLP natural language processing toolkit. In: Proceedings of 52nd Annual Meeting of the Association for Computational Linguistics: System Demonstrations, Baltimore, Maryland, pp. 55–60. Association for Computational Linguistics, June 2014
30. van Rijsbergen, C.J.: Information Retrieval. Butterworth, London (1979)
31. Zadeh, L.A.: The concept of a linguistic variable and its application to approximate reasoning - I. Inf. Sci. **8**(3), 199–249 (1975)
32. Hellendoorn, H., Thomas, C.: Defuzzification in fuzzy controllers. Intell. Fuzzy Syst. **1**, 109–123 (1993)
33. Dragoni, M., Tettamanzi, A., da Costa Pereira, C.: DRANZIERA: an evaluation protocol for multi-domain opinion mining. In: Calzolari, N., Choukri, K., Declerck, T., Goggi, S., Grobelnik, M., Maegaard, B., Mariani, J., Mazo, H., Moreno, A., Odijk, J., Piperidis, S. (eds.) Proceedings of the Tenth International Conference on Language Resources and Evaluation (LREC 2016), Paris, France. European Language Resources Association (ELRA), May 2016

A Knowledge-Based Approach for Aspect-Based Opinion Mining

Marco Federici[1] and Mauro Dragoni[2(✉)]

[1] Universitá di Trento, Trento, Italy
federici@fbk.eu
[2] Fondazione Bruno Kessler, Trento, Italy
dragoni@fbk.eu

Abstract. In the last decade, the focus of the Opinion Mining field moved to detection of the pairs "aspect-polarity" instead of limiting approaches in the computation of the general polarity of a text. In this work, we propose an aspect-based opinion mining system based on the use of semantic resources for the extraction of the aspects from a text and for the computation of their polarities. The proposed system participated at the third edition of the Semantic Sentiment Analysis (SSA) challenge took place during ESWC 2016 achieving the runner-up place in the Task #2 concerning the aspect-based sentiment analysis. Moreover, a further evaluation performed on the SemEval 2015 benchmarks demonstrated the feasibility of the proposed approach.

1 Introduction and Related Work

Opinion Mining is a natural language processing (NLP) task that aims to classify documents according to their opinion (polarity) on a given subject [1]. This task has created a considerable interest due to its wide applications in different domains: marketing, politics, social sciences, etc. Generally, the polarity of a document is computed by analyzing the expressions contained in the full text without distinguishing which are the subjects of each opinion. In the last decade, the research in the opinion mining field focused on the "aspect-based opinion mining" [2] consisting in the extraction of all subjects ("aspects") from documents and the opinions that are associated with them.

For clarification, let us consider the following example:

Yesterday, I bought a new smartphone.
The quality of the display is very good, but the buttery lasts too little.

In the sentence above, we may identify three aspects: "smartphone", "display", and "battery". As the reader may see, each aspect has a different opinion associated with it. The list below summarizes such associations:

- "display" → "very good"
- "battery" → "too little"

H. Sack et al. (Eds.): SemWebEval 2016, CCIS 641, pp. 141–152, 2016.
DOI: 10.1007/978-3-319-46565-4_11

– "smarthphone" → no explicit opinions, therefore polarity can be inferred by averaging the opinions associated with all other aspects.

The topic of aspect-based sentiment analysis has been explored under different perspectives. A comprehensive review of the last available systems can be found in the proceedings of SemEval 2015[1].

The paper is structured as follows. Section 2 briefly provides an overview of the aspect extraction task. Section 3 introduces the background knowledge used during the development of the system. Section 4 presents the underlying NLP layer upon which it has been developed the system described Sect. 5. Sections 6 and 7 shows the results obtained on the ESWC 2016 SSA challenge and on the SemEval 2015 benchmark, respectively. Finally, Sect. 8 provide a description about how the tasks of the challenge have been addressed and it concludes the paper.

2   Related Work

The topic of opinion mining has been studied extensively in the literature [3,4], where several techniques have been proposed and validated.

All the approaches presented so far operate at the document-level [5,6]; while, for improving the accuracy of the opinion classification, a more fine-grained analysis of the text, i.e., the opinion classification of every single sentence has to be performed [7,8]. In the literature, we may find approaches ranging from the use of fuzzy logic [9,10] to the use of aggregation techniques [11] for computing the score aggregation of opinion words. In the case of sentence-level opinion classification, two different sub-tasks have to be addressed: (i) to determine if the sentence is subjective or objective, and (ii) in the case that the sentence is subjective, to determine if the opinion expressed in the sentence is positive, negative, or neutral. The task of classifying a sentence as subjective or objective, called "subjectivity classification", has been widely discussed in the literature [7,8] and systems implementing the capabilities of identifying opinion's holder, target, and polarity have been presented [12].

The growth of online product reviews was the perfect floor for using opinion mining techniques in marketing activities. The issue of detecting the different opinions concerning the same product expressed in the same review emerged as a challenging problem. Such a task has been faced by introducing *aspect* extraction approaches aiming to extract, from each sentence, which is the aspect the opinion refers to. In the literature, many approaches have been proposed: conditional random fields (CRF) [13,14], hidden Markov models (HMM) [15–17], sequential rule mining [18], dependency tree kernels [19], clustering [20], and genetic algorithms [21]. In [22,23], a method was proposed to extract both opinion words and aspects simultaneously by exploiting some syntactic relations of opinion words and aspects.

---

[1] http://alt.qcri.org/semeval2015/cdrom/pdf/SemEval082.pdf.

At the same time, the social dimension of the Web opens up the opportunity to combine computer science and social sciences to better recognize, interpret, and process opinions and sentiments expressed over it. Such multi-disciplinary approach has been called *sentic computing* [24].

Above, we mentioned approaches that do not consider the domain analyzed documents belong to. The use of domain adaptation demonstrated that opinion classification is highly sensitive to the domain from which the training data is extracted. A classifier trained using opinionated documents from one domain often performs poorly when applied or tested on opinionated documents from another domain. The reason is that using the same words and even the same language constructs can carry different opinions, depending on the domain.

The classic scenario is when the same word in one domain may have positive connotations, but in another domain may have negative one; therefore, domain adaptation is needed. In the literature, different approaches related to the Multi-Domain sentiment analysis have been proposed. Briefly, two main categories may be identified: (i) the transfer of learned classifiers across different domains [25–28], and (ii) the use of propagation of labels through graph structures [9,29,30].

While on one side such approaches demonstrated their effectiveness in working in a multi-domain environment, on the other one, they suffer by the limitation in abstracting their usage within any domain different from the ones used for building the model.

3   Preliminaries

The system is implemented on top of a background knowledge used for representing the linguistic connections between "concepts" described in several resources. Below, it is possible to find the list of such resources and the links where further information about them may be found.

WordNet[2] [31] is one of the most important resource available to researchers in the field of text analysis, computational linguistics, and many related areas. In the implemented system, WordNet has been used as starting point for the construction of the semantic graph used by the system (see Sect. 5). However, due to some coverage limitations occurring in WordNet, it has been extended by linking further terms coming from the Roget's Thesaurus [32].

SenticNet[3] [33] is a publicly available resource for opinion mining that exploits both Artificial Intelligence and Semantic Web techniques to infer the polarity associated with common-sense concepts and represent it in a semantic-aware format. In particular, SenticNet uses dimensionality reduction to calculate the affective valence of a set of Open Mind concepts and represent it in a machine-accessible and machine-processable format.

General Inquirer dictionary[4] [34] is an English-language dictionary containing almost 12,000 elements associated with their polarity in different contexts.

---

[2] https://wordnet.princeton.edu/.

[3] http://sentic.net/.

[4] http://www.wjh.harvard.edu/~inquirer/spreadsheet_guide.htm.

Such dictionary is the result of the integration between the "Harvard" and the "Lasswell" general-purpose dictionaries as well as a dictionary of categories define by the dictionary creators. When necessary, for ambiguous words, specific polarity for each sense is specified.

4   The Underlying NLP Layer

The presented system has been implemented on top of existing Natural Language Processing libraries. In particular, it uses different functionalities offered by the Stanford NLP Library.

WordNet[5] [31] resource is used together with Stanford's part of speech annotation to detect compound nouns. Lists of consecutive nouns and word sequences contained in Wordnet compound nouns vocabulary are merged into a single word in order to force Stanford library to consider them as a single unit during the following phases. The entire text is then fed to the co-reference resolution module to compute pronoun references which are stored in an index-reference map. Details about the textual analysis are provided in Sect. 5.

The next operation consists in detecting which word expresses polarity within each sentence. To achieve this task *SenticNet, General Inquirer dictionary* and *MPQA* sentiment lexicons have been used.

While SenticNet expresses polarity values in the continuous range from -1 to 1, the other two resources been normalized: the General Inquirer words have positive values of polarity if they belong to the "Positiv" class while negative if they belong to "Negativ" one, zero otherwise, similarly, MPQA "polarity" labels are used to infer a numerical values. Only words with a non-zero polarity value in at least one resource are considered as opinion words (e.g. word "third" is not present in MPQA and SenticNet and has a 0 value according to General Inquirer, consequently, it is not a valid opinion word; on the other hand, word "huge" has a positive 0.069 value according to SenticNet, a negative value in MPQA and 0 value according to General Inquirer, therefore, it is a possible opinion word even if lexicons express contrasting values). Every noun (single or complex) is considered an aspect as long as it's connected to at least one opinion and it's not in the stopword list. This list has been created starting from the "Onix" text retrieval engine stopwords list[6] and it contains words without a specific meaning (such as "thing") and special characters.

Opinions associated with pronouns are connected to the aspect they are referring to; instead, if pronouns reference can't be resolved, they are both discarded.

The main task of the system is, then, represented by connecting opinions with possible aspects. Two different approaches have been tested with a few variants. The first one relies on the syntactic tree while the second one is based on grammar dependencies.

---

[5] https://wordnet.princeton.edu/.

[6] The used stopwords list is available at http://www.lextek.com/manuals/onix/stopwords1.html.

The sentence "I enjoyed the screen resolution, it's amazing for such a cheap laptop." has been used to underline differences in connection techniques.

The preliminary phase merges words "screen" and "resolution" into a single word "Screenresolution" because they are consecutive nouns. Co-reference resolution module extracts a relation between "it" and "Screenresolution". This relation is stored so that every possible opinion that would be connected to "it" will be connected to "Screenresolution" instead. Figure 1 shows the syntax tree while Fig. 2 represents the grammar relation graph generated starting from the example sentence. Both structures have been computed using Stanford NLP modules ("parse", "depparse").

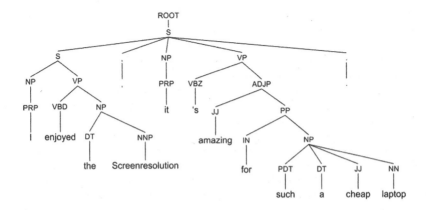

Fig. 1. Example of syntax tree.

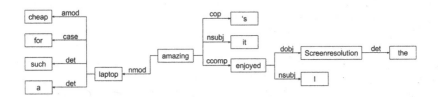

Fig. 2. Example of the grammar relations graph.

5   The Implemented System

The aspect extraction component is based on a six-phases approach as presented below.

Phase 1. Sentences are given as input to the Stanford NLP Library[7] and they are annotated with part of speech (POS) tags in order to detect nouns, adjectives, and pronouns.

Phase 2. Tokens annotated as adjectives are considered for computing opinion scores, while sequences of one or more consecutive nouns (for example "support" tagged as "NN" followed by "team" tagged "NN" as well) and complex linguistic structures recognized through the use of Wordnet (for example "hard" annotated as "JJ" and "disk" annotated as "NN") are aggregated and marked as potential aspects. This step is shown in Fig. 3.

Phase 3. Co-reference resolution is applied for resolving pronouns co-references between nouns. Example about how co-reference is applied is shown in Fig. 3.

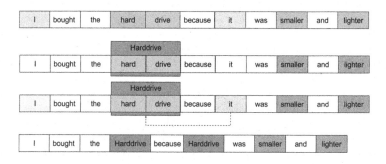

Fig. 3. Example of noun aggregation and co-reference resolution.

Phase 4. After the aggregation of compound names, we changed the sentences by replacing compound names with single tokens to ensure that they are considered as single entities during the opinion resolution phase. This way, it will be possible to exchange each pronoun with the corresponding label of the aspect they are referring to.

Phase 5. Stanford Parser is used for generating a syntax tree that is exploited in the last phase for associating opinions with aspects. Concerning the definition of the associations between aspects and opinions, during preliminary testing activities, we tried different approaches. Among them:

– each aspect has been connected with each opinion contained in the same sentence, where as sentence delimiters, we used the markers "S", "SBAR", and "FRAG" detected in the parsed tree;
– if an opinion is expressed in a sentence without nouns, such an opinion has been associated with the aspects belonging to the same noun phrase only.

Example of the generated parsed tree is shown in Fig. 4.

---

[7] http://stanfordnlp.github.io/CoreNLP/index.html.

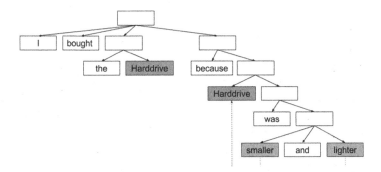

Fig. 4. Example of generated parse tree.

Phase 6. Finally, aspects without associated opinions are discarded, while remaining ones are stored. Example is shown in Fig. 5.

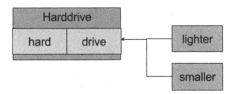

Fig. 5. Example of opinion association.

6 ESWC-2016 SSA Challenge Task #2: Aspect-Based Sentiment Analysis

The expected output of this task was a set of aspects of the reviewed product and a binary polarity value associated to each of such aspects. So, for example, while for classic binary polarity inference task an overall polarity (positive or negative) was expected for a review about a mobile phone, this task required a set of aspects (such as *speaker, touchscreen, camera,* etc.) and a polarity value (positive or negative) associated with each of such aspects. Engines were assessed according to both aspect extraction and aspect polarity detection using precision, recall, f-measure, and accuracy similarly as performed during the first edition of the Concept-Level Sentiment Analysis challenge held during ESWC2014 and re-proposed at SemEval 2015 Task 12[8]. Please refer to SemEval 2016 Task 5[9] for details on the precision-recall analysis. Figure 6 shows an example of the output schema for Task #2.

---

[8] http://www.alt.qcri.org/semeval2015/task12/.

[9] http://alt.qcri.org/semeval2016/task5/.

```xml
<?xml version="1.0" encoding="UTF-8" standalone="yes"?>
<Review rid="1">
    <sentences>
        <sentence id="348:0">
            <text>Most everything is fine with this machine: speed, capacity, build.</text>
                <Opinions>
                    <Opinion aspect="MACHINE" polarity="positive"/>
                </Opinions>
        </sentence>
        <sentence id="348:1">
            <text>The only thing I don't understand is that the resolution of the
              screen isn't high enough for some pages, such as Yahoo!Mail.
            </text>
                <Opinions>
                    <Opinion aspect="SCREEN" polarity="negative"/>
                </Opinions>
        </sentence>
        <sentence id="277:2">
            <text>The screen takes some getting use to, because it is smaller
              than the laptop.</text>
                <Opinions>
                    <Opinion aspect="SCREEN" polarity="negative"/>
                </Opinions>
        </sentence>
    </sentences>
</Review>
```

Fig. 6. Task #2 output example. Input is the same without the opinion tag and its descendant nodes.

The training set was composed by 5,058 sentences coming from two different domains: "Laptop" (3,048 sentences) and "Restaurant" (2,000 sentences). While, the test set was composed by 891 sentences coming from the "Laptop" (728 sentences) and "Hotels" (163 sentence). The reason for which we decided to use the "Hotels" domain in the test set with respect to the "Restaurant" one was to observe the capability of the participant systems to be general purpose with respect to the training set.

7   System Evaluation and Error Analysis

The system has been tested on two aspect-based sentiment analysis datasets by following the "Semi-Open" setting of the DRANZIERA protocol [35]:

D1 The SemEval 2015 Task 12 training set benchmark, consisting in sentences belonging to the "Laptop" and "Restaurant" domains.
D2 The ESWC2016 Benchmark on Semantic Sentiment Analysis test set, consisting in sentences belonging to the "Laptop" and "Hotels" domains.

To compute results, a notion of correctness has to be introduced: if the extracted aspects is equal, contained or contains the correct one, it's considered to be correct (for example if the extracted aspect is "screen", while the annotated one is "screen of the computer" or vice versa, the result of the system is considered to correct).

Table 1. Results obtained by the presented system on the SemEval 2015 Task 12 dataset and on the test set adopted for the challenge.

Dataset	Precision	Recall	F-Measure	Polarity accuracy
D1	0.39969	0.39478	0.39722	0.91720
D2	0.34820	0.35745	0.35276	0.84925

Table 2. Precision-recall analysis and winners for Task 2.

System	Precision	Recall	F-Measure	Accuracy
Soufian Jebbara and Philipp Cimiano **Aspect-Based Sentiment Analysis Using a Two-Step Neural Network Architectures**	0.41471	0.45196	0.43253	0.87356
Marco Federici and Mauro Dragoni • **A Knowledge-based Approach For Aspect-Based Opinion Mining**	0.34820	0.35745	0.35276	0.84925
Andi Rexha, Mark Kröll Mauro Dragoni and Roman Kern **Exploiting Propositions for Opinion Mining**	N/A	N/A	N/A	N/A

Table 1 shows the results obtained on each dataset; while, Table 2 shows the full results of Task #2 participants.

Figure 7 shows an analysis of error cases. Values have been computed according to the first 100 sentences of the "Laptop" dataset.

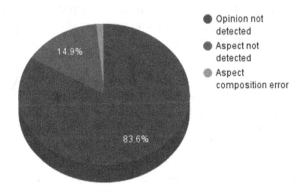

● Opinion not detected
● Aspect not detected
● Aspect composition error

14.9%

83.6%

Fig. 7. Overall error analysis.

The majority of false negatives are given by the impossibility to detect opinions expressed by verbs. For example, in the sentence "I generally like this place" or more complex expressions "tech support would not fix the problem unless I bought your plan for $150 plus".

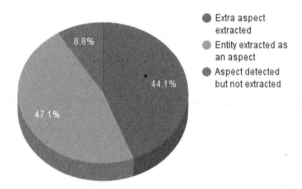

Fig. 8. Specific error analysis on aspect extraction.

Other issues are correlated to the association algorithm. Figure 8 shows the specific error analysis related to the extraction of aspects, always computed on the same 100 sentences of the "Laptop" dataset.

Even if the syntax-tree-based approach tends to produce a significant number of true positives, relationships are often imprecise. A relevant example is represented by the sentence "I was extremely happy with the OS itself." in the "Laptop" dataset. The approach connects the opinion adjective "happy" with the potential aspect "OS", correctly recognized as an aspect in the sentence.

A relevant part of false positives are generated due to the incapability of discriminating aspects from the entity itself. In facts, almost half of them consists in associations between opinion words and the entity reviewed that are correct. However, they must not be considered during the aspect extraction task (for example the aspect "laptop" in the example sentence should not be considered according to the definition of aspect).

8   Conclusions

In this paper, we presented the system submitted to the third edition of the Semantic Sentiment Analysis run during ESWC 2016. The system participated only at Task #2 and obtained the second place. The problem of detecting aspects in sentences is very relevant in the sentiment analysis community. Further work in this direction will be performed by starting from the analysis of the errors provided in the previous section.

References

1. Pang, B., Lee, L., Vaithyanathan, S.: Thumbs up? Sentiment classification using machine learning techniques. In: Proceedings of EMNLP, Philadelphia, pp. 79–86. Association for Computational Linguistics, July 2002

2. Hu, M., Liu, B.: Mining and summarizing customer reviews. In: Proceedings of the Tenth ACM SIGKDD International Conference on Knowledge Discovery and Data Mining, pp. 168–177. ACM (2004)
3. Pang, B., Lee, L.: Opinion mining and sentiment analysis. Found. Trends Inf. Retrieval **2**(1–2), 1–135 (2008)
4. Liu, B., Zhang, L.: A survey of opinion mining and sentiment analysis. In: Aggarwal, C.C., Zhai, C.X. (eds.) Mining Text Data, pp. 415–463. Springer, Berlin (2012)
5. Dragoni, M.: SHELLFBK: an information retrieval-based system for multi-domain sentiment analysis. In: Proceedings of the 9th International Workshop on Semantic Evaluation, SemEval 2015, Denver, Colorado, pp. 502–509. Association for Computational Linguistics, June 2015
6. Petrucci, G., Dragoni, M.: An information retrieval-based system for multi-domain sentiment analysis. In: Gandon, F., et al. (eds.) SemWebEval 2015. CCIS, vol. 548, pp. 234–243. Springer, Heidelberg (2015). doi:10.1007/978-3-319-25518-7_20
7. Riloff, E., Patwardhan, S., Wiebe, J.: Feature subsumption for opinion analysis. In: EMNLP, pp. 440–448 (2006)
8. Wilson, T., Wiebe, J., Hwa, R.: Recognizing strong and weak opinion clauses. Comput. Intell. **22**(2), 73–99 (2006)
9. Dragoni, M., Tettamanzi, A.G., da Costa Pereira, C.: Propagating and aggregating fuzzy polarities for concept-level sentiment analysis. Cogn. Comput. **7**(2), 186–197 (2015)
10. Dragoni, M., Tettamanzi, A.G.B., da Costa Pereira, C.: A fuzzy system for concept-level sentiment analysis. In: Presutti, V., et al. (eds.) SemWebEval 2014. CCIS, vol. 475, pp. 21–27. Springer, Heidelberg (2014)
11. da Costa Pereira, C., Dragoni, M., Pasi, G.: A prioritized "and" aggregation operator for multidimensional relevance assessment. In: Serra, R., Cucchiara, R. (eds.) AI*IA 2009. LNCS, vol. 5883, pp. 72–81. Springer, Heidelberg (2009)
12. Palmero Aprosio, A., Corcoglioniti, F., Dragoni, M., Rospocher, M.: Supervised opinion frames detection with RAID. In: Gandon, F., et al. (eds.) SemWebEval 2015. CCIS, vol. 548, pp. 251–263. Springer, Heidelberg (2015). doi:10.1007/978-3-319-25518-7_22
13. Jakob, N., Gurevych, I.: Extracting opinion targets in a single and cross-domain setting with conditional random fields. In: EMNLP, pp. 1035–1045 (2010)
14. Lafferty, J.D., McCallum, A., Pereira, F.C.N.: Conditional random fields: probabilistic models for segmenting and labeling sequence data. In: ICML, pp. 282–289 (2001)
15. Freitag, D., McCallum, A.: Information extraction with HMM structures learned by stochastic optimization. In: AAAI/IAAI, pp. 584–589 (2000)
16. Jin, W., Ho, H.H.: A novel lexicalized HMM-based learning framework for web opinion mining. In: Proceedings of the 26th Annual International Conference on Machine Learning, ICML 2009, pp. 465–472. ACM, New York (2009)
17. Jin, W., Ho, H.H., Srihari, R.K.: OpinionMiner: a novel machine learning system for web opinion mining and extraction. In: KDD, pp. 1195–1204 (2009)
18. Liu, B., Hu, M., Cheng, J.: Opinion observer: analyzing and comparing opinions on the web. In: WWW, pp. 342–351 (2005)
19. Wu, Y., Zhang, Q., Huang, X., Wu, L.: Phrase dependency parsing for opinion mining. In: EMNLP, pp. 1533–1541 (2009)
20. Su, Q., Xu, X., Guo, H., Guo, Z., Wu, X., Zhang, X., Swen, B., Su, Z.: Hidden sentiment association in chinese web opinion mining. In: WWW, pp. 959–968 (2008)

21. Dragoni, M., Azzini, A., Tettamanzi, A.G.B.: A novel similarity-based crossover for artificial neural network evolution. In: Schaefer, R., Cotta, C., Kołodziej, J., Rudolph, G. (eds.) PPSN XI. LNCS, vol. 6238, pp. 344–353. Springer, Heidelberg (2010)

22. Qiu, G., Liu, B., Bu, J., Chen, C.: Expanding domain sentiment lexicon through double propagation. In: IJCAI, pp. 1199–1204 (2009)

23. Qiu, G., Liu, B., Bu, J., Chen, C.: Opinion word expansion and target extraction through double propagation. Comput. Linguist. **37**(1), 9–27 (2011)

24. Cambria, E., Hussain, A.: Sentic Computing: Techniques, Tools, and Applications. SpringerBriefs in Cognitive Computation, vol. 2. Springer, Dordrecht (2012)

25. Blitzer, J., Dredze, M., Pereira, F.: Biographies, bollywood, boom-boxes and blenders: domain adaptation for sentiment classification. In: ACL, pp. 187–205 (2007)

26. Pan, S.J., Ni, X., Sun, J.T., Yang, Q., Chen, Z.: Cross-domain sentiment classification via spectral feature alignment. In: WWW, pp. 751–760 (2010)

27. Bollegala, D., Weir, D.J., Carroll, J.A.: Cross-domain sentiment classification using a sentiment sensitive thesaurus. IEEE Trans. Knowl. Data Eng. **25**(8), 1719–1731 (2013)

28. Yoshida, Y., Hirao, T., Iwata, T., Nagata, M., Matsumoto, Y.: Transfer learning for multiple-domain sentiment analysis–identifying domain dependent/independent word polarity. In: AAAI, pp. 1286–1291 (2011)

29. Ponomareva, N., Thelwall, M.: Semi-supervised vs. cross-domain graphs for sentiment analysis. In: RANLP, pp. 571–578 (2013)

30. Huang, S., Niu, Z., Shi, C.: Automatic construction of domain-specific sentiment lexicon based on constrained label propagation. Knowl.-Based Syst. **56**, 191–200 (2014)

31. Fellbaum, C.: WordNet: An Electronic Lexical Database. MIT Press, Cambridge (1998)

32. Kipfer, B.A.: Roget's 21st Century Thesaurus, 3rd edn. (2005)

33. Cambria, E., Speer, R., Havasi, C., Hussain, A.: SenticNet: a publicly available semantic resource for opinion mining. In: AAAI Fall Symposium: Commonsense Knowledge (2010)

34. Stone, P.J., Dunphy, D., Marshall, S.: The General Inquirer: A Computer Approach to Content Analysis. M.I.T. Press, Oxford (1966)

35. Dragoni, M., Tettamanzi, A., da Costa Pereira, C.: Dranziera: an evaluation protocol for multi-domain opinion mining. In: Calzolari, N., Choukri, K., Declerck, T., Goggi, S., Grobelnik, M., Maegaard, B., Mariani, J., Mazo, H., Moreno, A., Odijk, J., Piperidis, S. (eds.) Proceedings of the Tenth International Conference on Language Resources and Evaluation (LREC 2016), Paris, France. European Language Resources Association (ELRA), May 2016

Aspect-Based Sentiment Analysis Using a Two-Step Neural Network Architecture

Soufian Jebbara$^{(\boxtimes)}$ and Philipp Cimiano

Semantic Computing Group,
Cognitive Interaction Technology – Center of Excellence (CITEC),
Bielefeld University, Bielefeld, Germany
sjebbara@cit-ec.uni-bielefeld.de

Abstract. The World Wide Web holds a wealth of information in the form of unstructured texts such as customer reviews for products, events and more. By extracting and analyzing the expressed opinions in customer reviews in a fine-grained way, valuable opportunities and insights for customers and businesses can be gained.

We propose a neural network based system to address the task of Aspect-Based Sentiment Analysis to compete in Task 2 of the ESWC-2016 Challenge on Semantic Sentiment Analysis. Our proposed architecture divides the task in two subtasks: aspect term extraction and aspect-specific sentiment extraction. This approach is flexible in that it allows to address each subtask independently. As a first step, a recurrent neural network is used to extract aspects from a text by framing the problem as a sequence labeling task. In a second step, a recurrent network processes each extracted aspect with respect to its context and predicts a sentiment label. The system uses pretrained semantic word embedding features which we experimentally enhance with semantic knowledge extracted from WordNet. Further features extracted from SenticNet prove to be beneficial for the extraction of sentiment labels. As the best performing system in its category, our proposed system proves to be an effective approach for Aspect-Based Sentiment Analysis.

1 Introduction

The World Wide Web contains customer reviews for all kinds of topics and entities such as products, movies, events, restaurants and more. The wealth of information that is expressed in these reviews in the form of the writer's opinion offers valuable opportunities and insights for customers and businesses altogether. However, due to the vast amounts of customer reviews that are available in the Web, the manual extraction and analysis of these opinions is infeasible and thus requires automated tools. First attempts to extract opinions automatically have focused on extracting an overall polarity on a document or sentence level. This, however, is a too coarse-grained approach as it neglects huge amounts of information in these reviews.

In a more fine-grained way, sentiment analysis can be regarded as a relation extraction problem in which the sentiment of some opinion holder towards a

© Springer International Publishing Switzerland 2016
H. Sack et al. (Eds.): SemWebEval 2016, CCIS 641, pp. 153–167, 2016.
DOI: 10.1007/978-3-319-46565-4_12

certain aspect of a product needs to be extracted. The following example clearly shows that the mere extraction of an overall polarity for a sentence is not sufficient:

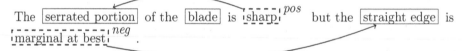

The serrated portion of the blade is sharp^{pos} but the straight edge is marginal at best^{neg} .

where aspect terms are outlined with solid boxes, opinion phrases with dashed ones, opinion polarities are displayed as superscripts, and aspect-opinion dependencies are depicted as arrows. Sentiment analysis needs to be regarded thus on a more fine-grained level that allows to assign sentiments to individual aspects in order to extract complex opinions more accurately.

In this work, we present a system that competes in the ESWC 2016 Challenge on Semantic Sentiment Analysis addressing the task of Aspect-Based Sentiment Analysis. The goal of this task is to extract a set of aspect terms with their respective binary polarities (**positive** and **negative**) from a given sentence. The sentences in the overall dataset are extracted from online reviews from different domains (restaurants, laptops and hotels). We approach the problem in two steps: (i) the extraction of aspect terms and (ii) the assignment of a polarity label to each extracted aspect term. Following this approach, we design a modular, neural network based architecture that is easy to extend.

In the following, we give a brief overview of related work in the field of aspect-based sentiment analysis. Afterwards, we present our overall system and describe its two main components and the features we employ. We further analyse the performance of our architecture on both subtasks and give insights into its predictive performance. Lastly, we conclude the paper and give suggestions for further improvements.

2   Related Work

Our work is inspired by different related approaches for sentiment analysis. Overall, our work is in line with the growing interest of providing more fine-grained, aspect-based sentiment analysis [14,15,23], going beyond a mere text classification or regression problem that aims at predicting an overall sentiment for a text.

San Vicente et al. [24] present a system that addresses opinion target extraction as a sequence labeling problem based on a perceptron algorithm with local features. The extraction of a sentiment polarity for an extracted opinion target is performed using an SVM. The approach uses a window of words around a given opinion target and classifies it based on a set of features such as word clusters, Part-of-Speech tags and polarity lexicon features.

Toh and Wang [32] propose a Conditional Random Field (CRF) as a sequence labeler that includes a variety of features such as POS tags and dependencies, word clusters and WordNet taxonomies. Additionally, the authors employ a logistic regression classifier to address aspect term polarity classification.

Jakob and Gurevych [11] follow a very similar approach that addresses opinion target extraction as a sequence labeling problem using CRFs. Their approach includes features derived from words, POS tags and dependency paths, and performs well in a single and cross-domain setting.

Klinger and Cimiano [13,14] have modeled the task of joint aspect and opinion term extraction using probabilistic graphical models and rely on Markov Chain Monte Carlo methods for inference. They have demonstrated the impact of a joint architecture on the task with a strong impact on the extraction of aspect terms, but less so for the extraction of opinion terms.

Lakkaraju et al. [15] present a recursive neural network architecture that is capable of extracting multiple aspect categories[1] and their respective sentiments jointly in one model or separately using two softmax classifiers. They show that the joint modeling of aspect categories and sentiments is beneficial for the predictive performance of their system.

Another way to address opinion extraction is the summarization of reviews. Hu and Liu [10] present an approach that summarizes reviews based on the product features for which an opinion is expressed using data mining and natural language processing techniques. Similarly, Titov and McDonald [30] describe a statistical model for joint aspect and sentiment modeling for the summarization of reviews. The method is based on Multi-Grain Latent Dirichlet Allocation which models global and local topics extended by a Multi-Aspect Sentiment Model.

Lastly, the general idea expressed in this paper to incorporate semantic web technologies in a machine learning framework for sentiment analysis is rooted in previous contributions of ESWC Challenges [1,4,7,27].

3   Aspect-Based Sentiment Analysis

We follow a two-step approach in designing a system that is capable of extracting a writer's sentiment towards certain aspects of an entity (such as a product or restaurant). As a first step, given a text, the system extracts explicitly expressed aspects[2] in this text. Secondly, each extracted aspect term is processed individually and a sentiment value is assigned given the context of the aspect term.

This two-step approach allows us to extract an arbitrary amount of aspects from a text. Additionally, by decoupling the aspect extraction from the sentiment extraction, the system is also applicable to settings where aspect terms are already given and only the individual sentiments towards these aspects need to be extracted. The following sections elaborate on our design and feature choices for our aspect and sentiment extraction components.

---

[1] Here, we distinguish between the terminologies of aspect *category* extraction and aspect *term* extraction: The set of possible aspect categories is predefined and rather small (e.g. Price, Battery, Accessories, Display, Portability, Camera), while aspect terms can take many shapes (e.g. *"sake menu"*, *"wine selection"* or *"French Onion soup"*).

[2] Parts of a sentence that refer to an aspect of the product, event, entity, etc.

3.1   Features

In this section, we describe the features we use to address aspect term extraction and aspect-specific sentiment extraction. For both sub tasks, we lowercase each input sentence and tokenize it using a simple regular expression in a preprocessing step. We do not remove punctuations or stopwords, but keep them intact.

Word Embeddings. The most important features that we use are pretrained word embeddings which have been successfully employed in numerous NLP tasks [6,16,20,22,25]. We use the skip-gram model [20] with negative sampling on a huge corpus of $\approx$83 million Amazon reviews [18,19] to compute 100 dimensional word embeddings. In total, our computation of word embeddings yields vectors for $\approx$1 million words. For this work, however, we reduce this vocabulary to only contain the 100,000 most frequent words. The resulting vocabulary is denoted as V.

In a pre-processing step, we replace rare words that appear less than 10 times in our dataset with a special token <UNK> and learn a *placeholder* vector for this token. At test time, we use this token as a replacement for Out-of-Vocabulary words. The sequence of word embedding vectors for a sentence with words[3] $1 \ldots N$ is denoted as:

$$[w]_1^N = \{w_1, \ldots, w_N\} \text{ with } w_i \in \mathbb{R}^{100}.$$

By using this domain-specific dataset we expect to obtain embeddings that capture the semantics of each word for our targeted domain more closely than embeddings trained on domain-independent data. A welcomed side effect of using this huge dataset of reviews is that we also obtain word embeddings for misspelled forms of a word that appear commonly in reviews. As shown in Table 1, the learned representation of a misspelled word is in many cases very close[4] to its correctly spelled counterpart.

Table 1. Three commonly used words in product reviews and their 3 nearest neighbors in the embedding space. Often, misspelled versions (*italic*) of the original word are among its closest neighbors.

Word	Speed	Quality	Display
Nearest neighbors	*spped*	*qualtiy*	displays
	speeds	*qualilty*	*diplay*
	speeed	*qulaity*	*dislay*

---

[3] For a more convenient notation, we use words and their respective indices interchangeably.

[4] We use the euclidean vector distance as a distance measure.

Although our approach technically works without any features apart from word embeddings, we are interested in improving its performance by means of semantic web technology. For that, we employ features derived from two graph-based semantic resources: WordNet and SenticNet.

Retrofitting Word Embeddings to WordNet. Although word embeddings have been shown to encode semantic and syntactic features of their respective words well [20, 21, 25], we try to enhance their encoded semantics by using a lexical resource. For this, we employ a technique called *retrofitting* [8]. The idea behind retrofitting is to iteratively adapt precomputed word vectors to better fit the (lexical) relations modeled in a given lexical resource. The graph-based algorithm gradually "moves" each word vector towards the word vectors of its neighboring nodes while still staying close to its original position.

Formally, following the notation by Faruqui et al. [8], let $V = \{v_1, \ldots, v_{N_V}\}$ be the considered vocabulary of size N_V and $\hat{W} = (\hat{w}_1, \ldots, \hat{w}_{|V|})$ with $w_i \in \mathbb{R}^d$ are their respective precomputed word vectors. $G = (V, E)$ is the graph of semantic relationships to which we want to fit the word vectors with $(v_i, v_j) \in E \subseteq V \times V$ denoting the edges between words. With $W = (w_1, \ldots, w_{|V|})$ being the fitted word vectors, the algorithm tries to minimize the following objective function:

$$\Psi(\hat{W}, W) = \sum_{i=1}^{|V|} \left[\alpha_i ||w_i - \hat{w}_i||^2 + \sum_{(i,j) \in E} \beta_{ij} ||w_i - w_j||^2 \right] \tag{1}$$

The online update rule for each w_i is then:

$$w_i = \frac{\sum_{j:(i,j) \in E} \beta_{ij} w_j + \alpha_i \hat{w}_i}{\sum_{j:(i,j) \in E} \beta_{ij} + \alpha_i} \tag{2}$$

where α and β are parameters of the retrofitting procedure.

In this work, we chose WordNet [9] as our lexico-semantic resource. We construct a subgraph of the WordNet relations that links each word in our vocabulary to all its synonyms (lemma names) in the WordNet graph. We set all $\alpha_i = 1$ and all $\beta_{ij} = 1/degree(i)$ and run the retrofitting algorithm for 10 iterations. The resulting embeddings are still very similar to their original embeddings, yet incorporate part of the semantics of WordNet. We investigate the benefit of using these retrofitted word embeddings in comparison to their original counterparts in Sect. 4.

SenticNet. SenticNet 3 [2] is a graph-based, concept-level resource for semantic and affective information. For each of the 30,000 concepts that are part of the knowledge graph, SenticNet 3 provides real-valued scores for 5 *sentics*: pleasantness, attention, sensitivity, aptitude, polarity.

We experimentally include the provided scores in our system as an additional input source that our networks can draw information from. Since these sentics encode information about the semantics and polarity of a concept, the aspect-specific sentiment extraction component is expected to benefit from the additional information in particular. For that, we construct a 5-dimensional feature vector s_c for each concept c that is represented in SenticNet 3. We refer to these vectors as *sentic vectors*.

Unfortunately, our system is not designed to process text on a concept level but only on a word level. Therefore, we omit all multi-word concepts (e.g. `notice_problem` or `beautiful_music`) in SenticNet 3 and only keep single-word concepts (e.g. `experience` or `improvement`) that are part of our vocabulary V. Doing that, we can treat the sentic vector s_i as an additional word vector for the word i. To account for Out-of-Vocabulary words during test time, we provide a default vector $s_{unk} = \mathbf{0}$. The sequence of sentic vectors for a sentence with words $1 \ldots N$ is denoted as:

$$[s]_1^N = \{s_1, \ldots, s_N\} \text{ with } s_i \in \mathbb{R}^5.$$

Part-of-Speech Tags. Apart from these word embeddings and sentic vectors, our system can incorporate other features as well. For each word in a text, Part-of-Speech (POS) tags can be provided that might aid both the aspect extraction and aspect-specific sentiment extraction components. When including POS tags, we employ a 1-of-K coding scheme that transforms each tag into a K-dimensional vector that represents this specific tag. Specifically, we use the Stanford POS Tagger [17] with a tag set of 45 tags. These vectors are then concatenated with their respective word vectors before being fed to the extraction components. The sequence of POS tag vectors for a sentence with words $1 \ldots N$ is denoted as:

$$[p]_1^N = \{p_1, \ldots, p_N\} \text{ with } p_i \in \mathbb{R}^{45}.$$

3.2 Aspect Term Extraction

Our first step in extracting aspect-based sentiment from a text is the extraction of mentioned aspect terms. We propose a system to extract an arbitrary number of aspect terms from a given text by framing the extraction as a sequence labeling problem. For this, we encode expressed aspect terms using the IOB2 tagging scheme [31]. According to this scheme, each word in our text receives one of 3 tags, namely **I**, **O** or **B** that indicate if the word is at the **B**eginning, **I**nside or **O**utside of an annotation:

The	sake	menu	should	not	be	overlooked	!
O	**B**	**I**	O	O	O	O	O

This tagging scheme allows us to encode multiple non-overlapping aspect terms at once. Ultimately, each tag is represented as a 1-of-K vector:

$$I = \begin{bmatrix} 1 \\ 0 \\ 0 \end{bmatrix}, O = \begin{bmatrix} 0 \\ 1 \\ 0 \end{bmatrix}, B = \begin{bmatrix} 0 \\ 0 \\ 1 \end{bmatrix}.$$

We design a neural network based sequence tagger that reads in a sequence of words and predicts a sequence of corresponding IOB2 tags that encode the detected aspect terms. Figure 1 depicts the neural network component.

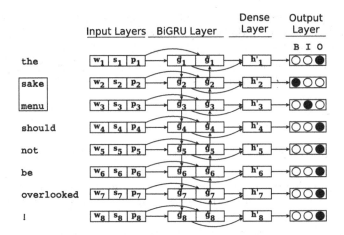

Fig. 1. The aspect term extraction component. The network processes the input sentence as a sequence of word vectors w_i, sentic vectors s_i and POS tags p_i using a bidirectional GRU layer and regular feed-forward layers. The output of the network is a predicted tag sequence in the IOB2 format. The aspect term that is to be predicted is outlined in the input sentence.

Neural Network Sequence Tagger. The procedure to generate a tag sequence for a given word sequence can be described as follows: First, the sequence of words is mapped to a sequence of word embedding vectors $[w]_1^N = \{w_1, \ldots, w_N\}$, sentic vectors $[s]_1^N = \{s_1, \ldots, s_N\}$ and POS tag vectors $[p]_1^N = \{p_1, \ldots, p_N\}$ using the resources described in Sect. 3.1. We concatenate each word vector with its corresponding sentic vector and POS tag vector to receive the sequence:

$$[u]_1^N = \{u_1, \ldots, u_N\} = \{(w_1, s_1, p_1)^T, \ldots, (w_N, s_N, p_N)^T\} \text{ with } u_i \in \mathbb{R}^{100+5+45}.$$

The resulting sequence is passed to a bidirectional layer [28] of Gated Recurrent Units (GRU, [3]) that produces an output sequence of recurrent states:

$$[g]_1^N = \mathrm{BiGRU}([u]_1^N) = \{g_1, \ldots, g_N\} \text{ with } g_i \in \mathbb{R}^{50},$$

using a combination of update and reset gates in each recurrent hidden unit. Despite its simpler architecture and less demanding computations, the GRU is shown to be a competitive alternative to the well-known Long Short-Term Memory [5]. In practice, we implement the bidirectional GRU layer as two separate GRU layers. One layer processes the input sequence in a forward direction (left-to-right) while the other processes it in reversed order (right-to-left).

The sequences of hidden states of each GRU layer are concatenated element wise in order to yield a single sequence of hidden states:

$$[g]_1^N = \{(\overrightarrow{g}_1, \overleftarrow{g}_1)^T, \ldots, (\overrightarrow{g}_N, \overleftarrow{g}_N)^T\} \text{ with } \overrightarrow{g}_i, \overleftarrow{g}_i \in \mathbb{R}^{25},$$

where $\overrightarrow{g}_i$ and $\overleftarrow{g}_i$ are the hidden states for the forward and backward GRU layer, respectively. Each hidden state g_i is passed to a regular feed-forward layer that produces a further hidden representation $h_i' \in \mathbb{R}^{50}$ for that state. Lastly, a final layer in the network projects each h_i' of the previous layer to a probability distribution q_i over all possible output tags, namely I, O or B, using a softmax activation function:

$$[q]_1^N = \{q_1, \ldots q_N\} \text{ with } q_i \in \mathbb{R}^3.$$

For each word, we choose the tag with the highest probability as its predicted IOB2 tag.

Since the prediction of each tag can be interpreted as a classification, the network is trained to minimize the categorical cross-entropy between expected tag distribution p_i and predicted tag distribution q_i of each word i:

$$H(p_i, q_i) = -\sum_{t \in \mathcal{T}} p_i(t) \log(q_i(t)),$$

where $\mathcal{T} = \{I, O, B\}$ is the set of IOB2 tags, $p_i(t) \in \{0,1\}$ is the expected probability of tag t and $q(t) \in [0,1]$ the predicted probability. The network's parameters are optimized using the stochastic optimization technique *Adam* [12].

For further processing, a predicted tag sequence can be decoded into aspect term annotations using the IOB2 scheme in reverse. Note that we do not enforce the syntactic correctness of the predicted IOB2 scheme on a network-level. It is possible that the network produces a tag sequence that is not correct in terms of the employed IOB2 scheme. Thus, we post process each predicted tag sequence such that it constitutes a valid IOB2 tag sequence. Specifically, we replace each *I* tag that follows an *O* tag with a *B* in order to properly mark the beginning of an aspect term.

3.3 Aspect-Specific Sentiment Extraction

The second step in our two-step architecture for aspect-based sentiment extraction is the prediction of a polarity label given a previously detected aspect term. We address this aspect-specific sentiment extraction using a recurrent neural network that is, in parts, very similar to the architecture for aspect term extraction in Sect. 3.2.

In order to predict a polarity label for a *specific* aspect term in a sentence, we need to mark the aspect term in question. For this, we apply a similar technique as has been done for relation extraction [33] and Semantic Role Labeling [6]. We tag each word in the input sentence with its relative distance to the aspect term, as follows:

Great	**service**	,	great	*food*
−1	0	1	2	3	4

where the bold word "**service**" is the aspect term for which we want to extract the polarity. The italic word "*food*" marks another aspect term. The relative distance to the selected aspect term is shown below each word. This sequence of relative distances implicitly encodes the position of the aspect term in question in the sentence. In theory, this strategy permits to incorporate long range information in the prediction process in contrast to cutting a fixed-sized (and usually small) window of words around the aspect term in the sentence. In practice, we do not use the raw distance values directly but represent them as 10 dimensional distance embedding vectors similar as in [26,29,33] and treat them as learnable parameters in our network. We further denote the sequence of distance embedding vectors for a sentence of N words as:

$$[d]_1^N = \{d_1, \ldots, d_N\} \text{ with } d_i \in \mathbb{R}^{10}.$$

Figure 2 depicts the neural network component.

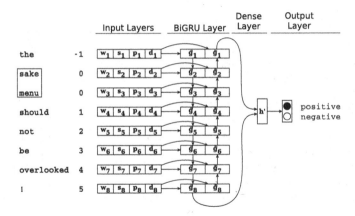

Fig. 2. The aspect-specific sentiment extraction component. The network processes the input sentence as a sequence of word vectors w_i, sentic vectors s_i, POS tags p_i and distance embeddings d_i using a bidirectional GRU layer and regular feed-forward layers. The output of the network is a single predicted polarity label for the aspect term of interest. The aspect term for which a polarity label is to be predicted is outlined in the input sentence.

Neural Network Polarity Extraction. The procedure for predicting a polarity label for an aspect term can be described as follows: Assume we have a sentence and an already extracted aspect term. We concatenate each word vector with its corresponding sentic vector, its POS tag vector and distance vector to receive the sequence:

$$[u]_1^N = \{(w_1, s_1, p_1, d_1)^T, \ldots, (w_N, s_N, p_N, d_N)^T\} \text{ with } u_i \in \mathbb{R}^{100+5+45+10}.$$

The resulting sequence is passed to a bidirectional GRU layer that produces an output sequence of recurrent states:

$$[g]_1^N = \text{BIGRU}([u]_1^N) = \{(\overrightarrow{g}_1, \overleftarrow{g}_1)^T, \ldots, (\overrightarrow{g}_N, \overleftarrow{g}_N)^T\} \text{ with } \overrightarrow{g}_i, \overleftarrow{g}_i \in \mathbb{R}^{25}.$$

We take the final hidden state $\overrightarrow{g}_N$ of the forward GRU and the final hidden state $\overleftarrow{g}_1$ of the backward GRU[5] and concatenate them to receive a fixed sized representation $h = (\overrightarrow{g}_N, \overleftarrow{g}_1)^T \in \mathbb{R}^{50}$ of the aspect term in the whole input sentence. Next, the network passes the hidden representation h of the aspect term through a densely connected feed-forward layer producing another hidden representation $h' \in \mathbb{R}^{50}$. As a last step, a final densely connected layer with a softmax activation function projects h' to a 2-dimensional vector $q \in \mathbb{R}^2$ representing a probability distribution over the two polarity labels positive and negative. We consider the label with the highest estimated probability to be the predicted polarity label for the given aspect term.

Again, we train the network to minimize the categorical cross-entropy between expected polarity label distribution p and predicted polarity label distribution q of each aspect term:

$$H(p, q) = - \sum_{l \in \mathcal{L}} p(l) \log(q(l)),$$

where $\mathcal{L} = \{\text{positive}, \text{negative}\}$ is the set of polarity labels and $p(l)$ and $q(l)$ the expected and predicted probability, respectively, for label l. As before, we apply the *Adam* technique to update network parameters.

4   Experiments and Evaluation

In order to see the performance of the overall system and the impact of the individual features, we perform an evaluation on the provided training data for the aspect-based sentiment analysis task. Based on that we select a final model configuration that is used in the actual challenge evaluation on additional test data.

Evaluation on Training Data. All experiments on the training data are performed as a 5-fold cross-validation. We evaluate the two steps of our approach separately to better see the individual performances of the two components.

Since we do not have access to official evaluation scripts, we evaluate aspect term extraction using Precision, Recall and F_1 score. We only take explicitly mentioned aspect terms into account[6] that have a polarity label of either positive or negative. Identical annotations i.e. annotations that target the same aspect term (in terms of character offsets) with the same polarity, are considered as one. Table 2 shows the results for aspect term extraction for different feature combinations. Here, WE denotes the usage of amazon review word embeddings, WE-Retro denotes the retrofitted embeddings, POS specifies additional POS tag features and Sentics indicates the usage of sentic vectors.

---

[5] Since this GRU processes the sequence in a reversed direction, the final hidden state is the hidden state for the first word.

[6] We exclude annotations with aspect = "NULL".

Table 2. Results of 5-fold cross-validation for aspect term extraction using different feature combinations.

Features	F_1	Precision	Recall
WE+POS	**0.684**	0.659	**0.710**
WE+POS+Sentics	0.679	**0.663**	0.697
WE-Retro+POS	0.678	0.651	0.708
WE-Retro+POS+Sentics	0.679	0.655	0.706

Comparing the models in Table 2, we can see that using the retrofitted embeddings seems to downgrade the performance of our system. Also, employing the sentic vectors for aspect term extraction degrades the networks performance. This is not completely unexpected, though, since the sentic vectors mainly encode sentiment information and aspect term extraction on its own is rather decoupled from the actual sentiment extraction. A more positive effect would be expected for the second step in our system, the prediction of polarity labels.

To evaluate the aspect-specific sentiment extraction, we extract polarity labels for all aspect terms of the ground truth annotations. By separating aspect term extraction and sentiment extraction, we can better evaluate the sentiment extraction in isolation. Again, we only consider unique aspect terms that are either labeled with a `positive` or `negative` polarity. We report the performance of our sentiment extraction in terms of the accuracy of the system for different feature combinations. Table 3 shows the results for the 5-fold cross-validation on the training data. `WE`, `WE-Retro`, `POS` and `Sentics` are defined as before, while `Dist` denotes the obligatory distance embedding features.

While the retrofitted embeddings do not contribute positively to the performance for sentiment extraction either, a notable gain is achieved using the sentic vectors in our component for aspect-specific sentiment extraction. Here, we observe a gain of 3.5% points accuracy compared to using only word embeddings, distance embeddings and POS tags. Apart from that, the usage of sentic vectors drastically reduces the training time needed to achieve these results. The best results for the `WE+POS+Dist` and `WE-Retro+POS+Dist` model were achieved with 102 iterations over the training portion of the data, while the `WE+POS+Dist+Sentic` and `WE-Retro+POS+Dist+Sentic` model reached their best performances for only 12 and 9 iterations, respectively. See Fig. 3 for a visualization of the system's accuracy with respect to the employed features and the iteration over the training data.

Evaluation on Test Data. Apart from our custom evaluation, each participating system is evaluated on a separate test set of customer reviews as part of the Sentiment Analysis Challenge. While the annotated training data covers the domains *laptops* and *restaurants*, the data for the test set is obtained from the domains *restaurants* and *hotels* in order to test the systems on a previously unseen review domain. For comparability, the predicted results for each system

Table 3. Results of 5-fold cross-validation for aspect-specific sentiment extraction using different feature combinations.

Features	Accuracy
WE+POS+Dist	0.776
WE+POS+Dist+Sentics	**0.811**
WE-Retro+POS+Dist	0.776
WE-Retro+POS+Dist+Sentics	0.809

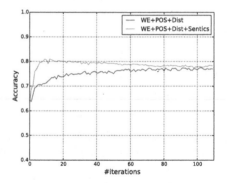

Fig. 3. Visualization of the performance gain of using sentic vectors with respect to the number of iterations over the training data. By using additional sentic vectors we achieve better results with less training needed.

are evaluated by the organizers. Aspect term extraction is evaluated with precision, recall and F_1 score regarding exact matches. Polarity extraction is evaluated with the accuracy of the predicted polarity label with respect to the subset of correctly extracted aspect terms from the previous step.

For this evaluation, we train final models for our two architectural components using knowledge gained from our preliminary results on the training data. The aspect term extract model WE+POS is trained on all training samples for 5 epochs and the polarity extraction model WE+POS+Dist+Sentics for 10 epochs. The official evaluation on the test data shows an F_1 score of 0.433 with a precision of 0.415 and a recall of 0.452 for the aspect term extraction in separation. The extraction of aspect-specific polarity labels for correctly identified aspect terms results in an accuracy of 0.874. With these results, the proposed system achieves the highest scores of the 2016th ESWC fine-grained sentiment analysis challenge.

5   Conclusion

With this work we propose a two-step approach for aspect-based sentiment analysis. We decouple the extraction of aspects and sentiment labels in order to obtain a flexibly applicable system. By using a recurrent neural network, we present a novel neural network based approach to tackle aspect extraction as a sequence labeling task. Furthermore, we present a novel way to address aspect-specific sentiment extraction using a recurrent neural network architecture with distance embedding features. This model is able to extract sentiments expressed towards a specific aspect that is mentioned in the text and is thus able to detect multiple opinions in a single sentence.

Both components of our overall sentiment analysis system incorporate additional semantic knowledge by using pretrained word vectors that are *retrofitted* to a semantic lexicon as well as semantic and sentiment-related features obtained from SenticNet. Although our first experiments could not show a benefit in using the retrofitted embeddings, the sentics obtained from SenticNet proved to be a valuable feature for extracting aspect-based polarity labels that increased accuracy and shortened training time considerably.

For this work, we could only incorporate single-word concepts from SenticNet as additional features. For the future, we plan to modify our architecture to permit incorporation of all concepts from SenticNet, thus moving the system to concept-level sentiment analysis even further.

Acknowledgements. This work was supported by the Cluster of Excellence Cognitive Interaction Technology 'CITEC' (EXC 277) at Bielefeld University, which is funded by the German Research Foundation (DFG).

References

1. Aprosio, A.P., Corcoglioniti, F., Dragoni, M., Rospocher, M.: Supervised opinion frames detection with RAID. In: Gandon, F., Cabrio, E., Stankovic, M., Zimmermann, A. (eds.) SemWebEval 2015. CCIS, vol. 548, pp. 251–263. Springer, Heidelberg (2015). doi:10.1007/978-3-319-25518-7_22

2. Cambria, E., Olsher, D., Rajagopal, D.: SenticNet 3: a common and commonsense knowledge base for cognition-driven sentiment analysis. In: Proceedings of the Twenty-Eighth AAAI Conference on Artificial Intelligence, pp. 1515–1521 (2014)

3. Cho, K., Van Merriënboer, B., Gülçehre, Ç., Bahdanau, D., Bougares, F., Schwenk, H., Bengio, Y.: Learning phrase representations using RNN encoder-decoder for statistical machine translation. In: Proceedings of the 2014 Conference on Empirical Methods in Natural Language Processing (EMNLP), Doha, Qatar, pp. 1724–1734. Association for Computational Linguistics, October 2014

4. Chung, J.K.-C., Wu, C.-E., Tsai, R.T.-H.: Polarity detection of online reviews using sentiment concepts: NCU IISR team at ESWC-14 challenge on concept-level sentiment analysis. In: Presutti, V., et al. (eds.) SemWebEval 2014. CCIS, vol. 475, pp. 53–58. Springer, Heidelberg (2014)

5. Chung, J., Gülçehre, Ç., Cho, K., Bengio, Y.: Empirical evaluation of gated recurrent neural networks on sequence modeling. In: NIPS Deep Learning Workshop (2014)

6. Collobert, R., Weston, J., Bottou, L., Karlen, M., Kavukcuoglu, K., Kuksa, P.: Natural language processing (almost) from scratch. J. Mach. Learn. Res. **12**, 2493–2537 (2011)

7. Dragoni, M., Tettamanzi, A.G.B., da Costa Pereira, C.: A fuzzy system for concept-level sentiment analysis. In: Presutti, V., et al. (eds.) SemWebEval 2014. CCIS, vol. 475, pp. 21–27. Springer, Heidelberg (2014)

8. Faruqui, M., Dodge, J., Jauhar, S.K., Dyer, C., Hovy, E., Smith, N.A.: Retrofitting word vectors to semantic lexicons. In: Proceedings of Human Language Technologies: The 2015 Annual Conference of the North American Chapter of the ACL, pp. 1606–1615 (2015)

9. Fellbaum, C.: WordNet and Wordnets. In: Brown, K. (ed.) Encyclopedia of Language and Linguistics, pp. 665–670. Elsevier, Oxford (2005)
10. Hu, M., Liu, B.: Mining and summarizing customer reviews. In: Proceedings of the Tenth ACM SIGKDD International Conference on Knowledge Discovery and Data Mining (KDD 2004), pp. 168–177. ACM, New York (2004)
11. Jakob, N., Gurevych, I.: Extracting opinion targets in a single- and cross-domain setting with conditional random fields. In: Proceedings of the Conference on Empirical Methods in Natural Language Processing, pp. 1035–1045, October 2010
12. Kingma, D., Ba, J.: Adam: a method for stochastic optimization. In: International Conference on Learning Representations (2015)
13. Klinger, R., Cimiano, P.: Bi-directional inter-dependencies of subjective expressions and targets and their value for a joint model. In: Proceedings of the 51st Annual Meeting of the Association for Computational Linguistics (ACL), Short Papers, vol. 2, pp. 848–854, August 2013
14. Klinger, R., Cimiano, P.: Joint and pipeline probabilistic models for fine-grained sentiment analysis: extracting aspects, subjective phrases and their relations. In: Proceedings of the 13th IEEE International Conference on Data Mining Workshops (ICDM), pp. 937–944, December 2013 .
15. Lakkaraju, H., Socher, R., Manning, C.: Aspect specific sentiment analysis using hierarchical deep learning. In: Proceedings of the NIPS Workshop on Deep Learning and Representation Learning (2014)
16. Le, Q., Mikolov, T.: Distributed representations of sentences and documents. ICML **32**, 1188–1196 (2014)
17. Manning, C., Surdeanu, M., Bauer, J., Finkel, J., Bethard, S., McClosky, D.: The Stanford CoreNLP natural language processing toolkit. In: Proceedings of 52nd Annual Meeting of the Association for Computational Linguistics: System Demonstrations, pp. 55–60 (2014)
18. McAuley, J., Pandey, R., Leskovec, J.: Inferring networks of substitutable and complementary products. In: Proceedings of the 21th ACM SIGKDD International Conference on Knowledge Discovery and Data Mining (KDD 2015), pp. 785–794. ACM, New York (2015)
19. McAuley, J., Targett, C., Shi, Q., van den Hengel, A.: Image-based recommendations on styles and substitutes. In: Proceedings of the 38th International ACM SIGIR Conference on Research and Development in Information Retrieval, pp. 43–52. ACM (2015)
20. Mikolov, T., Corrado, G., Chen, K., Dean, J.: Efficient estimation of word representations in vector space. In: Proceedings of the International Conference on Learning Representations (2013)
21. Mikolov, T., Sutskever, I., Chen, K., Corrado, G.S., Dean, J.: Distributed representations of words and phrases and their compositionality. In: Advances in Neural Information Processing Systems, pp. 3111–3119 (2013)
22. Pennington, J., Socher, R., Manning, C.: GloVe: global vectors for word representation. In: Proceedings of the 2014 Conference on Empirical Methods in Natural Language Processing (EMNLP), Doha, Qatar, pp. 1532–1543. Association for Computational Linguistics (2014)
23. Pontiki, M., Galanis, D., Papageorgiou, H., Manandhar, S., Androutsopoulos, I.: Semeval-2015 task 12: aspect based sentiment analysis. In: Proceedings of the 9th International Workshop on Semantic Evaluation, Denver, Colorado, pp. 486–495. Association for Computational Linguistics, June 2015

24. San Vicente, I., Saralegi, X., Agerri, R.: EliXa: a modular and flexible ABSA platform. In: Proceedings of the 9th International Workshop on Semantic Evaluation, Denver, Colorado, pp. 748–752. Association for Computational Linguistics, June 2015

25. dos Santos, C., Zadrozny, B.: Learning character-level representations for part-of-speech tagging. In: Proceedings of the 31st International Conference on Machine Learning, pp. 1818–1826 (2014)

26. dos Santos, C.N., Xiang, B., Zhou, B.: Classifying relations by ranking with convolutional neural networks. In: Proceedings of the 53rd Annual Meeting of the Association for Computational Linguistics and the 7th International Joint Conference on Natural Language Processing, vol. 1, pp. 626–634 (2015)

27. Schouten, K., Frasincar, F.: The benefit of concept-based features for sentiment analysis. In: Gandon, F., Cabrio, E., Stankovic, M., Zimmermann, A. (eds.) SemWebEval 2015. CCIS, vol. 548, pp. 223–233. Springer, Heidelberg (2015). doi:10.1007/978-3-319-25518-7_19

28. Schuster, M., Paliwal, K.K.: Bidirectional recurrent neural networks. IEEE Trans. Sig. Process. **45**(11), 2673–2681 (1997)

29. Sun, Y., Lin, L., Tang, D., Yang, N., Ji, Z., Wang, X.: Modeling mention, context and entity with neural networks for entity disambiguation. In: Proceedings of the 24th International Conference on Artificial Intelligence (IJCAI), pp. 1333–1339. AAAI Press (2015)

30. Titov, I., Mcdonald, R.: A joint model of text and aspect ratings for sentiment summarization. In: Proceedings of Annual Meeting of the Association for Computational Linguistics (ACL), pp. 308–316 (2008)

31. Tjong Kim Sang, E.F., Veenstra, J.: Representing text chunks. In: Proceedings of European Chapter of the ACL (EACL), Bergen, Norway, pp. 173–179 (1999)

32. Toh, Z., Wang, W.: DLIREC: aspect term extraction and term polarity classification system. In: Proceedings of the 8th International Workshop on Semantic Evaluation, pp. 235–240 (2014)

33. Zeng, D., Liu, K., Lai, S., Zhou, G., Zhao, J.: Relation classification via convolutional deep neural network. In: Proceedings of the 25th International Conference on Computational Linguistics (COLING), pp. 2335–2344 (2014)

Challenge on Question Answering
over Linked Data

6th Open Challenge on Question Answering over Linked Data (QALD-6)

Christina Unger[1]([✉]), Axel-Cyrille Ngonga Ngomo[2], and Elena Cabrio[3]

[1] CITEC, Bielefeld University, Bielefeld, Germany
cunger@cit-ec.uni-bielefeld.de
[2] AKSW, University of Leipzig, Leipzig, Germany
ngonga@informatik.uni-leipzig.de
[3] CNRS, Inria, I3S, Université Côte dAzur, Rocquencourt, France
elena.cabrio@unice.fr

1 Introduction

The past years have seen a growing amount of research on question answering over Semantic Web data (for an overview see [1]), shaping an interaction paradigm that allows end users to profit from the expressive power of Semantic Web standards while at the same time hiding their complexity behind an intuitive and easy-to-use interface. The *Question Answering over Linked Data*[1] challenge (QALD) provides up-to-date benchmarks for assessing and comparing systems that mediate between a user, expressing his or her information need in natural language, and RDF data. It thus targets all researchers and practitioners working on querying linked data, natural language processing for question answering, multilingual information retrieval and related topics.

The key challenge for question answering over linked data is to translate a user's information need into a form that can be evaluated using standard Semantic Web query processing and inferencing techniques. To focus on specific aspects and involved challenges, QALD proposes three tasks: *(i)* multilingual question answering over DBpedia, *(ii)* hybrid question answering over both RDF and free text data, and *(iii)* question answering over statistical data in RDF data cubes. The main goal is to gain insights into the strengths and shortcomings of different approaches and into possible solutions for coping with the heterogeneous and distributed nature of Semantic Web data.

QALD-6 is the sixth installment of the QALD challenge and focuses on multilingual and hybrid questions over DBpedia as well as on statistical questions over RDF datacubes.

2 Task Description

The main task of QALD is to retrieve answers to natural language questions or keywords from a given RDF dataset. In order to focus on specific aspects and challenges involved, QALD-6 comprises three tasks.

---

[1] http://www.sc.cit-ec.uni-bielefeld.de/qald.

© Springer International Publishing Switzerland 2016
H. Sack et al. (Eds.): SemWebEval 2016, CCIS 641, pp. 171–177, 2016.
DOI: 10.1007/978-3-319-46565-4_13

2.1   Task 1: Multilingual Question Answering over RDF Data

Given the diversity of languages used on the web, there is an impeding need to facilitate multilingual access to semantic data. The core task of QALD is thus to retrieve answers from an RDF data repository given an information need expressed in a variety of natural languages.

The underlying RDF dataset is DBpedia, a community effort to extract structured information from Wikipedia and to make this information available as RDF data. The RDF dataset relevant for the challenge is the official DBpedia 2015 dataset for English, including links, most importantly to YAGO[2] categories. The dataset can either be downloaded[3] or accessed from the official SPARQL endpoint[4]. The training data consists of the 350 questions compiled from previous instantiations of the challenge, available in eight different languages: English, Spanish, German, Italian, French, Dutch, Romanian, and Farsi. The test dataset consists of 100 similar questions in the same languages, compiled from existing question and query logs so as to provide unbiased questions expressing real-world information needs.

The questions are general, open-domain factual questions, such as *Which book has the most pages?*. They vary with respect to their complexity, including questions with counts (e.g., *How many children does Eddie Murphy have?*), superlatives (e.g., *Which museum in New York has the most visitors?*), comparatives (e.g., *Is Lake Baikal bigger than the Great Bear Lake?*), and temporal aggregators (e.g., *How many companies were founded in the same year as Google?*). Each question is annotated with keywords in all eight languages, a manually specified SPARQL query and the answers that this query retrieved from the official SPARQL endpoint. As an additional challenge, some of the questions are out of scope, i.e., they cannot be answered with respect to the dataset. One example is *Give me all animal species that live in the Amazon rainforest*. Statistics about the distribution of question and answer types are reported in Table 1.

2.2   Task 2: Hybrid Question Answering over RDF and Free Text Data

A lot of information is still available only in textual form, both on the web and in the form of labels and abstracts in linked data sources. Therefore, approaches are needed that can deal with the specific character of structured data as well as with finding information in several sources, processing both structured and unstructured information, and combining such gathered information into one answer.

QALD therefore includes a task on hybrid question answering, asking systems to retrieve answers for questions that require the integration of data both from RDF and from textual sources. The task builds on DBpedia 2015 as RDF

---

[2] For detailed information on the YAGO class hierarchy, please see http://www. mpi-inf.mpg.de/yago-naga/yago/.

[3] http://downloads.dbpedia.org/2015-10/core-i18n/en/.

[4] http://dbpedia.org/sparql/.

Table 1. The distribution of question and answer types in Task 1: multilingual question answering over DBpedia 2015.

	Train	Test
Total	350	100
Question type		
Aggregation	72 (21 %)	12
Other namespaces	108 (31 %)	27
`dbpedia.org/property/`	81 (23 %)	23
YAGO	20 (6 %)	2
FOAF	2 (1 %)	0
Out of scope	15 (4 %)	4
Answer type		
`resource`	256 (73 %)	81
`number`	44 (13 %)	7
`boolean`	28 (8 %)	3
`date`	15 (4 %)	3
`string`	7 (2 %)	6

knowledge base, together with free text abstracts contained in DBpedia as well as, optionally, the English Wikipedia as textual data source.

The training data comprises 50 English questions and builds on the questions compiled for last year's challenge (partly based on questions used in the INEX Linked Data track[5]). The test question set contains 50 similar questions. The questions are annotated with answers as well as a pseudo query that indicates which information can be obtained from RDF data and which from free text. The pseudo query is like an RDF query but can contain free text as subject, property, or object of a triple. An example is the question *Who is the front man of the band that wrote Coffee &TV?*, with the following corresponding pseudo query containing three triples, two RDF triples and a triple containing free text as property and object:

```
SELECT DISTINCT ?uri
WHERE {
    <http://dbpedia.org/resource/Coffee_&_TV>
    <http://dbpedia.org/ontology/musicalArtist> ?x .
    ?x <http://dbpedia.org/ontology/bandMember> ?uri .
    ?uri text:''is'' text:''frontman'' .
}
```

One way to answer the question is to first retrieve the band members of the musical artist associated with the song Coffee & TV from the RDF data using the first two triples, and then check the abstract of the returned URIs for the

---

[5] http://inex.mmci.uni-saarland.de/tracks/dc/index.html.

information whether they are the frontman of the band. In this case, the abstract of Damon Albarn contains the following sentence:

```
He is best known for being the frontman of the Britpop/alternative
rock band Blur [...]
```

All queries are designed in a way that they require both RDF data and free text to be answered. The main goal is not to take into account the vast amount of data available and problems arising from noisy, duplicate and conflicting information, but rather to enable a controlled and fair evaluation, given that hybrid question answering is a still very young line of research.

2.3   Task 3: Statistical Question Answering over RDF Data Cubes

With this new task, QALD aims to stimulate the development of approaches that can handle multi-dimensional, statistical data modelled using the RDF data cube vocabulary. The provided data consists of a selection of 50 data cubes from the LinkedSpending[6] government spending knowledge base. It could be either downloaded[7] or accessed through a SPARQL endpoint.[8] Question answering over RDF data cubes poses challenges that are different from general, open-domain question answering as represented by the above described two tasks, such as the different data structure, explicit or implied aggregations and intervals. An example is the question *How much did the Philippines receive in 2007?* with the following SPARQL query:

```
select SUM(xsd:decimal(?v1))
{
  ?o a qb:Observation.
  ?o :recipient-country :ph.
  ?o qb:dataSet :finland-aid.
  ?o :refYear ?v0.
  filter(year(?v0)=2007).
  ?o :finland-aid-amount ?v1.
}
```

This question involves the following challenges:

- *Implied aggregation:* The expected sum total is not directly mentioned.
- *Lexical gap:* The knowledge that an *amount* can be *received* is not given, so the measure *amount* can only be indirectly ascertained as the only measure with the correct answer type through the question word (*How much*).
- *Ambiguity:* Data cubes can contain large amounts of numerical values, impeding the conclusion that *2007* references a time period.

---

[6] http://linkedspending.aksw.org.

[7] http://linkedspending.aksw.org/extensions/page/page/export/qbench2datasets. zip.

[8] http://cubeqa.aksw.org/sparql.

– *Data cube identification:* There is no reference to the description of the correct data cube (*Finland Aid Data*) but it can be matched to both the Phillipines and the year of 2007.

The training question set consists of the 100 question benchmark compiled in the CubeQA project,[9] annotated with SPARQL queries, answers and the correct data cube for each question. The test question set contains 50 additional questions, each answerable using a single data cube.

3   Evaluation Measures

The results submitted by participating systems were automatically compared to the gold standard results and evaluated with respect to precision and recall. For each question q, precision, recall and F-measure were computed as follows:[10]

$$Recall(q) = \frac{\text{number of correct system answers for } q}{\text{number of gold standard answers for } q}$$

$$Precision(q) = \frac{\text{number of correct system answers for } q}{\text{number of system answers for } q}$$

$$\textit{F-Measure}(q) = \frac{2 * Precision(q) \times Recall(q)}{Precision(q) + Recall(q)}$$

On the basis of these measures, both macro and micro precision, recall and F-measure were computed, both globally and locally, i.e. for all questions, and only those questions for which the system provides answers.

4   Results

This year, QALD received 13 systems submissions from 6 countries: China, France, Italy, Germany, Iran, USA. For the first year, systems have been working on questions in languages other than English, more specifically in Farsi and Spanish. Table 2 reports the results for Task 1 (Multilingual QA), Table 3 for Task 2 (Hybrid QA), and Table 4 for Task 3 (Statistical QA over RDF datacubes). In each table, the first column specifies the system name and the language it processed. Systems marked with a * constitute late submissions (i.e. after the official end of the test phase). *Processed* states for how many of the questions the system provided an answer. *Recall*, *Precision* and *F-1* report the macro-measures with respect to the number of processed questions. *F-1 Global* in addition reports the macro F-1 measure with respect to the total number of questions.

---

[9] http://aksw.org/Projects/CubeQA.html.
[10] In the case of out-of-scope questions, an empty answer set counts as precision and recall 1, while a non-empty answer set counts as precision and recall 0.

Table 2. Results for Task 1: multilingual question answering over DBpedia, sorted by global macro F-1 score (over all 100 questions).

		Processed	Recall	Precision	F-1	F-1 global
CANaLI	(en)	100	**0.89**	**0.89**	**0.89**	**0.89**
UTQA	(en)	100	0.69	0.82	0.75	0.75
KWGAnswer	(en)	100	0.59	0.85	0.70	0.70
UTQA	(es)	100	0.62	0.76	0.68	0.68
UTQA	(fa)	100	0.61	0.70	0.65	0.65
NbFramework	(en)	63	0.85	0.87	0.86	0.54
SemGraphQA	(en)	100	0.25	0.70	0.37	0.37
PersianQA*	(fa)	100	0.19	0.91	0.31	0.31
UIQA (with manual)	(en)	44	0.63	0.54	0.58	0.25
UIQA (without manual)	(en)	36	0.53	0.43	0.48	0.17

Table 3. Results for Task 2: hybrid question answering, sorted by global macro F-1 score (over all 25 questions).

	Processed	Recall	Precision	F-1	F-1 global
Xser	23	**0.38**	0.43	**0.41**	**0.39**
AskDBpedia	21	0.33	0.51	0.40	0.35
FirstRun LIMSI	24	0.04	**0.61**	0.08	0.08

Table 4. Results for Task 3: statistical question answering over RDF datacubes, sorted by global macro F-1 score (over all 50 questions).

	Processed	Recall	Precision	F-1	F-1 global
SPARKLIS (expert)	50	**0.94**	**0.96**	**0.95**	**0.95**
SPARKLIS (beginner)	50	0.76	0.88	0.82	0.82
QA^3	44	0.62	0.59	0.60	0.53
CubeQA	49	0.41	0.49	0.45	0.44

Both the results for multilingual question answering and for hybrid question answering show a significant improvement over last year's challenge, with an average F-measure of 0.53 and 0.27, respectively. In the past two years [2,3], the average F-measure for multilingual question answering was 0.43 and 0.33, for hybrid question answering it was below 0.20. The fact that there is still a big difference in F-measure between the first and last system in all tasks illustrates that systems employ a wide variety of different approaches, methods and tools, which achieve quite different results. In fact, there is hardly any question that was not (at least partially) successfully tackled by at least one system. Similar to earlier challenges, the biggest challenge is still the matching of natural language expressions to correct vocabulary elements.

5   Future Perspectives

The QALD challenge continues to attract a growing number of participants working on question answering over RDF data, and it is successful in providing a benchmark that many systems use for evaluation also independent of the yearly challenge.

In addition to the core task, multilingual question answering over an RDF repository, QALD always includes other, often changing tasks like hybrid question answering (now running for the third time) and statistical question answering over RDF datacubes (included for the first time). These additional tasks are usually provided in collaboration with the community. We aim to intensify this collaboration, in order to disseminate ideas from researchers as well as requirements from industry.

The task of hybrid question answering is currently still limited to rather artificial questions and does not yet take into account the available amount of free text, or the redundancies and inconsistencies found in structured and unstructured data. In order to be able to evaluate and compare systems on a real-world hybrid question answering scenario, bigger data sources, other questions, and new evaluation methods are needed. These should be developed in collaboration with the Information Retrieval community, also in order to bring both fields closer together.

In general, the goal of QALD is to further contribute to developments in question answering over linked data, in particular by working with the community towards evolving shared tasks, a shared format for benchmarks and shared evaluation measures and frameworks. In addition, we aim to move towards benchmarking the scalability of frameworks by measuring their answer time within a controlled environment that will simulate users querying the systems. Here, a collaboration with the HOBBIT project[11] has already been initiated.

References

1. Höffner, K., Walter, S., Marx, E., Usbeck, R., Lehmann, J., Ngomo, A.-C.N.: Survey on challenges of question answering in the semantic web. Semant. Web Interoper. Usability Appl. (2016, to appear)
2. Unger, C., Forascu, C., Lopez, V., Ngomo, A.-C.N., Cabrio, E., Cimiano, P., Walter, S.: Question answering over linked data (QALD-4). In: CLEF Working Notes (2014)
3. Unger, C., Forascu, C., Lopez, V., Ngomo, A.-C.N., Cabrio, E., Cimiano, P., Walter, S.: Question answering over linked data (QALD-5). In: CLEF Working Notes (2015)

---

[11] http://project-hobbit.eu.

SPARKLIS on QALD-6 Statistical Questions

Sébastien Ferré[✉]

IRISA, Université de Rennes 1, Campus de Beaulieu, 35042 Rennes Cedex, France
ferre@irisa.fr

Abstract. This work focuses on the statistical questions introduced by
the QALD-6 challenge. With the growing amout of semantic data, including numerical data, the need for RDF analytics beyond semantic search
becomes a key issue of the Semantic Web. We have extended SPARKLIS
from semantic search to RDF analytics by covering the computation features of SPARQL (expressions, aggregations and groupings). We could
therefore participate to the new task on statistical questions, and we
report the achieved performance of SPARKLIS. Compared to other participants, SPARKLIS does not translate spontaneous questions by users,
but instead guide users in the construction of a question. Guidance is
based on the actual RDF data so as to ensure that built questions are
well-formed, non-ambiguous, and inhabited with answers. We show that
SPARKLIS enables superior results for both an expert user (94 % correct) and a beginner user (76 % correct).

1 Introduction

Question Answering (QA) systems [7] have so far rarely considered analytical
questions [5]. The best they can generally do is counting, e.g. *How many films
were directed by Tim Burton?*, or recognizing some predefined question templates [5]. Interestingly, the QALD challenge has released this year a new task,
"Statistical question answering over RDF datacubes", to foster research on RDF
analytics for end-users. The target dataset is a collection of datacubes, similar
to those used in the OLAP approach [1], but directly represented in RDF.

SPARQL 1.1 [10] is a valuable target formal language for RDF analytics, at least in terms of expressivity. It is even more expressive than OLAP
tools. It supports multiple aggregations, nested aggregations with sub-queries,
expressions and bindings on numbers, dates, and strings to derive new dimen-
sions/measures. Those features can be mixed with graph patterns with a lot
of flexibility. SPARQL engines are not as efficient as OLAP engines for data
analytics but they are continuously improving, and are already usable as our
participation to the challenge shows. The main reason that prevents end-users
to use SPARQL for RDF analytics is the difficulty to write queries. What is
already difficult for search is even more difficult for data analytics.

SPARKLIS [3] is a semantic search tool whose user interface and interaction
combines a query builder, faceted search [9] for guidance, and a Natural Language Interface (NLI) for readability of built queries. It can be used to explore

H. Sack et al. (Eds.): SemWebEval 2016, CCIS 641, pp. 178–187, 2016.
DOI: 10.1007/978-3-319-46565-4_14

and query SPARQL endpoints without the need for any preparation because data and its schema are discovered on the fly. We have extended SPARKLIS from semantic search to RDF analytics by covering the computation features of SPARQL (expressions, aggregations and groupings). We took the opportunity of the new QALD task on statistical questions to evaluate the performance and usability of the extended SPARKLIS for RDF analytics.

Section 2 recalls the principles and workings of SPARKLIS, and Sect. 3 explains how it has been extended to RDF analytics. Section 4 assesses the expressivity and scalability of the extended SPARKLIS. Section 5 reports on the participation of SPARKLIS at QALD-6. Section 6 concludes.

2   SPARKLIS: A Guided NLI for Semantic Search

SPARKLIS [3] reconciles expressivity and usability in semantic search by tightly combining a Query Builder (QB), a Natural Language Interface (NLI) [6], and a Faceted Search (FS) system [9]. As a QB it lets users build complex queries by composing elementary queries in an incremental fashion. An elementary query can be a class (e.g., ``a film''), a property (e.g., ``that has a director''), a RDF node (e.g., ``Tim Burton''), a reference to another node (e.g., ``the film''), or an operator (e.g., ``not'', ``or'', ``highest-to-lowest''). As a FS system, at every step, the query under construction is well-formed, query results are computed and displayed, and the suggested query elements are derived from actual data – not only schema – so as to prevent the construction of non-sensical or empty queries. The display of results and data-relevant suggestions at every step provides constant and acurate feedback to users during the construction process. This has been proved to support exploratory search, serendipity, and confidence about final results [9]. We have formalized the combination of query builders and faceted search as *Query-based Faceted Search* [4], for which we have proved the safeness and completeness of guided query construction over a large subset of SPARQL including basic graph patterns, cycles, UNION, OPTIONAL, and NOT EXISTS. Finally, as a NLI, everything presented to users – queries, suggested query elements, and results – are verbalized in natural language, completely hiding formal languages behind the user interface. Compared to Query Answering (QA) systems [7], the hard problem of spontaneous NL understanding (i.e., ambiguities, ellipsis, lexical and syntactic approximations) is avoided by controlling query formulation through guided query construction, and replaced by the simpler problem of NL generation.

Figure 1 shows a screenshot of SPARKLIS. Its user interface has three main parts, from top to bottom: the query, the suggested query elements organized into three lists, and the table of results. The query retrieves the list of "films starring an actor born since 2000, along with their director, director's nationality, the actor and his/her birth date, and their budget in decreasing order". An essential ingredient is the *query focus*, seen as the highlighted part of the query, and recalled above the suggestion lists: ``the nationality''. That focus determines what the suggestions are about, and where the suggestion chosen by the user will be inserted in the query. It can be changed

Fig. 1. Screenshot of SPARKLIS

by clicking on the relevant parts of the query (e.g., on ``starring`` to set the focus on actors). The sequence of selections that has led to this query is the following: ``every film``, ``that has a director``, ``that has a nationality``, focus on ``film``, ``that has a starring``, ``that has a birth date``, ``after 2000``, focus on ``film``, ``that has a budget``, ``highest-to-lowest``. Only the value ``2000`` has been written by the user, and with immediate feedback to show that the constraint ``after 2000`` does lead to answers.

SPARKLIS is available on-line[1] as a client-side application that works on top of SPARQL endpoints. A few endpoints are proposed (e.g., DBpedia, Mondial, Nobel Prizes) but any endpoint can be explored by simply entering its URL. A few configuration options allow to adapt to different endpoints (e.g., GET/POST method, sending with credentials), and to specify labelling properties. SPARKLIS now includes the YASGUI editor [8] to let advanced users access and modify the SPARQL translation of the query. The application page links to a list of clickable examples including the 50 test questions from the QALD-4 challenge, and to screencasts on YouTube.

3   Extending SPARKLIS to RDF Analytics

SPARKLIS is implemented along the principles of the N<A>F design pattern to bridge the gap between natural languages (NL) and formal languages (FL) [2], with SPARQL 1.1 on the FL side. N<A>F stands for (NL ← AST → FL) to indicate that translation goes from an intermediate representation AST (Abstract

---

[1] http://www.irisa.fr/LIS/ferre/sparklis/.

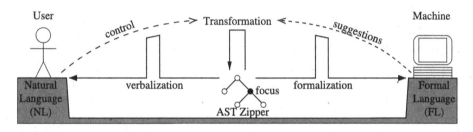

Fig. 2. Principle of zipper-based edition for bridging the gap between NL and FL

Syntax Tree) to both NL and FL, not from NL to FL. Figure 2 summarizes N<A>F. The user stands on the NL side, and does not understand the FL. The machine, here a SPARQL 1.1 endpoint, stands on the FL side, and does not understand the NL of the user. The central element of the bridge between the two sides is an *AST zipper*. In short, it is an Abstract Syntactic Tree (AST) representation of the user's question, split in two parts: the sub-tree at the *query focus* (see Sect. 2), and the rest of the tree called *query context*. Equivalently, the query focus can be seen as a distinguished node of the AST. The AST language is an intermediate between NL and FL, and is designed to make translations from the AST language to both NL (*verbalization*) and FL (*formalization*) as simple as possible. N<A>F includes the query builder approach, where the structure that is incrementally built is precisely the AST zipper. The AST zipper is initially the most simple query (''Give me everything''), and incrementally built by applying transformations to it. A transformation may insert or delete a query element, or move the focus. Transformations are suggested by the machine based on query semantics and actual data, and controlled by users after they have been verbalized in NL.

An advantage of N<A>F is the modularity of extending the coverage of a FL. Covering additional FL features requires first to add new constructs in AST zippers – plus new transformations to introduce them – and then to extend accordingly (1) verbalization, (2) formalization, and (3) suggestions. We explain in the following how we have extended SPARKLIS to RDF analytics by covering SPARQL 1.1 computations.

New AST Constructs. Before the extension, an AST was made of a single sentence representing essentially a graph pattern. After the extension, an AST is a sequence of sentences, where each sentence except the first refers to entities/values in previous sentences. There are three new kinds of sentences: aggregation sentences, binding sentences, and filter sentences. An aggregation sentence specificies what has to be aggregated (cube measure), the aggregation operator (e.g., count, sum), and the groupings (cube dimensions). Cube measure and dimensions are specified as references to entities/values in a previous sentence. A binding sentence introduces a new value defined by an expression combining operators, functions, and references to other entities/values. A filter sentence applies a constraint to be satisfied, and is defined by a Boolean expression.

New AST Transformations. In order to build the new kinds of sentences, new AST transformations have been defined. We recall that SPARKLIS suggests transformations according to the query focus and query results so as to ensure well-formed and sensical queries. To insert aggregation sentences, SPARKLIS suggests aggregation operators that are applicable to focus values (e.g., sum for numerical values, count for any kind of entities/values). When the focus is on an aggregation sentence, SPARKLIS suggests additional grouping dimensions, and it is also possible to remove a dimension. To build expressions in binding and filter sentences, SPARKLIS suggests operators (e.g., $+$), functions (e.g., `strlen`), and typed widgest for the input of literals (e.g., numbers, dates). Finally, a user name can be given to the result of an expression for more readable references.

Verbalization in English. Aggregations are verbalized with expressions like ''the number of department'' or ''the average amount''. Dimensions associated to an aggregation are introduced by ''for each''. The verbalization of expressions follow their nesting of operators and functions, which are verbalized either as mathematical operators for well-known operators (e.g., ''+'') or as plain English (e.g., ''the length of'' for function `strlen`). The following is an example of verbalization on geographical data that illustrates the four kinds of sentences.

```
give me every country that has a total GDP
                and that has a population
                and that has a continent
and give me GDP per capita = the total GDP * 1e6 / the population
and for each continent, give me the average GDP per capita
and where the GDP per capita > the average GDP per capita
```

That example query first retrieves the list of countries along with their total GDP, their population, and their continent (pattern sentence). Then it defines, for each country, GDP per capita as a function of total GDP and population (binding sentence). From there, it computes the average GDP per capita per continent (aggregation sentence with one dimension). Finally, it filters countries whose GDP per capita is above the average for their continent.

Formalization in SPARQL. The difficulties of the formalization do not lie at the level of individual sentences, for which the following templates are used, assuming some base pattern P.

> **aggregation** { SELECT ?d1 ... ?dn (g(?m) AS ?a)
>   WHERE { P } GROUP BY ?d1 ... ?dn }
> **binding** P BIND (expr AS ?name)
> **filter** P FILTER (expr)

We assume from the previous version of SPARKLIS that we can formalize a pattern sentence as a SPARQL graph pattern. The difficulties lie (a) in the identification of the base pattern P for each computation sentence, (b) in the dependencies between the sentences in a sequence, (c) on the impact of the focus position, and (d) on the computation of suggestions. We do not dive into the technical details here, but we provide the formalization of the above example on GDP per capita.

```
PREFIX xsd: <http://www.w3.org/2001/XMLSchema#>
PREFIX n1: <http://www.semwebtech.org/mondial/10/meta#>
SELECT DISTINCT ?Country_1 ?gdpTotal_34 ?population_37
                ?encompassed_259 ?expr_128 ?average_expr_167
WHERE { ?Country_1 a n1:Country .
        ?Country_1 n1:gdpTotal ?gdpTotal_34 .
        ?Country_1 n1:population ?population_37 .
        ?Country_1 n1:encompassed ?encompassed_259 .
        BIND (((?gdpTotal_34*"1e6"^^xsd:double)/?population_37)
             AS ?expr_128)
        { SELECT DISTINCT ?encompassed_259 (AVG(?expr_128)
                                        AS ?average_expr_167)
          WHERE { ?Country_1 a n1:Country .
                  ?Country_1 n1:gdpTotal ?gdpTotal_34 .
                  ?Country_1 n1:population ?population_37 .
                  ?Country_1 n1:encompassed ?encompassed_259 .
                  BIND (((?gdpTotal_34*"1e6"^^xsd:double)/?population_37)
                       AS ?expr_128) }
          GROUP BY ?encompassed_259
          LIMIT 200 }
        FILTER ( (?expr_128 > ?average_expr_167) ) }
LIMIT 200
```

Note the use of a subquery to compute the average GDP per capita for each continent, in order to compare it with the GDP per capita of each country. Note also that the formalization requires the duplication of the graph pattern about countries, which is transparent to the user who had only to specify it once.

Type-Based Suggestions. In the previous version of SPARKLIS, the suggestions are computed from the focus values by asking the SPARQL endpoint which classes and properties apply to those values. In the new version, the construction of well-formed expressions does not depend on data triples but on the *type* of values, aggregators, functions, and operators. Our set of types includes literal datatypes but also a type for URIs, a type for blank nodes, and a most general type for RDF nodes. We have organized types into a taxonomy to reflect their relative inclusions: e.g., integers are included in decimals, which are included in floats. We have implemented a *type inference* mechanism that determines the type of the query focus, depending on the query structure (e.g., surrounding operators), and on the actual values of the focus in the results. We account for complications such as focus values having different types, and the implicit conversions.

4   Assessment of Expressivity and Scalability

With the coverage of SPARQL computations, the source code of SPARKLIS has increased by 70 % from 5000 to 8500 lines of code, but this includes a few other improvements (e.g., YASGUI integration). That extension has significantly increased the expressivity of SPARKLIS and its coverage of SPARQL 1.1. The non-covered computations are a few technical functions like SHA256, the customization of the separator in the GROUP_CONCAT aggregator, and the application of COUNT to a row of variables (e.g., COUNT(*)). The other query features that are not yet covered are: GRAPH and SERVICE patterns, transitive closures of property paths, and CONSTRUCT/DESCRIBE forms of queries. About scalability, the extension has not much impact, apart from the evaluation of aggregations that are intrinsically costly, in particular in complex combinations. Bindings and filters do not impact subtantially efficiency as they are evaluated row-wise in answers. Type-based suggestions are cheap because they can be computed locally without requests to the endpoint.

5   Evaluation on QALD-6 Statistical Questions

The new version of SPARKLIS has been evaluated on the QALD-6 challenge[2] (Question Answering over Linked Data). It introduced a new task (Task 3) on "Statistical question answering over RDF datacubes". The dataset contains about 4 million transactions on government spendings all over the world, organized into 50 datacubes. Apart from a few properties (e.g., reference year), the properties used to represent the dimensions and measures of datacubes are specific to each datacube. There are about 16M triples in total. For the evaluation, we used SPARKLIS with the following configuration: http://cubeqa.aksw.org/ sparql as endpoint, and rdfs:label as labelling property for entities, classes and properties.

Range of Questions. All 100 training questions and 50 test questions are simple aggregations possibly as top-K queries, except for two questions (training Q23 and test Q23) that are more complex combinations requiring comparisons of two aggregations. As a consequence, all questions can be answered with SPARKLIS, and the Example page[3] provides the solutions to the 50 test questions as "Open" links. As an illustration, we here list the verbalization of a few test questions.

Q1. *How much was spent on public safety by the Town of Cary in 2010?*

```
it is true that 'Town of Cary, NC expenditures'
  is the data set of an Observation
    whose reference year is 2010
    and whose Class is 'Public Safety'
    and that has an Amount
  and give me the total Amount
```

---

[2] http://qald.sebastianwalter.org/index.php?x=home&q=6.

[3] http://www.irisa.fr/LIS/ferre/sparklis/examples.html.

Q4. *Which class achieved the highest revenue for the Town of Cary?*

```
it is true that 'Town of Cary, NC revenues'
  is the data set of an Observation
    that has a Class
    and that has an Amount
and for each Class
    give me the highest-to-lowest total Amount
```

Q10. *What was the average Uganda health budget amount in Namutumba District?*

```
it is true that 'Uganda Health Budget 2014-15'
  is the data set of an Observation
    whose To is 'Namutumba District'
    and that has an Amount
and give me the average Amount
```

Methodology. The fact that a question can be answered with SPARKLIS does not imply that a user manages to do it. There are two sources of difficulties: the tool and the data. Two users with different profiles were involved to evaluate those difficulties. The first user, the author of this paper, is an *expert* user of the tool. The second user, here called the *beginner*, has no IT background, has done business studies, and works on commercial analysis in a company. She works mostly with Excel, and had never used the tool before the experiment. Both users were totally unaware of the data before the experiment. They first became acquainted with the data – and the tool for the beginner! – by going through the training questions, and then went through the test questions before submitting their answers to the QALD evaluation engine. During the test phase, we measured the clock time including question reading and understanding by the user, user actions in the tool, and computations by the tool and the endpoint.

Table 1. Performance on QALD-6 statistical questions by two different users

User	Answered questions	Correct answers	Median time	Min time	Max time
Expert	49/50 = 0.98	47/50 = 0.94	1'30"	31"	6'20"
Beginner	44/50 = 0.88	38/50 = 0.76	4'	1'	10'

Results and Interpretation. Table 1 compares the performances of the expert and beginner users. The performance of the expert shows that virtually all questions (94 %) can be answered accurately and efficiently (half questions answered in less than 1'30"), once the user is fluent in the use of the tool. Although the median time of the beginner is nearly three times higher (4') than for the expert, the success rate of 76 % is very satisfying for a beginner given the complexity of the data, and the novelty of the tool compared to Excel. To compare with other challenge participants, their success rate is 50 % for QA[3] ($F_1 = 0.53$), and 38 %

for CubeQA ($F_1 = 0.44$). Also, the time range per question does not differ drastically between the two users. The difference in success rate is also mitigated when looking at the cause of errors. Among the 6 beginner errors, 1 can be explained by a lack of attention because the missed action was performed in several other questions, 3 can be explained by a lack of exposure to a similar case during the training phase ("When" questions, numeric filters like "more than 10000000"), and 2 can be explained by the difficulty to find the right property and value in a datacube. The 2 expert errors come from ambiguity in questions Q35 and Q42 that admit several equally plausible answers (e.g., several URIs having a same label). The explanation for unanswered questions is generally that the datacube is not explicit in the question. This requires to catch a first entity (e.g., "Research into Infrastructure" in Q33) by string matching, which appeared to be inefficient and unreliable on the SPARQL endpoint.

Discussion. The interactive data-centered approach of SPARKLIS has the advantage to provide a richer user experience than the automatic approach of classical QA. We wish to share here interesting observations about that user experience. The most efficient strategy to answer a question was to first select all datasets, and then to identify the right one by using question keywords. From there, the main difficulty was to identify the right properties to use as selectors or cube dimensions. Trial-and-error and experience was generally enough, but in the case of many possible properties, we used the construct `'has a relation to'` to cross all properties at once. When the dataset was not explicit, we used string matching to locate a first entity from which the datacube could easily be identified because URIs are hardly shared between datacubes. On one or two occasions, we needed to Google a name to relate it to a datacube: e.g., Nangarhar as part of Afghanistan, for which there is a datacube. Sometimes, there are several datacubes for a same location, and we could choose the right one thanks to a constraint on year, even if years do not appear in the datacube names. We remarked a few (intentional?) typos in questions but as humans we hardly noticed them. SPARKLIS often returns more results than required. For example, if the department with highest revenue is requested, the result contains not only that department but also its revenue, and also the department with the 2nd highest revenue, etc. We think it helps to build confidence in the system, and to detect errors early. In a number of questions we have observed that the application of some selectors is not necessary because they are redundant with other selectors, and that the application of some aggregations are not necessary because there is a single item. As additional support of our claim, note that a few errors were found by the expert user in the manually crafted golden standard of the training questions, and reported to the organizers.

6   Conclusion

We have demonstrated with SPARKLIS that guided construction of questions is a powerful alternative to Question Answering (QA) techniques based on NLP (Natural Language Processing). We have shown that all QALD-6 statistical questions can be answered correctly with the tool, and that the main cause for errors lies in the complexity of data. A difficulty is the need to learn how to use the tool, but we have shown that a non-IT beginner achieved a score of 76 % correct answers, still higher than the score achieved by QA approaches. SPARKLIS is freely available online, and can be used on top of any SPARQL endpoint conforming with standards.

Acknowledgement. I wish to thank QALD organizers for the datacube task, Pierre-Antoine Champin for valuable feedback and suggestions, and Eléonore Jouffe for her participation to the experiment.

References

1. Codd, E., Codd, S., Salley, C.: Providing OLAP (On-line Analytical Processing) to User-Analysts: An IT Mandate. Codd & Date Inc., San Jose (1993)
2. Ferré, S.: Bridging the gap between formal languages and natural languages with zippers. In: Sack, H., Blomqvist, E., d'Aquin, M., Ghidini, C., Ponzetto, S.P., Lange, C. (eds.) ESWC 2016. LNCS, vol. 9678, pp. 269–284. Springer, Heidelberg (2016). doi:10.1007/978-3-319-34129-3_17
3. Ferré, S.: Sparklis: an expressive query builder for SPARQL endpoints with guidance in natural language. Semant. Web: Interoper. Usability Appl. (2016, to appear)
4. Ferré, S., Hermann, A.: Reconciling faceted search and query languages for the Semant web. Int. J. Metadata Semant. Ontol. **7**(1), 37–54 (2012)
5. Höffner, K., Lehmann, J.: Towards question answering on statistical linked data. In: Semantic Systems. p. 4 (2014)
6. Kaufmann, E., Bernstein, A.: Evaluating the usability of natural language query languages and interfaces to semantic web knowledge bases. J. Web Semant. **8**(4), 377–393 (2010)
7. Lopez, V., Uren, V.S., Sabou, M., Motta, E.: Is question answering fit for the semantic web? A survey. Semant. Web **2**(2), 125–155 (2011)
8. Rietveld, L., Hoekstra, R.: YASGUI: not just another SPARQL client. In: Cimiano, P., Fernández, M., Lopez, V., Schlobach, S., Völker, J. (eds.) ESWC 2013. LNCS, vol. 7955, pp. 78–86. Springer, Heidelberg (2013)
9. Sacco, G.M., Tzitzikas, Y. (eds.): Dynamic Taxonomies and Faceted Search. IRS. Springer, Heidelberg (2009)
10. SPARQL 1.1 query language. W3C Recommendation (2012). http://www.w3.org/TR/sparql11-query/

Top-k Shortest Paths in Large Typed RDF Datasets Challenge

Top-K Shortest Paths in Large Typed RDF Datasets Challenge

Ioannis Papadakis[1]([⊠]), Michalis Stefanidakis[2], Phivos Mylonas[2],
Brigitte Endres Niggemeyer[3], and Spyridon Kazanas[1]

[1] School of Information Science and Informatics, Ionian University,
Ioannou Theotoki 72, 49100 Corfu, Greece
papadakis@ionio.gr, s.kazanas@gmail.com
[2] Department of Informatics, Ionian University,
7, Tsirigoti Square, 49100 Corfu, Greece
{mistral,fmylonas}@ionio.gr
[3] Rheinstr. 1a, 30519 Hannover, Germany
brigitteen@googlemail.com

Abstract. Perhaps the most widely appreciated linked data principle instructs linked data providers to provide useful information using the standards (i.e., RDF and SPARQL). Such information corresponds to patterns expressed as SPARQL queries that are matched against the RDF graph. Until recently, patterns had to specify the exact path that would match against the underlying graph. The advent of the SPARQL 1.1 Recommendation introduced property paths as a new graph matching paradigm that allows the employment of Kleene star * (and its variant Kleene plus +) unary operators to build SPARQL queries that are agnostic of the underlying RDF graph structure. In this paper, we present the Top-k Shortest Paths in large typed RDF Datasets Challenge. It highlights the key aspects of property path queries that employ the Kleene star operator, presenting three widely different approaches.

Keywords: SPARQL 1.1 · Property paths · Navigational queries · Kleene star · ESWC 2016

1 Introduction

In this paper, we present the Top-k Shortest Paths in large typed RDF Datasets Challenge that took place in Heraklion, Crete, from May, 31st to June, 2nd 2016 as part of the Extended Semantic Web Conference - ESWC 2016. The goal of this challenge was to highlight the key features of this special kind of SPARQL queries and accordingly promote research on this field.

Three groups submitted their solutions to the challenge: Hassan and colleagues [5], Hertling et al. [6], de Vocht et al. [7]. The award (an annual subscription to the Commercial License of Blazegraph's graph database with GPU acceleration) went to the Hertling team. Congratulations!

© Springer International Publishing Switzerland 2016
H. Sack et al. (Eds.): SemWebEval 2016, CCIS 641, pp. 191–199, 2016.
DOI: 10.1007/978-3-319-46565-4_15

Top-k shortest paths are at home in graph searching. Two nodes in a graph may be connected or not. In many occasions, if connections between them exist, the shortest or most appropriate ones are preferred.

Approaches for finding the top-k shortest paths may or may not accept loops inside paths. The Eppstein top-k path algorithm [9] applied by the winning team does not. Other solutions [e.g. 8, 12, 13] may index the paths through the graph.

The Challenge focuses on navigational queries expressed as property-path queries employing the Kleene star * (and it's variant +) unary operator. The advent of SPARQL 1.1 introduced property paths as a new graph matching paradigm that allows the employment of Kleene star * (and it's variant +) unary operators to build SPARQL queries that are agnostic of the underlying RDF graph structure.

For example, the graph pattern:

:Mike (foaf:knows)+ ?person

asks for all the persons that know somebody that knows somebody (and so on) that :Mike knows.

Property path queries extend the traditional graph pattern matching expressiveness of SPARQL 1.0 with navigational expressiveness. More specifically, navigational queries is a well-known type of queries on graph databases, where the corresponding results contain binary relations over the nodes of the graph [1].

Property paths in SPARQL 1.1 stirred up considerable discussion on their usability. They were considered as intrinsically untreatable or too hard to implement in a functional system. Voices in this chorus were [2, 3, 4, 11]. The Recommendation (www. w3.org/TR/sparql11-property-paths) reacted to criticism by replacing the initial *bag semantics* (i.e., duplicates within results are to be retained) underpinning property paths with the more agile *set semantics* (i.e., duplicates within results are dropped).

The ability to express path patterns that are agnostic of the underlying graph structure is certainly a step forward. However, one would also like to retrieve the actual paths through a property path SPARQL query. For this purpose navigational queries must ask for paths of arbitrary length within the RDF graph with concurrently specifying the actual graph patterns that constitute the path at query execution time.

Moreover, it is reasonable to assume that an RDF graph contains numerous paths of varying length between two arbitrary nodes. Therefore the results of a property path query should be ranked according to their path length. Shorter paths are assumed to be more meaningful.

The main part of this paper explains the setup of the challenge with its tasks, the training material and the assessment process. After that, the submissions with their widely different approaches to the challenge tasks are presented, followed by what we learned.

2   Structure of the Challenge

Challengers were confronted with a double set of tasks (see Fig. 1). In Task 1 the predicates of the property paths are free, whereas in Task 2, either the outgoing or the incoming predicate is prescribed. Both tasks include two subtasks. The first one is

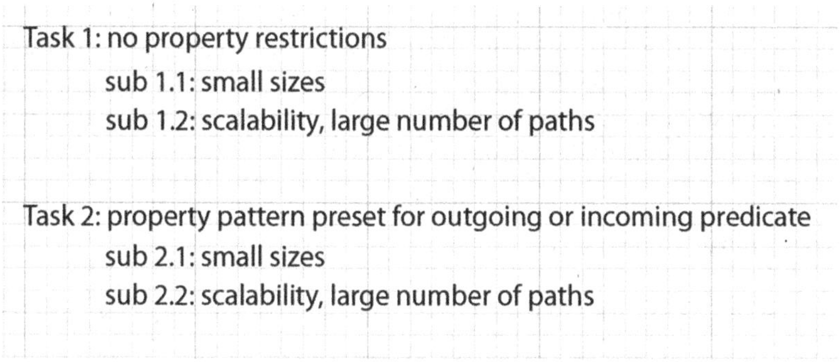

Task 1: no property restrictions
    sub 1.1: small sizes
    sub 1.2: scalability, large number of paths

Task 2: property pattern preset for outgoing or incoming predicate
    sub 2.1: small sizes
    sub 2.2: scalability, large number of paths

Fig. 1. Basic structure of the challenge

basic, the second huge in order to test the scalability of the approach. For all subtasks, the input data configuration was comprising the following items:

1. Start node: The first node of every path that will be returned.
2. End node: The last node of every path that will be returned.
3. k: The required number of paths.
4. ppath expression: A property path expression describing the pattern of the required paths.
5. Dataset: The RDF graph containing the start node and end node.

Challengers were expected to provide k paths between the Start node and the End node of the Dataset ordered by their length. A path is defined as a sequence of nodes and edges within the RDF graph:

(A,P1,U1,P2,U2,P3,U3,..,Un-1,Pn,B)

A corresponds to the Start node (i.e., subject), B to the End node (i.e., object), Pi to an edge (i.e., predicate) and Ui to an intermediate node. The path length equals to the corresponding number of edges (e.g., the above path length equals to n). The paths should comply with the given ppath expression.

All systems that participate in the Challenge should be able to provide ranked lists of paths pertaining to the requirements of each Task. In the **Annex** readers find a small practical example of the tasks and the expected results.

3   Datasets and Training Material

The training dataset was published not long after the announcement of the Challenge (available at: https://bitbucket.org/ipapadakis/eswc2016-challenge/downloads) and was accompanied with a set of input parameters together with the expected paths.

The evaluation dataset (together with the actual input data of each Task) was provided to all participants when the challengers had voted in.

The training and the evaluation datasets correspond to transformations of the 10 % dataset and the 100 % dataset of the DBpedia SPARQL Benchmark (available at:

http://aksw.org/Projects/DBPSB.html) respectively. The training dataset consists of 9,996,907 triples, while the evaluation dataset consists of 110,621,287 triples so that it is about 11 times larger. The datasets were slightly refurbished for easier treatment.

The training dataset was accompanied by four queries for each Task together with their corresponding results, 38 files of ranked paths.

4 Assessment Process

Each one of the three Challengers had to submit a paper describing the overall approach. Then, after the publication of the evaluation dataset, Challengers had to submit a set of four ranked path lists pertaining to the four sets of input parameters (two for each Task) that also had been published on the Challenge's website.

Each paper was assigned for reviewing to three members of the Program Committee. Then, the Challengers had to address the requirements of each Task by uploading the corresponding ranked path lists. The reviewers considered paper and system quality.

Overall paper quality (50 %) included presentation and organization of the paper; scientific/technical quality; English style; writing quality; figures and tables; innovation of approach; efficiency of approach (e.g., by measuring the complexity of an algorithm); consumption of resources; scalability.

System quality (50 %) was assessed by comparing results. The submitted sets were compared against each other in terms of the paths they contained. More specifically, the set that contained the highest number of shortest paths was ranked first, the set that contained the second highest number of shortest paths was ranked second, etc. In case of a tie, the tied sets were compared against the second shortest paths. If the sets remained tied, they were compared against the third shortest paths, and so on. All tasks were of equal weight.

5 Submissions

The three participating groups all managed to deliver paths, but not all requirements could be met, as the results table (Table 1) shows. Group 3 did not succeed beyond the first subtask of Task 1, so they were excluded from further assessment. Group 2 extracted all the requested paths, so that their technical approach scaled best. Their paper performed best as well. So that all in all Group 2 won the award.

Table 1. Challengers scoring.

Challenger	Task 1, part 1 (no. of paths)	Task 1, part 2 (no. of paths)	Task 2, part 1 (no. of paths)	Task 2, part 2 (no. of paths)	Paper review
1	377	450	370	4201	4
2	377	53008	374	52664	7
3	10				−2

The Challenger IDs at the first column refer to the following groups:

1. Laurens De Vocht, Ruben Verborgh, Erik Mannens and Rik Van de Walle. Using Triple Pattern Fragments To Enable Streaming of Top-k Shortest Paths via the Web
2. Sven Hertling, Markus Schröder, Christian Jilek and Andreas Dengel. Top-k Shortest Paths in Directed, Labeled Multigraphs
3. Zohaib Hassan, Mohammad Abdul Qadir, Muhammad Arshad Islam, Umer Shahzad and Nadeem Akhter. Modified MinG Algorithm to Find Top-K Shortest Paths from large RDF Graphs

Finding top-k shortest paths in an RDF graph cannot be easily solved by employing generic graph algorithms.

All contestants early recognized that and imposed additional constraints to their paths, such as allowing at most one loop or corresponding to a specific pattern. Moreover, the solutions exhibited in the challenge revealed that any classical graph algorithm may be overcharged by the sheer size of the usually huge and strongly interconnected linked data graphs, which may contain billions -or even trillions- of triples.

Graphs of that size pose also implementation problems: there is no guarantee that data will always fit in memory. For all these reasons, the proposed solutions used well-known approaches only as a base, tweaking them or building on top of them.

Modified MinG Algorithm to Find Top-K Shortest Paths from Large RDF Graphs. When preparing a query-answering solution, one can choose to move part of the computational burden to off-line processing that runs once in order to build some kind of index. This pre-built index is consulted later on every query, lowering thus the response time of the answering system. So do the authors of [5]. They build a modified MinG algorithm inspired from the ρ-index algorithm [8]. A hierarchical clustering schema is representing the graph interconnection information divided in levels of graph segments. This schema records the node interconnections in intra-segment Path Type Matrices (PTMs), as well as inter-segment Inverted Files (INFs) for various levels of segmentation. It is built in an off-line preprocessing step.

The same hierarchical structure is used during querying: beginning from both start and end of the requested path and iterating through the levels of segment information, all valid solutions are accumulated. A final step removes duplicate triples and sorts by path length, in order to achieve the final top-k solution. Due to compaction techniques the proposed implementation fares better for specific graph patterns in terms of indexing space and time compared to the algorithms it is based upon. The authors manage to keep their storage space to a reasonable size.

Top-K Shortest Paths in Directed, Labeled Multigraphs. In a more graph-theoretical based approach, [6] extends the classical Eppstein's top-k shortest path algorithm [9], which uses a bounded-degree graph to output k paths using breadth-first search. The authors implement a totally in-line (query-time) processing scheme, which computes a single source shortest path tree T for the target node of any query. They augment the classical algorithm by pruning any invalid paths, excluding them from further processing. A further modification allows for the discovery of paths starting or ending with a particular predicate, as requested in challenge terms. The algorithm is symmetrical to any end of the path; in order to find paths ending in a particular predicate the authors invert the graph and treat the ending node as the beginning node.

As other contestants, the authors had to employ memory compaction techniques, such as representing URIs with integer values, in order to make the totality of graph data fit in memory during query processing. They comply with the additional requirements of the challenge with a time complexity which is only moderately above the one of Eppstein's generic algorithm. The implementation results show that, in general, the overhead increases with the number of loops in the queried graph. This is why pruning invalid paths is an essential characteristic of the proposed solution.

Using Triple Pattern Fragments To Enable Streaming of Top-K Shortest Paths via the Web. The last but not least challenger team follows a more application-oriented approach [7]. The authors build on an existing platform for automated story-telling [10] that reduces the number of arbitrary resources revealed in interconnecting paths. The main characteristic of their proposal is a computationally inexpensive and asynchronous server-side interface that decomposes queries into simple and small parts: Triple Pattern Fragments. The application clients are in charge of querying by requesting and accumulating shortest paths ordered by length, while checking at the same time that the extracted paths fulfill the requirements of the challenge, i.e., they constitute unique-triple solutions. The implementation favors the querying of shortest paths first, which is beneficial to the control of candidate path explosion. The proposed system is decentralized: decomposing the initial query into small parts allows for the employment of multiple data endpoints without modification of the algorithm, coping also with the memory limitations imposed by a single centralized linked data graph. While more data needs to be buffered and streamed, the implementation results testify that the proposed solution scales very well with the size of input graph.

6   Conclusions – Lessons Learned

The *Top-k Shortest Paths in large typed RDF Datasets Challenge* was organised for the first time under the auspices of the 13th ESWC 2016 conference. Its primal goal was to raise the awareness of the semantic web community to property paths that have been recently introduced as part of the SPARQL1.1 Recommendations. A property path query that employs the Kleene star (*) or its variant Kleene plus (+) operators deserves much attention, also because of its scalability features that are directly proportional to the size of the underlying RDF graph.

Although the participation to the Challenge might have been greater, the Challenge illustrated that many scientific ways lead to top-k shortest paths. The three qualified Challengers presented completely different approaches:

- The first Challenger employed a triple pattern fragments technology to break the original query down to sequences of triple pattern fragment requests that were accordingly dispatched to clients for asynchronous processing.
- The second Challenger modified a well-known path finding algorithm to fit the requirements of the Challenge.

- The third Challenger aimed in creating a custom-made index structure that was specifically designed to serve navigational queries in RDF graphs.

All three Challengers realized that the bigger the RDF graph, the more difficult it is to retrieve the requested results. Thus, scalability issues are identified as a significant issue of this special kind of property path queries. Along these lines, future work is targeted towards systems capable of efficiently handling such queries in large RDF graphs.

Annex

Practical example. The image below (Fig. 2) depicts the example RDF dataset D1. Each node is a subject or object while each edge is a predicate.

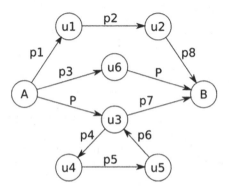

Fig. 2. Tiny example dataset

D1 contains exactly 4 paths from A to B:

9. (A,P,u3,p7,B)
10. (A,p3,u6,P,B)
11. (A,p1,u1,p2,u2,p8,B)
12. (A,P,u3,p4,u4,p5,u5,p6,u3,p7,B)

At this point, it should be mentioned that a path is valid only if it contains unique triples. For example, the path:

(A,P,u3,p4,u4,p5,u5,p6,u3,p4,u4,p5,u5,p6,u3,p7,B)

is not valid, since the triple: *u3,p4,u4* exists more than once.

Task 1. If task1 requires 3 paths between A and B within D1, the expected results should be the top-3 shortest paths from A to B:

1. (A,P,u3,p7,B)
2. (A,p3,u6,P,B)
3. (A,p1,u1,p2,u2,p8,B)

Note that the first two paths have the same length (i.e., 2), since they both contain two edges). The third path has a length that equals to 3, so it comes after the first two paths.

Task 2. If Task2 requires 2 paths between A and B within D1, the expected results should be the top-2 shortest paths from A to B that have P as their first or last edge:

4. (A,P,u3,p7,B)
5. (A,p3,u6,P,B)

If Task2 requires 3 paths between A and B within D1, the expected results should be the top-3 shortest paths from A to B that have P as their first or last edge:

6. (A,P,u3,p7,B)
7. (A,p3,u6,P,B)
8. (A,P,u3,p4,u4,p5,u5,p6,u3,p7,B)

Note that the path (A,p1,u1,p2,u2,p8,B) is omitted since P is neither the first nor the last edge of the path, so it does not fit into the requirements of the path pattern.

References

1. Zhang, X., Van den Bussche, J.: On the power of SPARQL in expressing navigational queries. Comput. J. **58**(11), 2841–2851 (2014)
2. Gubichev, A., Bedathur, S.J., Seufert, S.: Sparqling kleene: fast property paths in RDF-3X. In: First International Workshop on Graph Data Management Experiences and Systems, GRADES 2013, pp. 14:1–14:7. ACM, New York (2013)
3. Arenas, M., Conca, S., Perez, J.: Counting beyond a yottabyte, or how SPARQL 1.1 property paths will prevent adoption of the standard. In: Proceedings of the 21st International Conference on World Wide Web, WWW 2012, pp. 629–638. ACM, New York (2012)
4. Arenas, M., Gutierrez, C., Miranker, D.P., Perez, J., Sequeda, J.F.: Querying semantic data on the web? ACM SIGMOD Rec. **41**(4), 6–17 (2013)
5. Hassan, Z., Qadir, M. A., Islam, M. A., Shahzad, U., Akhter, N.: Modified MinG Algorithm to Find Top-K Shortest Paths from Large RDF Graphs. In: ESWC 2016
6. Hertling, S., Schroeder, M., Jilek, C., Dengel, A.: Top-K Shortest Paths in Directed, Labeled Multigraphs. In: ESWC 2016
7. De Vocht, L., Verborgh, R., Mannens, E, Van de Walle, R.: Using Triple Pattern Fragments to Enable Streaming of Top-K Shortest Paths via the Web. In: ESWC 2016
8. Barton, S.: Indexing graph structured data. Ph.D. thesis, Masaryk University, Brno, Czech Republic (2007)
9. Eppstein, D.: Finding the k shortest paths. SIAM J. Comput. **28**(2), 652–673 (1998)
10. De Vocht, L., Beecks, C., Verborgh, R., Seidl, T., Mannens, E., Van de Walle, R.: Improving semantic relatedness in paths for storytelling with linked data on the web. In: Gandon, F., Guéret, C., Villata, S., Breslin, J., Faron-Zucker, C., Zimmermann, A. (eds.) ESWC 2015. LNCS, vol. 9341, pp. 31–35. Springer, Heidelberg (2015). doi:10.1007/978-3-319-25639-9_6

11. Salvadores, M., Horridge, M., Alexander, P.R., Fergerson, R.W., Musen, M.A., Noy, N.F.: Using SPARQL to Query BioPortal Ontologies and Metadata
12. Tarjan, R.E.: Fast algorithms for solving path problems. J. ACM **28**(3), 594–614 (1981)
13. Dietz, P.F.: Maintaining order in a linked list. In: STOC 1982: Proceedings of the Fourteenth Annual ACM Symposium on Theory of Computing, pp. 122–127. ACM Press, New York (1982)

Top-k Shortest Paths in Directed Labeled Multigraphs

Sven Hertling[1]([✉]), Markus Schröder[1], Christian Jilek[1], and Andreas Dengel[1,2]

[1] Knowledge Management Group,
German Research Center for Artificial Intelligence (DFKI) GmbH,
Trippstadter Straße 122, Kaiserslautern, Germany
{sven.hertling,markus.schroeder,christian.jilek,andreas.dengel}@dfki.de
[2] Knowledge-Based Systems Group, Department of Computer Science,
University of Kaiserslautern, P.O. Box 3049, 67653 Kaiserslautern, Germany

Abstract. A top-k shortest path algorithm finds the k shortest paths of a given graph ordered by length. Interpreting graphs as RDF may lead to additional constraints, such as special loop restrictions or path patterns. Thus, traditional algorithms such as the ones by Dijkstra, Yen or Eppstein cannot be applied without further ado. We therefore implemented a solution method based on Eppstein's algorithm which is thoroughly discussed in this paper. Using this method we were able to solve all tasks of the ESWC 2016 Top-k Shortest Path Challenge while achieving only moderate overhead compared to the original version. However, we also identified some potential for improvements. Additionally, a concept for embedding our algorithm into a SPARQL endpoint is provided.

Keywords: Top-k shortest paths · Loop restrictions · Eppstein's algorithm

1 Introduction

A common approach for representing knowledge is the usage of a knowledge graph. Such graphs are well-known for their ability to describe relationships between concepts. Usually, a shortest path explains an obvious link between them, whereas subsequent shortest paths could reveal even more interesting relationships. A general way to compute such paths is the usage of a top-k shortest path algorithm which outputs shortest paths ordered by weight or length, respectively. In the area of semantic web, in which knowledge graphs are typically represented using RDF, such algorithms were already applied [3,7]. Interpreting graphs as RDF may lead to additional constraints, such as special loop restrictions or path patterns. In our case we will conceive an RDF graph as a directed labeled multigraph having RDF resources as vertices and RDF properties as edges. We further want to disallow paths including the same RDF triple (vertex - edge - vertex) multiple times. Apart from that, we tolerate multiply visited vertices and edges. That is why common algorithms cannot be applied without

© Springer International Publishing Switzerland 2016
H. Sack et al. (Eds.): SemWebEval 2016, CCIS 641, pp. 200–212, 2016.
DOI: 10.1007/978-3-319-46565-4_16

further ado. Additionally, even suitable modifications may result in unacceptable time and memory consumption in practice. For example, a slightly modified Dijkstra algorithm is not practical due to the large size of typical RDF graphs. Considering the constraints mentioned above, this paper proposes a solution method based on Eppstein's top-k shortest path algorithm [2].

The paper is structured as follows: In Sect. 2 other papers mainly about finding shortest paths are presented. Section 3 explains our approach which is evaluated and discussed in Sects. 4 and 5. Our algorithm's results on the evaluation sets of the ESWC 2016 Top-k Shortest Path Challenge are separately addressed in Sect. 6. Last, we shortly introduce our SPARQL endpoint for shortest paths (Sect. 7) and give an outlook on possible future work (Sect. 8).

2 Related Work

Approaches of finding top-k shortest paths can be divided in two categories based on whether their resulting paths are loopless or not. The latter is a less complex task.

One well-known representative of this category is Eppstein's algorithm [2]. Its core is a bounded-degree graph which outputs k paths using breadth-first search. Given a graph with n vertices and m edges this results in a time complexity of $O(m + n \log n + k)$. There are various modified versions: The algorithm by Jiménez et al. [5] lazily builds data structures for practical improvements whereas Aljazzar and Leue [1] allow the usage of a heuristic function to guide through the search space.

Yen's algorithm [12] is an example of the other category finding loopless paths. It has a time complexity of $O(kn(m + n \log n))$ which results from calling Dijkstra's algorithm multiple times to compute shortest paths.

Examining the complexity of evaluating property paths defined in SPARQL [6,8] and retrieving paths from large datasets [3,11] are other related research areas of interest.

More details will be given in the next section which is about our approach.

3 Approach

While Eppstein's top-k shortest path algorithm [2] finds paths with loops, our solution ensures that every path additionally complies to a unique triple condition. In particular, a path and the unique triple condition are defined as follows:

Definition 1. *A path is a sequence of alternating RDF resources and properties* $(r_0, p_1, r_1, \ldots, p_n, r_n)$ *where* $n > 0$.

Definition 2. *The unique triple condition for a path* $(r_0, p_1, r_1, \ldots, p_n, r_n)$ *is true if and only if* $\forall\, 0 < i \leq n : \exists! (r_{i-1}, p_i, r_i) \in p$.

A path is valid if (and only if) the unique triple condition is true.

There are paths which comply to this condition but contain multiple visited vertices. Yen's algorithm [12] would consider these paths to contain loops. Thus, we decided against softening the loop restriction on the one hand and inserting the unique triple condition on the other.

Our contribution is based on Eppstein's algorithm [2] which is shortly explained in the following. (For a more detailed introduction we kindly refer the reader to the original paper.) The algorithm uses a bounded-degree graph $P(G)$ built in $O(m + n \log n)$ that outputs k paths in linear time using breadth-first search. For building $P(G)$ a single source shortest path tree T for target vertex t has to be computed. All edges which are not part of T are called sidetrack edges, thus $G - T$ is a sidetrack graph. $P(G)$ is built with the help of two kinds of heaps: (1) $H_T(v)$ forms a heap of all minimal sidetracks on the shortest path from v to t. (2) $H_{out}(v)$ forms a heap of all outgoing sidetracks of v except the minimal sidetrack. Merging $H_T(v)$ and $H_{out}(v)$ results in $H_G(v)$. All H_G are connected by so-called cross edges to form $P(G)$. A breadth-first search on $P(G)$ with a priority queue outputs the top-k shortest paths ordered by weight in $O(k)$.

Our approach is depicted as pseudocode in Algorithm 1. In general, the given RDF dataset is interpreted as an edge-label directed multigraph (line 1). Before running the algorithm's main part, all statements containing a literal are removed in a preprocessing step[1]. Lines 2–13 are directly adopted from Eppstein: Like stated above, the shortest path tree T, the sidetrack graph $G - T$, the edge weights $\delta(e)$ and the heaps H_T and H_{out} are computed and necessary data structures are initialized.

Three modifications to the original Eppstein algorithm are made: (a) Every path has to be built and/or checked in order to immediately check its validity (line 16), (b) only valid paths are added to the result list R (lines 17–19) and (c) $P(G)$ is pruned whenever p is invalid due to multiple sidetracks (line 20).

There are cases in which a path does not comply to the unique triple condition: It may contain multiply used (1) sidetracks and/or (2) shortest path edges.

To item (1): Using $P(G)$ Eppstein's algorithm keeps track of activated sidetracks, thus multiply used ones are easily detectable. In this case we can stop adding a cross edge, because all further paths would become invalid. Additionally, if k is greater than the number of possible valid paths in G, this pruning guarantees our algorithm's termination. However, the computation does not stop until every sidetrack combination is generated and checked which may result in significant overhead.

---

[1] This is done since providing a literal as a target node could be ambiguous (e.g. several persons having the same first name). Thus, to apply our algorithm in such use cases, this ambiguity has to be resolved first. For example, one could first search all resources associated with the given literal and then choose the one that is actually meant as the target node (or subsequently run the algorithm on all of them).

To item (2): Since $P(G)$ only contains sidetracks, each path has to be built in order to detect multiply used shortest path edges. In that case we have to extend $P(G)$ and add a cross edge, since there could be a sidetrack in $P(G)$ which allows bypassing this problematic shortest path edge in the future. It's not possible to remove these nodes in $H_G(v)$, since the algorithm would then miss cross edges to potential valid paths.

A path is built with a (possibly empty) queue of sidetracks S and the start node s (see (a) above and line 16). If S already contains duplicates (case 1, see above) building the path is obsolete. If no duplicates are detected, the system initiates the building process in which sidetracks (if suitable) are preferred over shortest path edges. While building the path, duplicate shortest path edges can be detected by keeping track of used edges. There are special cases in which the given end node t is visited multiple times. Thus, the building path algorithm stops at t only if all given sidetracks in the queue are used.

In the given algorithm, $P(G)$ is only extended and $H_G(v)$ only constructed if necessary (lines 21–23). This is possible due to the pruning condition (line 20). Next, a breadth-first traversal step is performed on $P(G)$ and weights are updated (lines 25–29). Finally, all found and valid paths in R are returned (line 32).

For more complex problems we would also like to consider a special property path condition which we state as follows:

Definition 3. *The special property path condition for a path* $(r_0, p_1, r_1, \ldots, p_n, r_n)$ *and a given property P is true if and only if* $p_1 \equiv P \vee p_n \equiv P$.

A naive approach would be to simply delete all $(s, \neg P, r)$ edges from G. However, this would prevent valid paths like $(s, P, \ldots, s, \neg P, \ldots)$ from being found. That's why our approach (see Algorithm 2) introduces a dummy vertex v which ensures that all considered paths contain P as the first property. First, the algorithm collects all P-edges that start in s (we denote this set as A). If none is found the algorithm terminates, leaving only P-edges which end in t to be checked. By inserting the dummy vertex v and connecting it to all $head(e)$ of A we ensure that all paths from v to t contain P as the first property. After each found path the algorithm has to substitute v with s which is valid, since there actually exist outgoing P-edges from s.

Finding paths containing P-edges that end in t can be reduced to the previously solved problem. We therefore invert all edges in G and call Algorithm 2 with interchanged s and t. The resulting paths have to be reversed. Finally, both result lists are merged and ordered by length also removing duplicates.

In order to apply this approach, the previously introduced build path algorithm had to be modified. Since every path starts with $(v, P, r, \ldots)$ the occurrence of (s, P, r) would not be recognized as a duplicate. This is solved by artificially adding (s, P, r) to the activated sidetrack queue S and to the set of already used edges.

How our method performs in practical scenarios is discussed in the next section.

Data: start node s, end node t, required number of paths k, dataset D
Result: k paths between s and t in D ordered by length complying to unique triple condition

1   interpret D as an edge-labeled directed multigraph G;
2   compute shortest path tree T starting from t using Dijkstra;
3   compute sidetrack graph $\widetilde{G} = G - T$;
4   **for** $e \in \widetilde{G}$ **do**
5   $\quad$ compute $\delta(e) = l(e) + d(head(e), t) - d(tail(e), t)$;
6   **end**
7   **for** $v \in T$ **do**
8   $\quad$ compute $H_T(v)$ and $H_{out}(v)$;
9   **end**
10   initialize initial edge $init = (\emptyset, \emptyset, s)$;
11   initialize heap $H = < init >$;
12   initialize graph $P(G) = (\{init\}, \emptyset)$;
13   initialize result list $R = <>$;
14   **while** H *is not empty* $\wedge\ |R| < k$ **do**
15   $\quad$ remove minimal element m from H;
16   $\quad$ build path p using m;
17   $\quad$ **if** p *is valid* **then**
18   $\quad\quad$ add p to R;
19   $\quad$ **end**
20   $\quad$ **if** p *is valid* $\vee$ p *contains multiple shortest path edges* **then**
21   $\quad\quad$ compute $H_G(head(m))$ and add to $P(G)$;
22   $\quad\quad$ let $r = root(H_G(head(m)))$;
23   $\quad\quad$ add cross edge (m, r) with weight $\delta(r)$ to $P(G)$;
24   $\quad$ **end**
25   $\quad$ **for** $e \in out(m)$ **do**
26   $\quad\quad$ let $n = head(e)$;
27   $\quad\quad$ compute $\delta(n) = \delta(m) + l(e)$;
28   $\quad\quad$ add n to H;
29   $\quad$ **end**
30   $\quad$ remove m from $P(G)$;
31   **end**
32   return R;

Algorithm 1. Modified Eppstein Algorithm

Data: start node s, end node t, required number of paths k, property P,
   dataset D

Result: k paths between s and t in D ordered by length complying to
    unique triple and special property path condition with P

1 interpret D as a edge-labeled multigraph G;
2 let $A = \{e \in G : tail(e) \equiv s \wedge label(e) \equiv P\}$;
3 **if** A *is empty* **then**
4 | return $\emptyset$;
5 **end**
6 construct an arbitrary vertex $v \notin G$;
7 add v to G;
8 **for** $e \in A$ **do**
9 | add edge $(v, head(e))$ to G;
10 **end**
11 run modified Eppstein with $s' = v$, $t' = t$, $k' = k$;

Algorithm 2. Reuse of modified Eppstein to retrieve paths complying to the
special property path condition

4   Evaluation

We evaluated our approach on the training set of the ESWC 2016 Top-k Shortest
Path Challenge. This challenge consists of two tasks: Besides finding the top-
k shortest paths complying to the unique triple condition (see Definition 2),
the second one additionally introduces a special property path restriction (see
Definition 3).

The training set corresponds to a 10 % dataset generated from the DBpedia
knowledge base [9,10]. Since the data is not generated artificially like the LUBM
dataset [4], it is very heterogeneous and there are no well structured classes.
The dataset is processed so that there are no blank nodes, untyped classes or
unparsable triples[2]. As a consequence, each resource r is not a blank node and
has at least got one statement (r rdf:type *type*) after the processing.

In total the training set consists of 9,996,907 triples, 7,598,913 of them being
literal statements. After their removal the resulting set contains 2,397,994 triples.
There are 181,702 duplicate statements which do not provide additional information. In fact, there are 13 statements which appear 4 times in the dataset. Our
algorithm has to deal with multigraphs since there are 394,085 multiple edges
(also including duplicate triples). Moreover, there are 407 reflexive edges which
means that also smallest loops have to be handled. The most important number
is the average out degree. In this set, the mean value is 6.03 with a standard
deviation of 4.51. The vertices' number of outgoing edges ranges from 0 to 106.

---

[2] In particular, we found unparsable datatypes like <*http://dbpedia.org/datatype/*
brake horsepower>. Our first solution removed them, in a later version they were
fixed by URL-encoding the blank.

In this section we analyze our approach in terms of performance, space and overhead compared to the original Eppstein algorithm. Table 1 shows our results of Task 1 in which four queries with increasing values of k were given. For each value we recorded six different measures which are explained in more detail in the following:

The first one is the number of iterations needed by our algorithm, i.e. the number of executions of the *while* loop in line 14. Additional iterations in comparison to Eppstein's algorithm are characterized by the relative overhead to k: $\frac{(iterations-k)}{k}$. Next, we captured time in milliseconds from the computation of the shortest path tree T (line 2) till the return of R (line 32). All experiments were conducted on a typical end user notebook with an Intel 4 Core i7-4712MQ 2.30 GHz CPU, 16 GB RAM and Java 1.8 installed. The next column contains the number of times $P(G)$ was not extended or in other words pruned. This is the case when multiple sidetracks are detected, thus the *if* condition in line 20 is skipped. To give an impression of our algorithm's memory consumption we additionally record $|V_{P(G)}|$ and $|H|$. $|V_{P(G)}|$ is the number of vertices in $P(G)$ (line 12) whereas $|H|$ is the number of entries in heap H (line 11) after termination.

Like stated before, Task 2 introduced the special property path condition (see Definition 3). This means all paths need to have P as their first (i.e. $(s, P, \dots)$) or last (i.e. $(\dots, P, t)$) property. The latter case does not appear in queries 1, 3 and 4. For query 2 we separated the cases accordingly. All measures are found in Table 2.

The relation between k and the overhead of all queries is depicted in Fig. 1. k is measured on a logarithmic scale whereas the overhead is given in percent on a linear scale. Additionally, we plotted a trendline obtained by applying linear regression on all query results.

Results are further discussed in the next section.

5   Discussion

Since the expected results of the challenge were provided beforehand we could verify that our algorithm performed correctly for all queries. The individual results of our evaluation (see Tables 1, 2 and Fig. 1) will be discussed in the following, starting with Table 1.

Obviously, a high k results in a high overhead since more potentially invalid paths are found. This effect is mitigated by pruning of $P(G)$. The time column shows that 5 to 6 s are required to initialize the necessary data structures of the training set (see Algorithm 1, lines 2–13). Additional time is needed to discover the shortest paths which is done in the while loop. For a maximum k query 4 needs about 40 s for this task having a rather low overhead of 10 %. Please note that all time measurements were done using a non-optimized version of our algorithm. During implementation we identified some potential for improvements which was not exploited for time reasons. Moreover, we discovered that the number of vertices in $P(G)$ is proportional to k.

Table 1. Results of task 1

Query	k	Iterations	Overhead	Time (ms)	Pruned	$|V_{P(G)}|$	$|H|$
1	8	8	0.00 %	6531	0	83	15
1	344	362	5.23 %	6663	6	4401	715
1	1068	1128	5.61 %	6833	12	16408	2296
1	20152	21663	7.49 %	12404	407	293527	43701
2	3	3	0.00 %	5528	0	64	6
2	4	4	0.00 %	5662	0	80	8
2	79	91	15.18 %	5783	4	1702	174
2	154	169	9.74 %	5908	4	3193	336
3	36	36	0.00 %	5785	0	64	6
3	336	336	0.00 %	5932	0	80	8
3	4866	5034	3.45 %	8147	4	1702	174
4	2	2	0.00 %	6137	0	78	4
4	16	16	0.00 %	6230	0	524	34
4	250	254	1.60 %	6416	0	7389	515
4	1906	1980	3.88 %	8393	34	58426	3984
4	20224	21858	8.07 %	13980	678	619359	42879
4	175560	192367	9.57 %	45785	7592	5628758	380839

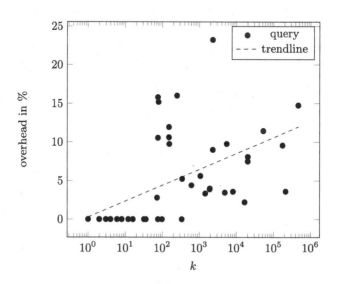

Fig. 1. Relation of k and corresponding overhead

Table 2. Results of task 2

Query	k	Iterations	Overhead	Time (ms)	Pruned	$|V_{P(G)}|$	$|H|$
1	32	32	0.00 %	6912	0	391	63
1	98	98	0.00 %	8353	0	1410	200
1	1914	1990	3.97 %	9119	28	27063	4030
1	16632	16999	2.20 %	12929	119	247948	34909
1	212988	220639	3.59 %	45968	2319	3178408	451439
2 $(s, P, \dots)$	3	3	0.00 %	6418	0	61	6
2 $(\dots, P, t)$	3	3	0.00 %	5297	0	99	5
2 $(s, P, \dots)$	4	4	0.00 %	5519	0	77	8
2 $(\dots, P, t)$	4	4	0.00 %	4560	0	127	8
2 $(s, P, \dots)$	76	88	15.78 %	5701	4	1635	167
2 $(\dots, P, t)$	76	84	10.52 %	4589	3	2818	170
2 $(s, P, \dots)$	151	167	10.59 %	5686	5	3149	328
2 $(\dots, P, t)$	151	169	11.92 %	4616	5	5572	338
2 $(s, P, \dots)$	2311	2848	23.23 %	7787	232	52309	5297
2 $(\dots, P, t)$	2311	2519	9.00 %	7988	83	89060	5362
3	12	12	0.00 %	5652	0	295	24
3	76	76	0.00 %	5554	0	1677	151
3	1440	1488	3.33 %	7259	16	32991	2869
3	8088	8377	3.57 %	10122	109	173570	16281
4	1	1	0.00 %	7827	0	34	1
4	6	6	0.00 %	8493	0	218	12
4	72	74	2.77 %	7664	0	2166	146
4	614	641	4.39 %	8025	12	19182	1280
4	5483	6018	9.75 %	10960	198	172989	11788
4	52649	58671	11.43 %	20077	2640	1707021	115534
4	471199	540815	14.77 %	71884	26962	15997816	1065488

Most findings also apply to Table 2. Although query 2 has high prune values
the overhead is the highest in comparison to all other queries. Our assumption is
that the execution of query 2 involves a high number of loops from s to t. That
is why it has got the highest overhead of 23 % for a maximum k of 2311.

Figure 1 gives a summarizing overview of the relation between k and the
overhead. Considering that k is plotted on a logarithmic scale we see that the
overhead only grows slowly. In particular, there are several queries having a k
less than 10^2 which do not cause any overhead. Most queries have an overhead
between 0 % to 15 % even if k approaches 500,000. However, for values greater
than 10^2 there are some outliers above 10 % though not reaching 25 %.

Thus, our algorithm achieves a time complexity which is moderately above the one of Eppstein's algorithm for the given training set. In general, our algorithm's overhead increases with the number of loops in the graph.

6 ESWC 2016 Challenge Evaluation

We also ran our algorithm on the evaluation set of the ESWC 2016 Top-k Shortest Path Challenge. It contains 1,551,041 URIs and about 13.6 million distinct triples, which is about 5.7 times the size of the training set used before. This led to memory problems with our first implementation. We therefore optimized our algorithm's memory efficiency by indexing all triples and storing them as an integer array during runtime. However, we still had to use a more powerful machine compared to the one used in Sect. 4 in order to complete all four evaluation tasks, in particular the fourth one. We used a virtual machine having 48 cores à 2.8 GHz[3] and 492 GiB of memory running openSUSE Linux 42.1 (64-bit) and Java OpenJDK 8. The achieved results, which are principally quite similar to those of the training set (see Sect. 4), are given in Table 3.

Table 3. Results of evaluation

Query	k	Iterations	Overhead	Time (ms)	Pruned	$|V_{P(G)}|$	$|H|$
1.1	377	389	3.18 %	31918	4	21115	739
1.2	53008	56231	6.08 %	35154	1589	2979808	106252
2.1 $(s, P, \dots)$	374	386	3.20 %	21283	4	20934	732
2.1 $(\dots, P, t)$	374	386	3.20 %	23349	2	48356	725
2.2 $(s, P, \dots)$	52664	55877	6.10 %	41206	1586	2961258	105579
2.2 $(\dots, P, t)$	52664	52766	0.19 %	271662	23	95446490	151819

Again, the overhead compared to the original Eppstein algorithm was moderate. Tasks 1.1 to 2.1 needed less than 10 GiB of RAM, whereas up to 80 GiB were required in order to complete Task 2.2. This results from Task 2.2 being quite complex: in the second part of the query $((\dots, P, t))$, the algorithm was only able to prune for 23 times. Thus, $P(G)$ had to be extended several times resulting in very high memory usage. A low overhead of 0.2 % indicates that this is not caused by our extensions but would also have been present when using the original Eppstein alone. Concerning runtime performance there was a noticeable overhead induced by the virtualization, especially also related to memory allocation. Please note that we therefore omitted freeing memory during the calculation for improved runtime performance.

To better use the algorithm in our daily work (e.g. our research about explanation aware computing), we embedded it into a SPARQL endpoint which is explained in the next section.

---

[3] Our algorithm is still implemented as a single thread application.

7 SPARQL Endpoint for Shortest Paths

In this section a modification to the SPARQL evaluation is provided to use the proposed algorithm for calculating k shortest paths. We extended the property path semantics to also include the additional shortest paths while keeping the SPARQL query language syntax and expressiveness. Therefore a SELECT query is used to retrieve the path components. The idea is to create a suitable number of hidden variables containing resources and properties alternately. They are numbered consecutively starting from zero. The path length is assigned to the ?length variable. Our approach is conform to SPARQL 1.1, i.e. FILTER and LIMIT instructions work as expected. We used the property path syntax to encode a shortest path semantic[4]. As an example, the first predicate can also be restricted to always be dbp:after with the following query:

```
SELECT *
WHERE {
    dbr:Felipe_Massa !:* dbr:Red_Bull.
    FILTER(?r1 = dbp:after).
}
LIMIT 3
```

Table 4. Example SPARQL response

?length	?r0	?r1	?r2	?r3	?r4	?r5	?r6
7	dbr:Felipe_Massa	dbp:after	dbr:Robert_Kŭbica	dbp:first Win	dbr:2008_Canadian_Grand_Prix	dbp:third Team	dbr:Red_Bull
7	dbr:Felipe_Massa	dbp:after	dbr:Robert_Kubica	dbo:first Win	dbr:2008_Canadian_Grand_Prix	dbp:third Team	dbr:Red_Bull
7	dbr:Felipe_Massa	dbp:after	dbr:Robert_Kubica	dbo:first Win	dbr:2008_Canadian_Grand_Prix	dbo:third Team	dbr:Red_Bull

A possible solution is depicted in Table 4. The downside of this approach is that all paths have to be computed to calculate the maximum amount of necessary variables. Another possibility is to name the variables directly in the

---

[4] Given two resources A and B, A as the source and B as the target, the corresponding SPARQL query is A !:* B. Technically, this searches for zero or more occurrences of properties (indicated by "*") between A and B not ("!") matching an introduced fake IRI ":".

query. With the help of the variable length, one can check if the amount of variables given in the query is enough. If this is not the case, the requester can create a new query with more variables. In the following, a brief overview of our SPARQL endpoint's implementation is shown. It uses the proposed path algorithm and responds to SELECT queries like in the given example above. We used FUSEKI SPARQL server based on Jena introducing a new Jena subsystem called GraphQuery. The QueryEngineMain of Jena is extended with an OpExecutor[5], making it possible to replace the evaluation of a specific operation. In this use case only the path operation is changed and all other operations are executed as usual. If the subject as well as the object are not variables, the proposed algorithm will be executed. We had to modify the evaluation of the query to prevent the removal of equalities in FILTER elements. Otherwise all variables which are not explicitly defined like our hidden ones, would be removed. As a consequence the FILTER statement would not have any effect. Creating or updating the actual dataset triggers the creation of the corresponding graph index. To create a running FUSEKI instance for it, we used the Jena Assembler specification.

In summary, we now have a running FUSEKI server which is able to retrieve all shortest path for a given SPARQL property path query.

8   Conclusion and Future Work

In this paper we presented several approaches for finding top-k shortest paths and pointed out why they are not directly applicable for our specified use case. Based on Eppstein's algorithm we therefore implemented our own solution method which induces only moderate overhead. We were able to solve all queries given in the ESWC 2016 Top-K Shortest Path Challenge and additionally provided further details on our algorithm's performance. One general problem was the overhead resulting from loops in the graph.

During the implementation we already identified some potential for improvements concerning time and memory consumption which was not exploited due to time reasons. For future work we could lazily build the different heaps and try to predict invalid paths earlier. Since our algorithm would not need to follow these paths the resulting overhead is far less.

Besides, a concept for embedding our algorithm into a SPARQL endpoint was provided.

Acknowledgement. This work was partially funded by the BMBF project Multimedia Opinion Mining (MOM: 01WI15002).

---

[5] https://jena.apache.org/documentation/query/arq-query-eval.html.

References

1. Aljazzar, H., Leue, S.: K*: a heuristic search algorithm for finding the k shortest paths. Artif. Intell. **175**(18), 2129–2154 (2011)
2. Eppstein, D.: Finding the k shortest paths. SIAM J. Comput. **28**(2), 652–673 (1998)
3. Gubichev, A., Neumann, T.: Path query processing on very large RDF graphs. In: WebDB. Citeseer (2011)
4. Guo, Y., Pan, Z., Heflin, J.: LUBM: a benchmark for owl knowledge base systems. Web Semant. Sci. Serv. Agents World Wide Web **3**(2) (2005)
5. Jiménez, V.M., Marzal, A.: A lazy version of Eppstein's K shortest paths algorithm. In: Jansen, K., Margraf, M., Mastrolli, M., Rolim, J.D.P. (eds.) WEA 2003. LNCS, vol. 2647, pp. 179–191. Springer, Heidelberg (2003)
6. Kostylev, E.V., Reutter, J.L., Romero, M., Vrgoč, D.: SPARQL with property paths. In: Arenas, M., et al. (eds.) ISWC 2015. LNCS, vol. 9366, pp. 3–18. Springer International Publishing, Switzerland (2015)
7. Lehmann, J., Schüppel, J., Auer, S.: Discovering unknown connections-the dbpedia relationship finder. In: CSSW, vol. 113, pp. 99–110 (2007)
8. Losemann, K., Martens, W.: The complexity of evaluating path expressions in SPARQL. In: Proceedings of the 31st Symposium on Principles of Database Systems, pp. 101–112. ACM (2012)
9. Morsey, M., Lehmann, J., Auer, S., Ngonga Ngomo, A.-C.: DBpedia SPARQL benchmark – performance assessment with real queries on real data. In: Aroyo, L., Welty, C., Alani, H., Taylor, J., Bernstein, A., Kagal, L., Noy, N., Blomqvist, E. (eds.) ISWC 2011, Part I. LNCS, vol. 7031, pp. 454–469. Springer, Heidelberg (2011)
10. Morsey, M., Lehmann, J., Auer, S., Ngonga Ngomo, A.-C.: Usage-centric benchmarking of RDF triple stores. In: Proceedings of the 26th AAAI Conference on Artificial Intelligence (AAAI 2012) (2012)
11. Przyjaciel-Zablocki, M., Schätzle, A., Hornung, T., Lausen, G.: RDFPath: path query processing on large RDF graphs with MapReduce. In: García-Castro, R., Fensel, D., Antoniou, G. (eds.) ESWC 2011. LNCS, vol. 7117, pp. 50–64. Springer, Heidelberg (2012)
12. Yen, J.Y.: Finding the k shortest loopless paths in a network. Manag. Sci. **17**(11), 712–716 (1971)

Modified MinG Algorithm to Find Top-K Shortest Paths from large RDF Graphs

Zohaib Hassan$^{(\boxtimes)}$, Mohammad Abdul Qadir,
Muhammad Arshad Islam$^{(\boxtimes)}$, Umer Shahzad, and Nadeem Akhter

Department of Computer Science,
Capital University of Science and Technology, Islamabad, Pakistan
zohaib200pk@hotmail.com,
{aqadir,arshad.islam}@cust.edu.pk,
umershah444@gmail.com, nadeemakhter82@gmail.com

Abstract. MinG algorithm indexes large RDF graphs in an efficient way and then uses the index to answer all path queries between two nodes of the graph. MinG reduces the computational and space complexity of indexing by not creating a special type of adjacency matrix called Path Type Matrix at each level of indexing. We only need Path Type Matrices at first and last level of indexing. MinG was modified to answer top-K shortest paths. The experiments were performed on specific case studies. Gain in the performance is significant due to reduction in the space to index a graph and also reduction in computation time to answer path queries.

Keywords: Top-K Shortest Paths · Graph Traversal · Graph Indexing · Algorithm · Graph Mining · Semantic Web

1 Introduction

We are witnessing 'big data' explosion both in terms of size as well as time. Rate of data production has reached to astonishing magnitude and data is originating from unprecedented variety of sources such as climate sensors, social media sites, digital pictures, transaction records and location based cell phone services[1]. Graphs are traditionally used to represent data that has strong relational characteristics. Such type of data includes online social networks, transportation networks, hyperlink graph of the World Wide Web and many more.

RDF (Resource Description Framework) is a widely considered World Wide Web Consortium standard that is used for data interchange[2]. RDF is represented as a directed labeled graph, where the edges represent the named link between two resources, represented by the graph nodes. Scalability challenges still persist both with respect to storing huge number of RDF triples as well as query them. Research community is already facing complexity issues to manage billions of RDF triples and it can be foreseen that this requirement will rise to trillions of triples in near future [1].

---

[1] http://www-01.ibm.com/software/data/bigdata/what-is-big-data.html
[2] http://www.w3.org/TR/rdf-primer, 2004

© Springer International Publishing Switzerland 2016
H. Sack et al. (Eds.): SemWebEval 2016, CCIS 641, pp. 213–227, 2016.
DOI: 10.1007/978-3-319-46565-4_17

Moreover, competitions are being organized to encourage researchers to build fast and efficient mechanism to store and query RDF graphs[3].

We can categories the mechanism of graph queries in two broad classes, i.e., on-the-fly approaches and indexing-based approaches. One approach is on the fly approach in which no pre-processing is done over the data. Query is processed by the system over the raw input graph to retrieve relevant results [2]. Indexing-based approach involves pre-processing over the input graph to reduce the computational time for answering queries. Usually the input graph is reduced to its features (path, tree, cycle etc.) that are further inserted into a particular data structure [3].

Shortest path queries are based on the fundamental concept in graph theory and is widely applied in the field like artificial intelligence and Web communities. In dense graphs, there are several shortest paths between two graph nodes and single shortest path queries are unable to exploit such underlying graph structures. As shown in Fig. 1, there are multiple shortest paths between the shaded node pair. To extract such structures in graphs, mechanisms for top-k shortest path are adapted. Intuitively, we may use variant of any shortest path algorithm such as Dijkstra's algorithm however, such methods are computationally very expensive $O((n+m)k)$. We can find several instances that compute top-k shortest paths such as Eppstein algorithm [4] with the complexity of $O(n+m+k)$ but implementation of this algorithm is prohibitively slow.

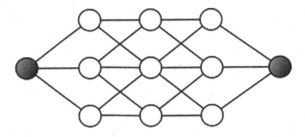

Fig. 1. Example depicting multiple shortest paths

In this paper, we have presented an index-based approach for top-k shortest paths in an RDF graph. The proposed approach is inspired from the work of Barton [5]. The author has proposed a ρ-index and a search algorithm to find all paths between any two vertices from a graph [6]. Modified MinG adapts this scheme to compute top-k shortest path and further reduces the cost of indexing with a faster logic to query the graph [17].

2   Related Work

There are numerous algorithms that can tell us the existence of a complex relationship between any two nodes from a graph. Such algorithms are known as graph reachability query answering algorithms. These approaches return a single Boolean value answering that a source vertex can reach a target vertex or not. These approaches

---

include the matrix-based approach [10], an interval based approach [11], the 2-Hop approach [12], the HOPI algorithm [13, 14], the HLSS approach [15], and an indexing scheme of Dual Labeling [16]. These approaches can answer the reachability queries in constant time. Although efficient, these approaches can be used to find only the existence of a relationship and not the actual relationships between any two vertices from a graph [17].

Several on-the-fly and indexing-based techniques for computing shortest paths have been proposed in the past. Tarjan's algorithm of a single source path expression problem [2] [7] is on the fly query-answering algorithm. Given a vertex A from a graph the algorithm finds for each vertex V a regular expression RE(A,V) that represents all paths between nodes A and V in the graph. The computational complexity of Tarjan's algorithm is $O(|E|)$, which takes a lot of time for strongly connected very large graphs.

When it comes to indexing approaches to find all paths between any two nodes from a graph, an indexing scheme of [11] uses a data structure known as suffix arrays. Suffix array is a well-known data structure used for full text search over the documents constructed on one-dimensional character strings. The algorithm extracts all path expressions from an RDF graph and to make query processing faster it creates a suffix array for all extracted path expressions. The problem with this approach is that it works only for DAGs.

Another indexing scheme proposed by [9] converts RDF graph into forest of trees. The algorithm assigns each node a signature, which includes a pre-order rank of that node, a post-order rank of that node, a pointer to first ancestor of the node and some other information. These signatures are then used to find all paths between any two nodes from graph.

3 Proposed algorithm

To find all paths between any two nodes of a graph the search algorithm of ρ-index uses intra segment Path Type Matrices (PTMs) and inter segment Inverted Files (INFs) after dividing a large graph into segments. Then the segments with INFs are considered as a graph with each segment as a node and the same approach to generate PTMs and INFs is applied. This will continue till the last graph of segments with a reasonable size is reached.

In this section we will present a search algorithm to find all paths between any two nodes from a graph, which is based on ρ-index. However, the difference is that our algorithm needs only the PTMs at the topmost and the bottommost segments. PTMs at any intermediate levels are not required. So while indexing, we will only create the PTMs at topmost and bottommost levels and not at any intermediate levels.

To answer query, instead of going from bottom to topmost level, identifying all segments involved and then building the paths from top to bottom [5], MinG starts from bottom, identify segments involved at bottommost level of indexing and builds the paths at current level. This will generate a set of possible paths at bottommost level. Then it moves on to next level of indexing, identifies the segments involved at that level and finds the paths at that level by using INFs. As there are no PTMs at any intermediate levels which requires $O(N^2)$ space to store the index and $O(N^3)$ time to multiply the graphs with N nodes.

3.1 PTMs and INFs Generation Approach

To explain the working of our algorithm, we will take a graph, which we will index up to three levels. Fig. 2 shows an example graph whose segments are created by including vertices of original graph into six different segments. The vertices are included into segments randomly.

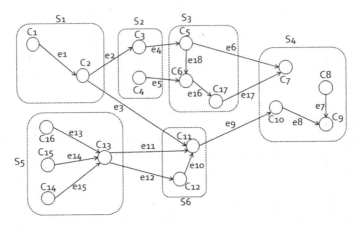

Fig. 2. An example graph and the first level of indexing

Fig. 3 shows the INFs of six segments created in Fig. 2. PTMs for the segments of Fig. 2 are shown in Fig. 4. This way a graph and segments of Fig. 2, INFs of Fig. 3 and PTMs of Fig. 4 form the first level of indexing. Fig. 5 shows a segment graph created from the segments in Fig. 2. This segment graph is further segmented in Fig. 5 to form the second level of indexing. Only the INFs will be created at this level, which are shown in Fig. 6.

S1	Out	S2	Out	S3	Out	S6	Out
C2	C3, C11	C3	C5	C5	C7	C11	C10
S4	**IN**	C4	C6	C17	C7		**IN**
C7	C5, C17		**IN**		**IN**	C11	C2, C13
C10	C11	C3	C2	C5	C3	C12	C13
				C6	C4		

S5	Out
C13	C11, C12

Fig. 3. Inverted Files at first level of indexing

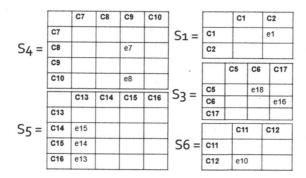

Fig. 4. Path Type Matrices at first level of indexing

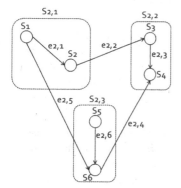

Fig. 5. Segment graph and the second level of indexing

S2,1	**Out**	**S2,3**	**Out**
S2	S3	S6	S4
			IN
S1	S6	S6	S1

S2,2	**IN**
S3	S2
S4	S6

Fig. 6. Inverted files at second level of indexing

Segment graph created from the segments of Fig. 5 is shown in Fig. 7. No further segmentation is performed for a graph of Fig. 7. As there is no segmentation at last level, the INFs cannot be created there. A segment graph in Fig. 7 is at highest level of indexing; therefore a PTM at this level will be created which is shown in Fig. 8.

3.2 Query Algorithm

Modified MinG algorithm to find K shortest paths between any two nodes from a graph performs the following steps.

1. Identify from first level the segment of Source and Target nodes.
2. At the first level, retrieve the entries from, PTM(Source), INF_EDGES_OUT (Source), INF_EDGES_IN(Target), PTM(Target). At any level other than the bottommost and the topmost, retrieve the entries from INF_EDGES_OUT(Source), INF_EDGES_IN(Target)
3. Generate a set of paths at level one by information we retrieved in step 2.

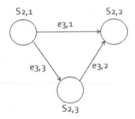

	S2,1	S2,2	S2,3
S2,1		e3,1 e3,3e3,2	e3,3
S2,2			
S2,3		e3,2	

Fig. 7. Segment graph at the third level of indexing   **Fig. 8.** Path Type Matrix at the last level of indexing

4. Identify that from which segment at next level a segment of source identified at step 1 belongs to. This newly identified segment is now Source. Update the Target by same procedure.
5. Remove the duplicate paths from a set of paths at step 4.
6. For each path in a set generated compute, INF_EDGES_OUT(Initial) U INF_EDGES_IN(Terminal) till cell entries related to first level of indexing are obtained and retrieve the following values that are distinct. PTM(Initial), EDGES_OUT (Initial), EDGES_IN(Terminal), PTM(Terminal)
7. Accumulate paths of all levels. As there is no segmentation on the topmost level so entries will only be retrieved from PTM at that level.
8. Once the processing of algorithm has been finished by reaching the topmost level, remove all those paths from the final set in which Initial I is not the Source S of user query or Terminal T is not the Target D of a user query.
9. The final set will contain all paths between S and D. For all those paths having $I = S$ and $T !=D$, find T in other paths of a set.
10. If T has been found in a path, join the partial path from T to terminal T' of that path with the path whose $T !=D$.
11. In the last step, while parsing the paths to find T, length K of paths can also be calculated in order to return only K length paths between source and destination nodes.

In step 2 and 7, set of paths is generated by comparing the terminal of one path with the initial element of the other paths. If terminal of one path is same as initial of other, merge that path. In order to reduce the comparisons, paths in a set can be sorted.

Now lets suppose that a user pose the query Paths (C1—C9). That is, find all paths between the nodes C1 and C9 from a graph shown in Fig. 2.

Vertex C1 is the source and vertex C9 is the target. The first step of algorithm is to check that from which segment the source and target vertex belongs to. The source vertex C1 belongs to the segment S1 and the target vertex C9 belongs to the segment S4.

The second step is to generate a set of paths at first level by retrieving entries from the PTMs and the INFs of segments identified at first step of algorithm. The entries from INFs will be retrieved in a way that outgoing paths from an INF of source vertex and incoming paths from an INF of target vertex will be queried. Firstly, the algorithm will retrieve entries from a PTM of S1 and outgoing paths from an INF of S1. The algorithm will then retrieve incoming paths from an INF of S4 and entries from a PTM of S4. Let us call this set of paths generated at first level as L0. A set L0 will initially contains the following paths,

- PTM(S1) = C1→C2
- EDGES_OUT(S1) = C2→C3, C2→C11
- EDGES_IN(S4) = C5→C7, C17→C7, C11→C10
- PTM(S4) = C8→C9, C10→C9

Note that we have written the incoming edges retrieved from an INF in a reverse direction. The INF of segment S4 in Fig. 2 shows that a vertex C7 has an incoming edge from C5 and C17. The direction of an arrow is towards C7. That is, it can be written as C7←C5 but we have written it as C5→C7.

After removing the duplicate paths if any, the algorithm will build the partial paths from the information it has in L0. It does this by merging those paths for which the terminal vertex of one path is same as the initial vertex of the other path. Starting from the paths of which the source vertex C1 is the initial, the algorithm compares the terminal vertex of that path with the initial vertices of the other paths in L0.

The path in L0 that starts from the source vertex is C1→C2. The terminal vertex C2 of that path is the initial vertex of two other paths in L0 that is, C2→C3 and C2→C11. By merging C2, two partial paths C1→C2→C3 and C1→C2→C11 will be formed. If paths in L0 are sorted based on the initials, the terminal vertex C2 of the path C1→C2 will be compared only with the initial vertices of the paths C2→C3 and C2→C11 and not with the initials of other paths in L0.

The terminal vertex C3 of the partial path C1→C2→C3 is not the initial of any path in L0. The terminal vertex C11 of the partial path C1→C2→C11 is the initial of the path C11→C10 in L0. Thus a new partial path C1→C2→C11→C10 will be formed. The terminal vertex C10 of this partial path matches the initial of the path C10→C9 in L0 thus C1→C2→C11→C10→C9 will be formed. The other paths that are still not processed in L0 are C5→C7, C17→C7 and C8→C9. The terminal vertices C7 and C9 of these paths are not the initials of any path in L0 thus the algorithm have finished building the paths in L0. The L0 will now contains the following information,

- C1→C2→C3
- C1→C2→C11→C10→C9
- C5→C7
- C17→C7
- C8→C9

The user query was Paths (C1—C9) and from L0 the algorithm has found the path from the source to the target which is C1→C2→C11→C10→C9. In this query the path has been found on the first level of indexing. Let us now consider a query in which path will not be found on the first level and the algorithm has to proceed to the second level where there are no PTMs.

Consider the user query Paths (C1—C7). The source vertex C1 is in S1 and thetarget vertex C7 is in S4. A set L0 will be same as that of the previous query Paths (C1—C9) as the segments identified in first step that is, S1 and S4 are same for both thequeries. By examining L0 no paths between C1 and C7 are found thus the algorithmneeds to proceed to the second level of indexing. It checks that from which segments at the next level the segments identified at the firstlevel belong to. The segments identified at first level were S1 and S4, the algorithmnow checks that from which segment S1 and S4 belongs to. Figure 5 shows that S1belongs to S2,1 and S4 belongs to S2,2.

To build a set L1 that is, a set of paths at level two. The algorithm retrieves the outgoing edges from an INF of S2,1 and incoming edges from an INF of S2,2.

- EDGES_OUT(S2,1) = S2→S3, S1→S6
- EDGES_IN(S2,2) = S2→S3, S6→S4

After removing the duplicate paths a set L1 will contains the following information,

- S2→S3
- S6→S4
- S1→S6

Now for each path in L1 the algorithm will retrieve the outgoing edges from an INF of the initial vertex of a path and the incoming edges from an INF of the terminal vertex of a path. The union of these two will be taken. For a Path S2→S3,

- EDGES_OUT(S2) = C3→C5, C4→C6
- EDGES_IN(S3) = C3→C5, C4→C6
- EDGES_OUT(S2) ∪ EDGES_IN(S3) = C3→C5, C4→C6.

Similarly for the paths S1→S6 and S6→S4

- EDGES_OUT(S1) = C2→C3, C2→C11
- EDGES_IN(S6) = C2→C11, C13→C11, C13→C12
- EDGES_OUT(S6) = C11→C10
- EDGES_IN(S4) = C5→C7, C17→C7, C11→C10
- EDGES_OUT(S1) ∪ EDGES_IN(S6) = C2→C3, C2→C11, C13→C11, C13→C12.
- EDGES_OUT(S6) ∪ EDGES_IN(S4) = C11→C10, C5→C7, C17→C7.

By doing all this, the algorithm has reached the bottommost level where we have the PTMs. Thus it will check the PTMs of the vertices in L1. If however the algorithm has not reached the bottommost level, it will keep on checking the INFs in the same fashion.

In L1 we have 5 vertices, S2, S3, S1, S6 and S4. Their PTMs are shown in Fig. 4. The algorithm will query only the PTMs of S2, S3 and S6 because when it was building the paths in L0 it has queried the PTMs of S1 and S4, so it will not query their PTMs at L1.

- PTM(S2) = Empty
- PTM(S3) = C5→C6, C6→C17
- PTM(S6) = C12→C11

The algorithm will now build set L1 with information it has collected. L1 will contain then

- C3→C5
- C2→C3
- C2→C11
- C13→C11
- C13→C12
- C11→C10

- C5→C7
- C17→C7
- C5→C6
- C6→C17
- C12→C11
- C4→C6

Each item in List is an instance of path having separate initial, terminal and intermediate nodes. We now have two sets of paths, L0 and L1. The algorithm will merge L0 and L1. The new set at this level will become,

- C3→C5
- C2→C3
- C2→C11
- C13→C11
- C13→C12
- C11→C10
- C1→C2→C3
- C1→C2→C11→C10→C9
- C5→C7

- C5→C7
- C17→C7
- C5→C6
- C6→C17
- C12→C11
- C4→C6
- C17→C7
- C8→C9

The paths are now build from this combined set in a same manner as the partial paths were built in individual sets. After building the paths, set L1 will contain

- C1→C2→C3→C5→C6→C17→C7
- C1→C2→C3→C5→C7
- C1→C2→C11→C10→C9
- C2→C3→C5→C7
- C2→C3→C5→C6→C17→C7
- C2→C11→C10
- C13→C11→C10
- C13→12→C11→C10
- C4→C6→C17→C7
- C8→C9

The user query was Paths(C1—C7). A set L1 contains all paths between source C1 and target C7 of a graph shown in Fig. 2. The required paths are

- C1→C2→C3→C5→C6→C17→C7
- C1→C2→C3→C5→C7

3.3 Cycle Management

Our technique uses the property of matrix algebra for the calculation of paths of certain length. The kth power of a graph G is a graph with the same set of vertices as G and an edge between two vertices if and only if there is a path of length at most k between them [18]. Since a path of length two between vertices *u* and *v* exists for every vertex *w* such that $\{u, w\}$ and $\{w, v\}$ are edges in G, the square of the adjacency matrix of G counts the number of such paths. Similarly, the (u, v)th element of the kth power of the adjacency matrix of G gives the number of paths of length k between vertices *u* and *v*.

When k^{th} power of an adjacency matrix is calculated, it includes the cycles of length as well. As the given challenge allows cycles to be included in the query results however the triple repetition is not permitted therefore, we calculate all paths and prune the repeated triples in the end.

4 Implementation

We have implemented our technique using C# and MySQL. The experiments have been carried out on a Windows 8.1 machine with 24GB RAM and an Intel Pentium CPU i5 2.30 GHz. The first implementation challenge we faced has been the reading the evaluation dataset into our system because of its huge size, i.e., approximately 19GB. We used the in memory hash-set to identify all unique subjects, predicates and objects. Later on, we used dictionary to convert this data from hash-set into integers to be imported into the database. The further process of implementation is divided into 6 modules as shown in Fig. 9.

1. **Segmentation module** generates segments of the input RDF graph. The maximum size of the segment is tuned at 40 nodes. This tuning factor has been selected keeping in view the guidelines of Barton's approach [8]. Our implementation reads the input in a sequential way and constructs segments arbitrarily. We save the *node URL* in hash table and use *hash-code* for further processing. Each segment is assigned a *segment ID*. While a set of nodes is selected to be part of a particular segment, all edges that do not lie in the same segment are saved in *INF* files.

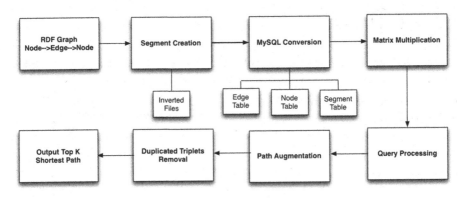

Fig. 9. Implementation Architecture

Once the segments for the complete input are generated, we rearrange nodes in segments that have more external edges as compared to internal edges. This way majority number of nodes have edges within that segment. As a result of this step, most of the segments are reduced to 39 nodes.

2. **MySQL conversion module** exports the data structures to MySQL for the sake of fast access. We have used MyISAM storage engine for this purpose. Segment structures are exported in the form of edge list into a MySQL database. Three tables are generated a) for nodes to segment relationship, b) for edge to node relationship, and c) for keeping segment level and its size as shown in Fig. 10.

3. **Path Type Matrices** for each segment at lowest level are generated using an adjacency matrix data structure and we calculate upto 40th power of the adjacency matrix. As described above, when two matrices are multiplied, the corresponding elements of rows and columns are concatenated to generate the path of that length. In case an empty matrix is generated before reaching till 40th power, then the process of matrix is terminated. All the matrices are summed at the end to obtain all paths of length 40 or less between any two nodes belonging to that segment. These *PTM* are also stored in MySQL database using a path list, which contains all path between any two nodes in the segment. The summary of the training data read through this mechanism is presented in Table 1.

Table 1. Parameter settings for index creation

Data Set	Feature	Value
Evaluation Data Set	Triples	110621287
	Unique Triple	46275613
	Unique Subjects	1457983
	Unique Objects	13751645
	Unique Predicates	21875
Training Data Set	Triples	9996907
	Unique Triple	9264607
	Unique Subjects	313036
	Unique Objects	13114
	Unique Predicates	3482755

4. **Query process module** reads data from the training dataset in the form of source and destination pairs along with value for k. The next step in the process is to consult the node table to check the segment for source and destination. If both source and destination lie in different segments then *INF* files are consulted to obtain the entire segment that have path from the source segments. If there are multiple outgoing edges from source segment then this would result in multiple paths. This process is iteratively continued till a segment containing the destination is extracted.

5. **Path Augmentation** accumulates all paths in the form of a list. When the destination segment is obtained, the list of path is consolidated according to the step explained above to extract the actual paths between source and destination. As this

technique produces all paths, therefore we have to sort them with respect to length and then the top k list of path is filtered.

6. **Duplicate Triplet** may be part of the augmented path list produced as a result of the previous steps because matrix multiplication process includes cycles in the paths. As stated earlier, the challenge problem allows the cycle to be included but no duplicate triplets are permissible therefore, this step removes any duplicated triplets.

5   Performance evaluation

Our initial testing shows that our implementation consumes less than 4 sec to compute the path between two nodes that generates approximately 20,000 items in step 5 of the implementation. During this step, 15K paths have been generated and 16MB space has been consumed. Our performance evaluation on the challenge dataset is yet to be completed and we will be reporting the complete evaluation on the challenge dataset in future. We are presenting the performance evaluation on our artificially created dataset below.

We calculate the time and space complexity based upon the PTMs only as the complexity of creating the INFs will be same for both approaches. Let us take five graphs with 100000, 200000, 300000, 400000 and 500000 nodes. These graphs are indexed up to five levels based upon the parameter settings shown in Table 2. Fifth level has only the segment graph and no further segmentation is performed at that level thus it is not shown in Table 2. Segment graph at fifth level contains 16, 32, 60, 8 and 10 nodes respectively based upon the parameter settings at previous levels. The index is created up to a user defined limit of 10, that is, transitive closure is computed up to length 10 of the paths.

Table 2. Parametersettingsforindexcreation

Graph Size	Level 1	Level 2	Level 3	Level 4
100000	5	5	5	10
200000	5	5	5	10
300000	5	5	5	10
400000	5	10	10	10
500000	5	10	10	10

The Computational complexity for a graph G100000 with parameter setting of Table 2 will be

- Level one (L1) nodes per segment = 5
- L1 total segments = 20000
- L1 cost per segment = matrix multiplication cost * weight limit
- L1 cost per segment = $(5)^3 * 10 = 1250$
- Total cost at L1 = Cost of one segment * number of segments
- Total cost at L1 = 1250 * 20000 = 25000000 memory locations

Calculating in similar manner, cost at level two, three, four and five will be 5000000, 1000000, 160000 and 40960 respectively. Total indexing cost of ρ-index for creating the PTMs will be the cost of L1+L2+L3+L4+L5 = 31200960. We have discussed in section III that our algorithm requires the PTMs only at first and last level of indexing. Thus the indexing cost of our algorithm for creating the PTMs will be the cost of L1+L5 = 25040960. Table 3 compares an indexing cost of our algorithm with ρ-index.

Table 3. Indexing cost

Graph Size	MinG	ρ-index
100000	25040960	31200960
200000	50327680	62727680
300000	77160000	101160000
400000	100005120	188805120
500000	125010000	236010000

Figure 10 shows an efficiency of our approach over the ρ-index. X-axis shows the graph size and to keep the scale of y-axis down, costs in Table 3 are scaled.

The number of times our algorithm uses less space than ρ-index is calculated in a same manner as we have calculated the number of times our algorithm is faster than ρ-index. Our experiments show that on average our algorithm uses 1.33 times less space than ρ-index. Search algorithm of ρ-index creates a special type of graph called a Transcription Graph to answer path queries. The vertices in a transcription graph can be processed strictly from left to right. MinG is designed with parallel processing in mind. It can build the paths from both directions at the same time. In MinG building of paths from the source and target vertex is independent of each other. This is not the case with the search algorithm of ρ-index.

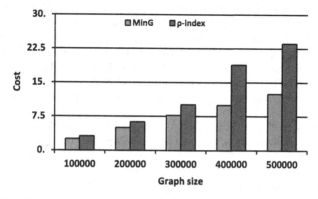

Fig. 10. Indexing cost of creating the PTMs for MinG and ρ-index

6 Conclusions

We have presented an algorithm to find top k shortest paths between any two vertices of a graph. The algorithm is based on indexing scheme of ρ-index. Our algorithm resulted 1.5 times faster and uses 1.3 times less memory than ρ-index on selected case studies. Our algorithm inherently supports parallel processing which is an additional advantage over the search algorithm of ρ-index. The experiments we have performed were based upon selected case studies; with some other scenarios the results may be different.

Reference

1. Wong, P.C., Haglin, D., Gillen, D., Chavarria, D., Castellana, V., Joslyn, C., Chappell, A., Zhang, S.: "A visual analytics paradigm enabling trillion-edge graph exploration", n IEEE 5th Symposium on Large Data Analysis and Visualization (LDAV 2015), October 25-26, pp. 57–64. Illinois, Chicago (2015)
2. Tarjan, R.E.: Fast algorithms for solving path problems. J. ACM **28**(3), 594–614 (1981)
3. Wang, J., Ntarmos, N., and Triantafillou, P. (2016) Indexing Query Graphs to Speed Up Graph Query Processing. In: EDBT: 19th International Conference on Extending Database Technology, Bordeaux, France, 15-18 March 2016
4. Eppstein, D.: Finding the *k* shortest paths. SIAM J. Computing **28**(2), 652–673 (1998)
5. S. Barton, "Indexing graph structured data," PhD Thesis, Masaryk University, Brno, Czech Republic, 2007
6. S. Barton, and P. Zezula, "Indexing structure for graph structured data," in *Mining Complex Data*, Springer Berlin / Heidelberg, 2009, pp. 167-188. Studies in Computational Intelligence, Vol. 165. ISBN 978-3-540-88066-
7. Tarjan, R.E.: A unified approach to path problems. J. ACM **28**(3), 577–593 (1981)
8. Matono, A., Amagasa, T., Yoshikawa, M., Uemura, S.: An indexing scheme for RDF and RDF Schema based on suffix arrays. *Proceedings of SWDB'03*, p. 2003. Co-located with VLDB, The first International Workshop on Semantic Web and Databases (2003)
9. S. Barton, "Indexing structure for discovering relationships in RDF graph recursively applying tree transformation", in *Proceedings of the Semantic Web Workshop at 27th Annual International ACM SIGIR Conference*, pp. 58–68, 2004
10. Agrawal, R., Dar, S., Jagadish, H.V.: Direct transitive closure algorithms: design and performance evaluation. ACM Transactions on Database Systems **15**(3), 427–458 (1990)
11. P.F. Dietz. "Maintaining order in a linked list," in *STOC'82: Proceedings of the fourteenth annual ACM symposium on Theory of computing*, pp. 122–127, New York, USA, 1982. ACM Press
12. E. Cohen, E. Halperin, H. Kaplan, and U. Zwick. "Reachability and distance queries via 2-hop labels," in *Proceedings of the 13th annual ACM-SIAM Symposium on Discrete algorithms*, pp. 937–946, 2002
13. Schenkel, Ralf, Theobald, Anja, Weikum, Gerhard: HOPI: An Efficient Connection Index for Complex XML Document Collections. In: Bertino, Elisa, Christodoulakis, Stavros, Plexousakis, Dimitris, Christophides, Vassilis, Koubarakis, Manolis, Böhm, Klemens (eds.) EDBT 2004. LNCS, vol. 2992, pp. 237–255. Springer, Heidelberg (2004)
14. R. Schenkel, A. Theobald, and G.Weikum. "Efficient creation and incremental maintenance of the HOPI index for complex xml document collections," In *ICDE*, 2005

15. H. He, H. Wang, J. Yang, and P.S. Yu, "Compact reachability labeling for graph-structured data," in *CIKM '05: Proceedings of the 14th ACM international conference on Information and knowledge management*, pp. 594–601, New York, USA, 2005. ACM Press.

16. H. Wang, H. He, J. Yang, P.S. Yu, and J.X. Yu, "Dual Labeling: Answering Graph Reachability Queries in Constant Time," in *Proceedings of the 22nd International Conference on Data Engineering (ICDE)*, pp. 75, 2006. IEEE Computer Society

17. S. Barton and P. Zezula, "rho Index – designing and evaluating an indexing structure for graph structured data." Technical Report FIMU-RS-2006-07, Faculty of Informatics, Masaryk University, 2006.

18. Skiena, S.: Implementing Discrete Mathematics: Combinatorics and Graph Theory with Mathematica. Addison-Wesley, Reading, MA (1990)

Using Triple Pattern Fragments to Enable Streaming of Top-k Shortest Paths via the Web

Laurens De Vocht[(✉)], Ruben Verborgh, and Erik Mannens

Data Science Lab, Ghent University, iMinds,
Sint-Pietersnieuwstraat 41, 9000 Ghent, Belgium
{laurens.devocht,ruben.verborgh,erik.mannens}@ugent.be

Abstract. Searching for relationships between Linked Data resources is typically interpreted as a pathfinding problem: looking for chains of intermediary nodes (hops) forming the connection or bridge between these resources in a single dataset or across multiple datasets. In many cases centralizing all needed linked data in a certain (specialized) repository or index to be able to run the algorithm is not possible or at least not desired. To address this, we propose an approach to top-k shortest pathfinding, which optimally translates a pathfinding query into sequences of triple pattern fragment requests. Triple Pattern Fragments were recently introduced as a solution to address the availability of data on the Web and the scalability of linked data client applications, preventing data processing bottlenecks on the server. The results are streamed to the client, thus allowing clients to do asynchronous processing of the top-k shortest paths. We explain how this approach behaves using a training dataset, a subset of DBpedia with 10 million triples, and show the trade-offs to a SPARQL approach where all the data is gathered in a single triple store on a single machine. Furthermore we investigate the scalability when increasing the size of the subset up to 110 million triples.

1 Introduction

A 'linked data' representation of data, as a graph with annotated edges, allows pathfinding algorithms to work on top of it. Applying such algorithms to linked data has the advantage that links between nodes are annotated, thus allowing interpreting the transitions between nodes and the meaning of a certain path. Unlike the 'generic' topic of pathfinding in graphs (e.g. in 2D or 3D spaces or for navigational purposes), pathfinding algorithms applied to linked data graphs, have been a less popular research topic so far. Pathfinding in large real-world linked data graphs can be a non-trivial task since such graphs typically exhibit small-world network properties. This means that most nodes are not neighbors of one another, but most nodes can be reached from every other node by a 'small' number steps. The centrality of graph-indexing and data pre-processing that many algorithms require, often turns to be an important bottleneck, which degrades the scalability.

© Springer International Publishing Switzerland 2016
H. Sack et al. (Eds.): SemWebEval 2016, CCIS 641, pp. 228–240, 2016.
DOI: 10.1007/978-3-319-46565-4_18

One particular type of pathfinding algorithms focuses on systems for top-k shortest pathfinding and make use of the following information to compute the paths:

- The *first* node of every path that will be returned and the *last* node of every path that will be returned.
- k, The required number of paths.
- A property path expression describing the pattern of the required paths.
- The RDF graph containing the start node and end node.

A top-k shortest path algorithm responds with a set of paths taking into account this information. More specifically, it orders all paths by their length.

Property paths in the RDF Query Language (SPARQL) version 1.1 introduced a pathfinding paradigm that uses unary operators to build SPARQL queries unaware of the dataset its structure. SPARQL queries like `:Einstein (:workedWith)+ ?scientist` include a pattern asking for all the scientists that worked with somebody that (etc.) worked with *:Einstein*. Nevertheless, it is tricky to retrieve a chain of relationships through such SPARQL queries.

In this paper we address the top-k challenge for datasets made available on the Web. Rather than building a single system where the data and the algorithm runs on the same machine, we opt for a streaming algorithm than can stream paths from a linked data server that can answer Triple Pattern Fragment (TPF) requests [11]. TPF provides a computationally inexpensive server-side interface that does not overload the server and guarantees high availability and instant responses. Basic triple patterns (i.e. *?s ?p ?o*) suffice to navigate across linked data graphs (no complex queries needed).

One could wonder why the use of TPFs is beneficial here in this case, given that each top-k query is quite compact and there is no way to make use of specialized indexes that are typically available in triple stores such as for example BlazeGraph[1]. The reason for this is threefold:

1. There is a low server cost where TPFs perform good in case of federation as well which is especially useful when centralization of the data is not possible or desired [12].
2. Fast execution is not always the goal, TPF allows shifting from pure speed optimization to other metrics. It would for example be possible to generate and pre-cache many of the fragments, leading to a better cost/performance ratio in the long term.
3. Show how versatily applicable TPFs are and to indicate where the performance trade-offs lie in different cases.

2   Related Work

The related work can be divided in approaches for: (i) finding paths and relationships in general, and (ii) specifically for top-k shortest paths. The former

---

[1] https://www.blazegraph.com/.

category considers approaches for retrieving semantic associations, with a par- †
ticular focus on finding paths, while the latter category considers methods to
find more than one path (top-k) in a graph.

The A* algorithm is often applied for revealing relations between resources.
In Linked Data it can be used to recombine data from multimedia archives and
social media for storytelling. For example, the implementation[2] of the "Everything is Connected Engine" (EiCE) [5] uses a distance metric based on the
Jaccard-distance for pathfinding. It applies the measure to estimate the similarity between two nodes and to assign a random-walk based weight, which ranks
more rare resources higher, thereby guaranteeing that paths between resources
prefer specific relations over general ones [9]. REX [7] is a system that takes like
the EiCE a pair of entities in a given knowledge base as input but while EiCE
makes heuristically optimizes the choice of relationship explanation, REX identifies a ranked list of relationships explanations. In contrast to the EiCE system,
which heuristically optimizes the choice of relationship explanations, the REX
system [7] identifies a ranked list of relationship explanations.

A slightly different approach with the same goal of association search is
Explass [3]. It provides a flat list (top-k) clusters and facet values for refocusing
and refining a search. The approach detects clusters by running pattern matches
on the datasets to compute frequent, informative and small overlapping patterns [3]. Similar to EiCE, there exist strategies to specifically the top-k shortest
path problem more efficiently, by working with an index and structural pruning [13]. The framework by Cedeno [2] is able to deal with weighted graphs
by enhancing RDF triples with a certain weight (cost) and introducing custom
query patterns to be able to retrieve the paths through SPARQL queries.

On a more theoretical level, Eppstein [6] described algorithms for top-k shortest path finding which are particularly suited when large number of paths needed
to be computed efficiently, and there exist a couple of implementations for it, for
example for the alignment of biological sequences [10]. Brander and Sinclair [1]
investigated four algorithms for a detailed study from over seventy papers written on the subject. These four were implemented in the C programming language and, on the basis of the results they made an assessment of their relative
performance in telecommunications networks. These implementations were not
reusable for semantic graphs due to their application specific implementation
and because in most cases the number of paths k the retrieved was much lower
(dozens up to hundreds) than what we aim for with this paper (up to thousands).

3   Approach

In this section we explain the *architecture* we set-up, the *algorithm* that we used
to compute the top-k shortest path and how it is *implemented*.

---

[2] http://demo.everythingisconnected.be/.

3.1 Architecture

Instead of running the pathfinding algorithm entirely on the server (the same machine as where the data is located), we choose to relocate CPU and memory intensive tasks to the another machine (client). The client translates the path queries into smaller, digestible fragments for the data endpoint. All optimizations and the execution of the algorithm are moved to the client. This has two benefits: (i) the CPU and memory bottleneck at server side is reduced; and (ii) the more complex data fragments to be translated stay on the server even though they do not require much CPU and memory resources, but they would introduce to many client-side requests.

3.2 Algorithm

The algorithm we use as basis was originally developed for automated story-telling. It reduces the number of arbitrary resources revealed in each path. The algorithm therefore added on top of an asynchronous implementation of the A* algorithm for Linked Data an additional resource pre-selection and a post-processing step to increases the semantic relatedness of resources and tweak the weights between links given a certain heuristic [4]. Preliminary evaluation results using the DBpedia dataset indicated that this algorithm succeeds in telling a story featuring better link estimation, especially in cases where other investi-gated algorithms did not make seemingly optimal choices of links. The advan-tage of this approach was that it, depending on the user preferences, generated a handful op to a dozen of paths within reasonable amount of time (a few seconds to a couple of minutes) but continued to stream additional paths until no more could be found.

However, With top-k shortest paths, we are interested - given a certain k, start, and destination node - in *all* the shortest paths ordered by length, not only those optimized to a specific query context. We therefore first retrieve all paths of a certain length before we score their relevance given the user input, rather than pre-processing the search domain and tweaking the search using weights and heuristics in the A* algorithm. This corresponds to the approach of iterative deepening depth first search. For each retrieved path, our algorithm makes sure that there are no loops: (i) start and destination node do not occur as intermediary nodes and (ii) there are no repetitions of combinations of the same predicate and object in a path. The dataset involved is identified by the URI of the Triple Pattern Fragments Server endpoint.

3.3 Implementation

The implementation of the top-k shortest path algorithm as an extension of the EiCE is a result of reverse engineering the original algorithm and redesigning the pipeline to be fit for streaming hundreds to thousands of paths and do any optimizations afterwards rather than pre-emptive delineating the search domain and heuristically tweaking which nodes should be inspected and in which order.

Figure 1 gives an overview on the relevant components of the EiCE involved. The implementation is published as an npm package[3].

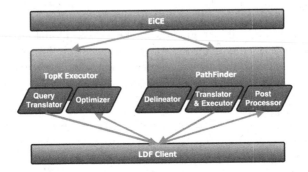

Fig. 1. The top-k shortest path algorithm is implemented as an extension of the EiCE, the *TopK Executor*, which stands side by side to the pathfinding algorithm at the core of the engine, the *Pathfinder*.

Each top k shortest path query is translated in the *Query Translator*. Each incoming combination of parameters (start, destination, required fixed predicate) is ordered with an increasing number of intermediary variable nodes and predicates. These intermediaries are interpretable as sequences of TPFs to be resolved.

For example for paths of length 1 this sequence with intermediary variables looks like `:Start PRED1 OBJ1 PRED2 :Dest`. `PRED1` or `PRED2` can be bound to a fixed predicate or not. It translates to the following TPFs (Table 1):

Table 1. For paths of length one only two triple patterns need to be resolved

Subject	Predicate	Object
:Start	*	*
*	*	:Dest

The generation of these sequences goes on until k paths are found and a new sequence is generated as soon as there are no more paths to be found, or TPFs received that contribute to resolving a sequence of a certain length. Each JOIN of patterns that end and start with a star (*) can lead to a very high number of possible combinations. To ensure that results start arriving instantly, we rely on the built-in optimization of the LDF Client to first bind the stars to matching TPFs with lower counts, to avoid an explosion of possible bindings for the other stars. Eventually if the TPFs with low counts are depleted, TPFs with larger counts and thus more joining possibilities will be considered.

---

[3] https://www.npmjs.com/package/everything_is_connected_engine.

4   Evaluation

To evaluate the pathfinding algorithm we participated in the Extended Semantic Web Conference (ESWC) 2016[4] "top k shortest path challenge", which consisted out of two tasks[5]. Each consisted of four queries $Q1$–$Q4$ with different number of results: k.

1. The first task $T1$ required a certain number of paths between two nodes of the dataset, ordered by their length.
2. The second task $T2$ differentiated from the first task by imposing a specific pattern to the required paths. More specifically, the second task required a certain number of paths between two nodes of the dataset, ordered by their length. Every path should have a particular predicate as the outgoing edge of the start node, or as the incoming edge of the destination node (Table 2).

Table 2. Overview of the training ($TxQx$) and the evaluation ($ExQx$) queries.

Query	Start	Destination	k	Predicate
$T1Q1$	Felipe_Massa	Red_Bull	20,152	
$T2Q1$	Felipe_Massa	Red_Bull	2,12,988	:firstWin
$T1Q2$	1952_Winter_Olympics	Elliot_Richardson	154	
$T2Q2$	1952_Winter_Olympics	Elliot_Richardson	2,311	:after
$T1Q3$	Karl_W._Hofmann	Elliot_Richardson	4,866	
$T2Q3$	Karl_W._Hofmann	Elliot_Richardson	8,088	:predecessor
$T1Q4$	James_K._Polk	Felix_Grundy	1,75,560	
$T2Q4$	James_K._Polk	Felix_Grundy	4,71,199	:president
$E1Q1$	1952_Winter_Olympics	Elliot_Richardson	377	
$E1Q2$	1952_Winter_Olympics	Elliot_Richardson	53,008	
$E2Q1$	1952_Winter_Olympics	Elliot_Richardson	374	:after
$E2Q2$	1952_Winter_Olympics	Elliot_Richardson	52,664	:after

We loaded the training dataset (a $\pm$ 10 M triples subset of DBpedia SPARQL Benchmark[6]) in Blazegraph 2.0.0 as N-Triples and into a Linked Data Fragments Server backed with a compressed Head Dictionary Triples (HDT) [8] index. The machine we used for testing had 8 GB RAM and 4 CPU cores (both client and server side). To validate the algorithm we measured the performance (execution times) and the quality of the results (precision and recall compared to the given training results) and looked into the streaming behavior of the results as time progresses.

---

[4] http://2016.eswc-conferences.org/.
[5] Details about the tasks can be found at https://bitbucket.org/ipapadakis/eswc2016-challenge.
[6] http://aksw.org/Projects/DBPSB.html.

As with the validation of the algorithm through the training tasks, we tested the scalability by using the evaluation dataset of the ESWC Top-k Challenge which consisted out of two tasks[7]. The evaluation dataset is a larger subset of DBpedia containing about 110 M triples. The first task consisted of two queries $E1Q1$ and $E1Q2$ with a different number of results required, without a given predicate. The second task consisted of two queries as well, $E2Q1$ and $E2Q2$, but this time it included a given predicate. The evaluation queries $E1$ are the same as $T1Q2$ and $E2$ the same as $T2Q2$. The difference is in the number of paths k required and the dataset size.

4.1   Expected Results

The training data included the expected max. number of results for each query. These are listed in Table 3.

Table 3. Expected results for the training queries with the expected maximum number of results (k) for a certain system

Task	1				2			
Query	1	2	3	4	1	2	3	4
Expected results (k)	20,152	154	4,866	1,75,560	2,12,988	2,311	8,088	4,71,199

4.2   Baseline

As a baseline, we executed each of each tasks as series of SPARQL queries against the Blazegraph SPARQL endpoint, with increasing path length (starting from 1 going up to the maximum path length in the training results). The term max depicts the highest path length, expressed as the number of intermediary nodes (hops), in the top k shortest path training results for each query - which differs (ranging from 5 to 7) but can be interpreted just alike.

Table 4 shows the precision, recall, runtimes and total results when executing SPARQL queries to retrieve the top-k shortest paths. The number of results retrieve is always higher than the expected training results, this is because the SPARQL results include loops. The top-k shortest paths according to the challenge specification were not supposed to include loops that include combinations of the same predicate and node. The recall at the maximum path length minus 1 in the training results is always 1.00 (complete).

---

[7] More details about the tasks can be found at https://bitbucket.org/ipapadakis/eswc2016-challenge/downloads/evaluation_input_data.txt and are included with the implementation as well.

Table 4. SPARQL Results for the training dataset with 10 M triples. Due to loops in the paths precision is never *1*. At the highest path length in task 2 for query 1 and 4, Blazegraph runs out-of-memory, and therefore the SPARQL query failed to produced sufficient paths.

Task	Query	Precision	Recall (length: $max - 1$)	Recall (length: max)	Runtime (s)	Results
1	1	0.86	1.00	1.00	4	23,448
1	2	0.91	1.00	1.00	2	169
1	3	0.97	1.00	1.00	2	5,034
1	4	0.88	1.00	1.00	29	2,00,246
2	1	<1.00	1.00	**0.08**	6	17,572
2	2	0.6	1.00	1.00	138	3,822
2	3	0.94	1.00	1.00	8	8,568
2	4	<1.00	1.00	**0.13**	47	61,125

4.3   Task 1: Retrieve k Paths Ordered by Length

We note in Table 5 on the one hand that the execution time for streaming paths is 10–100x slower when comparing to querying them to the SPARQL baseline. This is due to the additional checks and reordering that is executed each time a certain possible path is evaluated but also due to the overhead introduced by network traffic: instead of computing the paths on the server, all necessary fragments are transferred first to the client which then computes the paths based on the received fragments. On the other hand we see that the precision is greatly improved approaching or equal to *1* for all queries.

Table 5. The precision, recall and runtime performance for the highest successful value of k during the test for task 1.

Task	Query	Precision	Recall (length: $max - 1$)	Recall (length: max)	Runtime (s)	Results
1	1	0.97	1.00	0.98	1,160	20,152
1	2	1.00	1.00	0.99	2,069	153
1	3	0.99	1.00	0.10	510	518
1	4	0.97	1.00	0.86	1,100	1,57,098

Except for query 3 there is also a good recall for the queries. Figure 2 shows the streaming progression for this query. One of the reasons why execution halts at certain k as shown in the Fig. 2 is because the algorithm first looks for paths which shorter length and tries to retrieve those first before going on to paths with larger length. It might take some time, like in the case for query 3, for to algorithm to find the first chain of links between the start and destination after which the results start coming in. The server stopped delivering fragments for query 3 at $k = 518$ likely due to some internal time-out or queue overload fairly early on, in all the other cases the algorithm was able to progress longer than 500 s and deliver much more results.

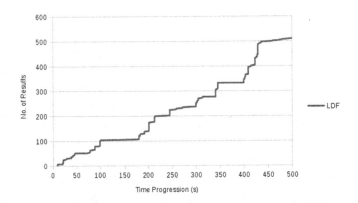

Fig. 2. Progression of the results streaming in task 1, query 3. At $k = 100$ and $300 < k < 350$ the streaming seems to pause and from 500 on the streaming speed severely decreases until halting at $k = 518$.

4.4   Task 2: Fixed Outgoing Edge Start Node or Ingoing Edge Destination Node

Streaming TPF's has a relatively low recall even at path length $max - 1$, which is with plain SPARQL queries as well at max path length. The runtime is here much lower than in all the other cases. At this point, the precision is still 1.0, which indicates that there are no incorrect results or loops among the found paths (Table 6).

Table 6. The precision, recall and runtime performance for the highest successful value of k during the test for task 2.

Task	Query	Precision	Recall (length: $max - 1$)	Recall (length: max)	Runtime (s)	Results
2	1	1.00	0.42	0.03	148	7,052
2	2	0.98	1.00	0.18	2,929	491
2	3	1.00	1.00	0.43	882	3,462
2	4	1.00	0.50	0.06	64	26,148

4.5 Streaming Behavior

For most queries SPARQL and TPF generate the results linear with regard to the time progression (except for the first few - shorter paths). This is clearly visible in Fig. 3 when plotting the results of, for example, query 1.

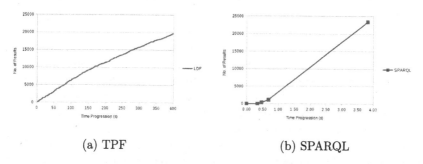

(a) TPF   (b) SPARQL

Fig. 3. The result progression is linear with TPF (reflecting the behavior with SPARQL) for most queries, here the results of query 1 are shown.

Figure 4 shows the progress of query 2. Query 2 initially produced many results very rapidly (shorter path lengths) but at some point when the path length became longer, the time to compute each next path increased. This behavior we noticed both with TPF and with SPARQL.

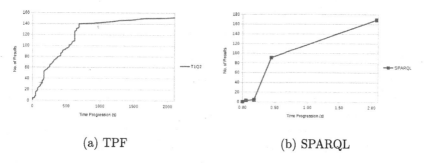

(a) TPF   (b) SPARQL

Fig. 4. For query 2, both SPARQL and TPF start generating the results fast, but at about 20 %–30 % of the total time result progression visibly slows down.

4.6 Scalability

We repeated the queries against SPARQL to see if there are any differences. As Table 7 shows, increasing dataset size has no remarkable impact of the dataset when using SPARQL. The speed of executing the queries is about the same in both cases.

Table 7. SPARQL Results for training and evaluation queries. No fixed predicate: *T1Q2* vs. *E1Q1*, fixed predicate: *T2Q2* vs. *E2Q2*

	Training			Evaluation		
	Results (#)	Time (s)	Results/s	Results (#)	Time (s)	Results/s
No fixed predicated	169	2.070	81	389	4.260	91
Fixed predicate	3,822	138.455	27	1,17,719	1,283.505	91

Figure 5 shows the progression of the result streaming over time for all evaluation queries compared to the two training queries using our algorithm. Both *E1Q1* and *E2Q1* succeeded in retrieving the first top-k paths 377 and 374 respectively. However, when the k is increased to higher numbers 53008 and 52664 respectively, the results indicate that – like with the training data – at some point the time to retrieve the next results increases significantly or leads to time-outs or buffer overflows. Nevertheless, the algorithm seems to hold stance in terms of scalability, there is no evidence that the increased dataset size (x10) has any impact on the performance. However, we note that with a high number *k* shortest paths requested, the query with fixed predicate in *E2Q2* is outperformed by the same query run against the training data *T2Q2*, on the other hand the query without fixed predicate *E1Q1* behaves more or less the same with the training data *T1Q2*. Both in the case of the training data, evaluation data, higher and lower values for *k*, the query with fixed predicate produces the results faster. This is due to most of the paths going through the given predicate anyway, but the algorithm does not need to determine this predicate, leading to a smaller search space.

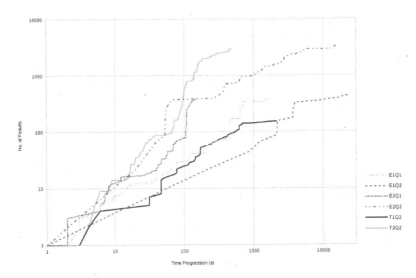

Fig. 5. The evaluation queries and the training queries behave similar the first 100 results and during the first 100 s, but then start diverging. In particular *T2Q2* outperforms *E2Q2* but *T1Q2* and *E1Q2* have similar performance. Plotted on a logarithmic scale on both axis.

5 Conclusions and Next Steps

We implemented a top-k shortest path algorithm by combining and applying two quite recent linked data technologies who were originally designed for different purposes than top-k shortest path finding. Nevertheless, the results with the data from the ESWC 2016 top-k shortest path challenge indicate good results for streaming paths. A larger dataset size does not seem to influence the query performance of the algorithm. The biggest impact, regardless of dataset size, is the number of paths requested k. The higher k (in particular when k is a number in thousands), the more data needs to be buffered and streamed. This led in a number of cases to time-outs from the server or long times to retrieve the next path. In future work, we will optimize the performance and implement ordering of shortest paths of equal length: further integrating the heuristic and weight features in the algorithm. It is also crucial to look into why the streaming for some queries stops early on - leading to lower recall, in particular when the first or the last predicate is given.

Acknowledgments. The research activities that have been described in this paper were funded by Ghent University, iMinds (Interdisciplinary institute for Technology) a research institute founded by the Flemish Government, the Institute for the Promotion of Innovation by Science and Technology in Flanders (IWT), the Fund for Scientific Research-Flanders (FWO-Flanders), and the European Union.

References

1. Brander, A.W., Sinclair, M.C.: A comparative study of k-shortest path algorithms. In: Merabti, M., Carew, M., Ball, F. (eds.) Performance Engineering of Computer and Telecommunications Systems, pp. 370–379. Springer, London (1996)
2. Cedeno, J.P.: A framework for top-K queries over weighted RDF graphs. Ph.D. thesis, Arizona State University (2010)
3. Cheng, G., Zhang, Y., Qu, Y.: Explass: exploring associations between entities via top-K ontological patterns and facets. In: Mika, P., et al. (eds.) ISWC 2014, Part II. LNCS, vol. 8797, pp. 422–437. Springer, Heidelberg (2014)
4. De Vocht, L., Beecks, C., Verborgh, R., Seidl, T., Mannens, E., Van de Walle, R.: Improving semantic relatedness in paths for storytelling with linked data on the web. In: Gandon, F., Guéret, C., Villata, S., Breslin, J., Faron-Zucker, C., Zimmermann, A. (eds.) ESWC 2015. LNCS, vol. 9341, pp. 31–35. Springer, Switzerland (2015). http://dx.doi.org/10.1007/978-3-319-25639-9_6
5. De Vocht, L., Coppens, S., Verborgh, R., Vander Sande, M., Mannens, E., Van de Walle, R.: Discovering meaningful connections between resources in the web of data. In: Proceedings of the 6th Workshop on Linked Data on the Web (LDOW2013) (2013)
6. Eppstein, D.: Finding the k shortest paths. SIAM J. Comput. **28**(2), 652–673 (1998)
7. Fang, L., Sarma, A.D., Yu, C., Bohannon, P.: REX: explaining relationships between entity pairs. Proc. VLDB Endow. **5**(3), 241–252 (2011)

8. Fernández, J.D., Martínez-Prieto, M.A., Gutierrez, C.: Compact representation of large RDF data sets for publishing and exchange. In: Patel-Schneider, P.F., et al. (eds.) ISWC 2010, Part I. LNCS, vol. 6496, pp. 193–208. Springer, Heidelberg (2010)

9. Moore, J.L., Steinke, F., Tresp, V.: A novel metric for information retrieval in semantic networks. In: García-Castro, R., Fensel, D., Antoniou, G. (eds.) ESWC 2011. LNCS, vol. 7117, pp. 65–79. Springer, Heidelberg (2012)

10. Shibuya, T., Imai, H.: New flexible approaches for multiple sequence alignment. J. Comput. Biol. **4**(3), 385–413 (1997)

11. Verborgh, R., et al.: Querying datasets on the web with high availability. In: Mika, P., et al. (eds.) ISWC 2014, Part I. LNCS, vol. 8796, pp. 180–196. Springer, Heidelberg (2014)

12. Verborgh, R., Vander Sande, M., Hartig, O., Van Herwegen, J., De Vocht, L., De Meester, B., Haesendonck, G., Colpaert, P.: Triple pattern fragments: a low-cost knowledge graph interface for the Web. J. Web Semant. **37–38**, 184–206 (2016)

13. Zhang, D.Y., Wu, W., Ouyang, C.F.: Top-k shortest path query processing in RDF graphs. Chin. J. Electron. **43**(8), 1531–1537 (2014)

Semantic Publishing Challenge

Semantic Publishing Challenge – Assessing the Quality of Scientific Output in Its Ecosystem

Anastasia Dimou[1(✉)], Angelo Di Iorio[2], Christoph Lange[3,4], and Sahar Vahdati[3]

[1] Ghent University - iMinds, Ghent, Belgium
anastasia.dimou@ugent.be
[2] Università di Bologna, Bologna, Italy
diiorio@cs.unibo.it
[3] University of Bonn, Bonn, Germany
vahdati@uni-bonn.de
[4] Fraunhofer IAIS, Sankt Augustin, Germany
math.semantic.web@gmail.com

Abstract. The Semantic Publishing Challenge aims to involve participants in extracting data from heterogeneous sources on scholarly publications, and produce Linked Data which can be exploited by the community itself. The 2014 edition was the first attempt to organize a challenge to enable the assessment of the quality of scientific output. The 2015 edition was more explicit regarding the potential techniques, i.e., information extraction and interlinking. The current 2016 edition focuses on the multiple dimensions of scientific quality and the great potential impact of producing Linked Data for this purpose. In this paper, we discuss the overall structure of the Semantic Publishing Challenge, as it is for the 2016 edition, as well as the submitted solutions and their evaluation.

Keywords: Linked data · Information extraction · Challenge

1 Introduction

Changes in technology, tools, funding and social aspects raise new challenges in scholarly publishing. On the other hand, a growing amount of research on publishing and consuming Linked Data, i.e., data represented and made available in a way that maximizes reusability, has facilitated Semantic Web adoption. In this context, the idea of *Semantic Publishing* emerged. Semantic Publishing is "the enhancement of scholarly publications by the use of modern Web standards to improve interactivity, openness and usability, including the use of ontologies to encode rich semantics in the form of machine-readable RDF metadata" [15,16]. It is expected that the semantic publishing advent will foster advanced services for scholars and non-expert users, such as semantic search, identification of research trends, discovery of connections between research works, people and institutions, and so on. However, to achieve such services, richer datasets represented as Linked Data are required. Nevertheless, even though several tools have been

© Springer International Publishing Switzerland 2016
H. Sack et al. (Eds.): SemWebEval 2016, CCIS 641, pp. 243–254, 2016.
DOI: 10.1007/978-3-319-46565-4_19

developed to generate Linked Data about scholarly publications, the procedure is still cumbersome and most available datasets still have some limitations.

The Semantic Publishing Challenge series aims at investigating novel approaches for improving scholarly publishing using Linked Data technology [3,7]. The first editions focused on extracting information both from non-semantic as well as semantic data sources: these sources range from proceeding volumes and their papers (as published in PDF) to semantically enhanced datasets (with information derived from different data sources).

In this paper we present the 2016 edition of the Semantic Publishing Challenge, giving an overview of the tasks, together with a short introduction of the proposed solutions and their final results.

The remaining of the paper is structured as follows. Section 2 introduces the Challenge, Sect. 3 presents the different tasks, Sect. 4 the training and evaluation datasets considered to accomplish the tasks and Sect. 5 the queries to be addressed. Section 6 presents the different solutions which were submitted and Sect. 7 their evaluation and the corresponding results.

2   Semantic Publishing Challenge

In 2014, we considered defining a challenge about Semantic Publishing which would act as an enabler for evaluating different solutions producing corresponding Linked Data [7]. The tasks are defined in a way that such solutions can generate richer Linked Data for Semantic Publishing compared to existing datasets. Datasets existing at that time focused on basic bibliographic metadata, or domain specific data, rendering more advanced applications, especially in respect to assessing the scientific output, difficult, if not impossible.

The 2014 edition was designed to produce an initial dataset which would be useful for future challenges and the community could experiment on it. However, the two information extraction tasks with an objective evaluation had received few submissions. Thus, an open task with a subjective evaluation was also introduced. For the 2015 edition, one of the 2014 edition's tasks, **Task 1**, remained the same because the results were encouraging and it was intended to give another opportunity and incentive to the 2014 edition's participants to improve their tools and participate again, though without excluding new participants.

The other task with an objective evaluation, namely **Task 2**, which was focused on extracting information from the papers' full text, remained also the same, but the underlying data source was changed. In 2014, papers encoded in XML JATS[1], a language for encoding journal articles derived from the NLM Archiving and Interchange DTD, and its TaxPub extension for taxonomic treatments, served as the data source. In 2015, the same data source as for Task 1 was considered as the data source for Task 2: the CEUR-WS.org open access computer science workshop proceedings were considered as the input dataset for Task 2 too, aiming to foster synergies between the two tasks and to encourage

---

[1] http://jats.nlm.nih.gov/.

participants to compete in both tasks. For this edition in 2016, both **Task 1** and **Task 2** remained the same.

Nevertheless, in 2015 there was only one team competing for both tasks (cf. the overview of the 2015 challenge [3]). Therefore, aligning Tasks 1 and 2 became a priority for the 2016 edition which was expected to be achieved relying on **Task 3**. Task 3 was radically changed from the 2014 edition which was an open task with a subjective evaluation to the 2015 edition whose Task 3 was formed in a way that allowed an objective evaluation. For 2015, it aimed at interlinking CEUR-WS.org dataset with other existing Linked Data. For the 2016 edition, Task 3 was designed to be more focused on promoting synergies and aligning Tasks 1 and 2. The dataset of the 2015 winning solutions both for Tasks 1 and 2 were considered as the data sources to be aligned, while aligning the resulting dataset with external Linked Data were of subsequent priority.

3 Tasks and Motivation

In this section, we outline the different tasks defined, or how existing were adjusted for the 2016 edition of the challenge, and we describe the underlying motivation for defining or modifying each one of them.

3.1 Task 1

Task 1 was designed to assess the ability to extract data from a full body of HTML documents. Task 1 is an extension of the 2015 edition. All quality indicators from the previous edition are reconsidered, some are defined more precisely, while one was completely new.

To be more precise, the input dataset for Task 1 consists of HTML documents at different levels of encoding quality and semantics. Therefore, Task 1 mainly requires to employ information extraction and semantic annotation techniques. Participants are asked to extract information from a set of HTML tables of contents published in the CEUR-WS.org workshop proceedings. The extracted information enables describing data which might act as means for assessing the quality of these workshops, for instance by measuring growth, longevity, etc.

Motivation. Common questions related to the quality of a scientific venue include whether a researcher should submit a paper to it or accept an invitation to its programme committee, or whether a publisher should publish its proceedings, and whether a company should sponsor it [2]. Being aware of the quality of an event helps to assess the quality of the papers accepted there.

3.2 Task 2

Task 2 was designed to assess the ability to extract data from the papers full text, namely their PDF corpus. It follows the last two editions' Tasks 2, which were focused on extracting information from citations in the first place, as well as affiliations and funding since the 2015 edition. For the 2016 third edition, the aforementioned are still in scope (apart from citations), but extracting information regarding the internal structure, e.g., tables and figures, comes also in context.

To be more precise, the input dataset for Task 2 consists of PDF documents of papers published with CEUR-WS.org. Therefore, Task 2 mainly requires PDF mining techniques and some natural language processing. The extracted information describes the organisation of the paper and provides a deeper understanding of the context in which it was written. The extracted information should describe, on the one hand, the internal structure of sections, tables, figures and, on the other hand, the authors affiliations and research institutions, and funding sources.

Motivation. Scientific papers are not isolated units. Common questions related to the quality of a scientific contributions include factors that directly or indirectly contribute to the origin and development of a paper include citations, affiliations, funding agencies or even the venue where the paper was presented. The internal organisation and the structural components of a paper are also good indicators of its quality and potential impact.

3.3   Task 3

Task 3 was designed to assess the ability to generate cross-datasets links. It follows the previous, 2015 edition. However, the 2016 edition narrows down the task's scope to a smaller number of external datasets, whereas cross-task links between the previous edition's Tasks 1 and 2 datasets are now also explored.

To be more precise, the input dataset for Task 3 consists of Linked Datasets. Therefore, Task 3 mainly requires entity interlinking techniques and some natural language processing. Participants are asked to interlink the CEUR-WS.org dataset with relevant datasets already existing in the Linked Open Data cloud. In particular, they are expected to interlink persons, papers, events, organisations and publications. All these entities are identified, disambiguated and interlinked to their correspondences in other datasets.

Motivation. Scientific papers and venues are not isolated units and should not be considered separately from each other. They belong in a broader context of scientific contributions, offering complementary information when it is associated with prior existing information, else it remains incomplete.

4   Input Dataset

In this section, we describe the input dataset considered for the different tasks. A summary of statistics related to the training and evaluation dataset for each task is summarized in Table 1.

Table 1. Training and evaluation datasets

	Training dataset	Evaluation dataset
Task 1		
Workshops	118	50
Task 2		
Papers	45	40
Task 3		
Datasets	5	5

4.1   Task 1 Dataset

To support the evolution of extraction tools, the 2016 training dataset is largely the same as the union of the 2015 training and evaluation dataset, with a few additions. To be more precise, the Task 1 training dataset consists of:

- one HTML index page linking to all CEUR-WS.org workshop proceedings volumes[2] (invalid but still uniformly structured HTML 4);
- the volumes' HTML tables of contents[3], which link to the individual workshop papers. Their format is largely uniform but has changed over time. In more details, the training dataset consists of:
  - valid HTML5 pages with microformats and sometimes RDFa,
  - valid and invalid HTML 4.01 with or without microformats.

4.2   Task 2 Dataset

To support the evolution of extraction tools, but also to align with Task 1, the 2016 training dataset is largely the same as the one of the 2015 edition, taken from some of the workshops which are also analyzed in Task 1[4]. The selected papers use different formats and styles (ACM, LNCS, IEEE) and different rules for bibliographic references, headers, affiliations and acknowledgments.

The training and evaluation datasets were totally disjoint (differently formed compared to the past editions) and shared the same internal structure, with the same distribution of styles. Papers were clustered according to their similarities and randomly selected within each cluster.

4.3   Task 3 Dataset

To align with Task 1 and Task 2, Task 3 considered the previous 2015 edition output from Tasks 1 and 2, besides the external datasets that already exist in the Linked Data cloud and were considered also in the 2015 edition. In total 5

---

[2] http://ceur-ws.org/.

[3] https://github.com/ceurws/lod/wiki/SemPub16_Task1.

[4] https://github.com/ceurws/lod/wiki/SemPub16_Task2.

different dataset were considered for the training and evaluation dataset. First of all, the CEUR-WS.org proceedings dataset as it was formed by the solutions that performed best for Task 1 [8][5] and Task 2 [17][6] in 2015. Then, the COLINDA[7], the DBLP[8] and the Springer LD[9] datasets were also considered as input datasets.

5   Queries

In this section, we describe the queries which the different solutions should be able to answer. Based on the results of those queries, the solutions were evaluated for their capacity to address the different tasks.

5.1   Task 1

The submitted solutions are required to produce a dataset for Task 1 against which the following queries can be answered, roughly ordered by increasing difficulty:

- **Q1.1**: List the full names of all editors of the proceedings of workshop W.
- **Q1.2**: Count the number of papers in workshop W.
- **Q1.3**: List the full names of all authors who have (co-)authored a paper in workshop W.
- **Q1.4**: Identify the full names of those chairs of workshop W who are affiliated in the same country in which the workshop took place.
- **Q1.5**: Compute the average length (in numbers of pages) of a paper in workshop W.
- **Q1.6**: Find out whether the proceedings of workshop W were published on CEUR-WS.org before the workshop took place.
- **Q1.7**: Identify all editions that the workshop series titled T has published with CEUR-WS.org.
- **Q1.8**: Identify the full names of those chairs of the workshop series titled T that have so far been a chair in every edition of the workshop that was published with CEUR-WS.org.
- **Q1.9**: Identify all CEUR-WS.org proceedings volumes in which papers of workshops of conference C in year Y were published.
- **Q1.10**: Identify those papers of workshop W that were (co-)authored by at least one chair of the workshop.
- **Q1.11**: List the full names of all authors of invited papers in workshop W.
- **Q1.12**: Determine the number of editions that the workshop series titled T has had, regardless of whether published with CEUR-WS.org.

---

[5] http://rml.io/data/SPC2016/CEUR-WS/CEUR-WStask1.rdf.gz.
[6] http://rml.io/data/SPC2016/CEUR-WS/CEUR-WStask2.rdf.gz.
[7] http://www.colinda.org/.
[8] http://dblp.l3s.de/dblp++.php.
[9] http://lod.springer.com/.

- **Q1.13**: Determine the title (without year) that workshop W had in its first edition.
- **Q1.14**: Of the workshops of conference C in year Y, identify those that did not publish with CEUR-WS.org in the following year (and that therefore probably no longer took place).
- **Q1.15**: Identify the papers of the workshop titled T (which was published in a joint volume V with other workshops).
- **Q1.16**: List the full names of all editors of the proceedings of the workshop titled T (which was published in a joint volume V with other workshops).
- **Q1.17**: Of the workshops that had editions at conference C both in year Y and $Y + 1$, identify the workshop(s) with the biggest percentage of growth.
- **Q1.18**: Return the acronyms of those workshops of conference C in year Y whose previous edition was co-located with a different conference series.
- **Q1.19**: Of the workshop series titled T, identify those editions that took place more than two months later/earlier than the previous edition that was published with CEUR-WS.org.
- **Q1.20**: Identify the affiliations and countries of all editors of the proceedings of workshop W. Use DBpedia resources for the countries.
- **Q1.21**: Identify the full names of those authors of papers in the workshop series titled T that have so far been a (co-)author of a paper in every edition of the workshop that was published with CEUR-WS.org.

5.2   Task 2

The submitted solutions are required to produce a dataset for Task 2, against which the following queries can be answered:

- **Q 2.1**: Affiliations in a paper:
  Identify the affiliations of the authors of the paper X.
- **Q 2.2**: Countries in affiliations:
  Identify the countries of the affiliations of the authors in the paper X.
- **Q 2.3**: Supplementary material:
  Identify the supplementary material(s) for the paper X.
- **Q 2.4**: Sections:
  Identify the titles of the first-level sections of the paper X.
- **Q 2.5**: Tables:
  Identify the captions of the tables in the paper X
- **Q 2.6**: Figures:
  Identify the captions of the figures in the paper X.
- **Q 2.7**: Funding agencies:
  Identify the funding agencies that funded the research presented in the paper X (or part of it).
- **Q 2.8**: EU projects:
  Identify the EU project(s) that supported the research presented in the paper X (or part of it).

5.3   Task 3

The submitted solutions are required to produce a dataset for Task 3 answering the following queries, roughly ordered by increasing difficulty:

- **Q 3.1**: Same person – Multiple URIs:
  Identify and interlink same entities that represent the same editor and/or author but appear with different URIs within the CEUR dataset of Task 1.
- **Q 3.2**: Same conference – Multiple URIs:
  Identify and interlink same entities that represent the same conference but appear with different URIs within the CEUR dataset of Task 1.
- **Q 3.3**: Same cited paper – Multiple URIs:
  Identify and interlink same entities that represent the same cited paper but appear with different URIs within the CEUR dataset of Task 2.
- **Q 3.4**: Same people – Different URIs in CEUR-WS subsets:
  Identify and interlink same entities that represent the same editor and/or author but appear with different URIs within the CEUR dataset of Task 1 and Task 2.
- **Q 3.5**: Same workshops in the CEUR-WS and COLINDA datasets:
  Identify and interlink same entities that represent the same workshop but appear with different URIs within the CEUR dataset of Task 1 and the COLINDA dataset.
- **Q 3.6**: Same workshops in the CEUR-WS and DBLP datasets:
  Identify and interlink same entities that represent the same workshop but appear with different URIs within the CEUR dataset of Task 1 and the DBLP dataset.
- **Q 3.7**: Same people in the CEUR-WS and DBLP datasets:
  Identify and interlink same entities that represent the same person but appear with different URIs within the CEUR dataset of Task 1 and the DBLP dataset.
- **Q 3.8**: Cited papers in CEUR dataset presented at conferences described in Springer dataset:
  Identify and interlink same entities that represent the same conference but appear with different URIs within the CEUR dataset of Task 2 and the Springer dataset.

6   Solutions

Five solutions were submitted and accepted for Task 2, while there were no solutions at all submitted neither for Task 1 nor for Task 3.

6.1   Solution 1

Solution 1 by Ahmad et al. [1] proposed a heuristic-based approach that uses a fruitful combination of tag-based and plain-text-based information extraction techniques which is not frequently encountered in bibliography. Their approach identifies patterns and rules from integrated formats which are stored in knowledge bases. The PDF extraction occurs using the PDFX library[10], while the PDFbox Java library[11] is considered to extract the supportive material links because the former extracts them as plain text, whereas the later as links. Besides the PDF parsers (both PDF-to-XML and PDF-to-Text), the entire solution is modular consisting additionally of the following modules: content pre-processing, rule identifier, information extraction and triplification. The information extraction module, in its own turn, consists of the following sub-extractors: (i) authors extractor, (ii) section heading extractor, (iii) table extractor, (iv) figure extractor, (v) supplementary material extractor, and (vi) funding extractor.

6.2   Solution 2

Solution 2 by Klampfl and Kern [6] extended their approach for the 2015 edition [5]. They implemented a processing pipeline that analyzes a PDF document structure incorporating a diverse set of machine learning techniques, unsupervised to extract text blocks and supervised to classify blocks into different meta-data categories. Heuristics are applied to detect the reference section and sequence classification to categorize the tokens of individual references strings. Last, Named Entity Recognition (NER) is used to extract references to grants, funding agencies and EU projects. In 2016, they changed or improved some parts of their solution. They employed different processing steps of their tool which were not used in the previous edition. To be more precise, the current solution processes section headings, hierarchy and captions, but it also introduces novel aspects for extracting links from supplementary material. Its modular structure allows separate training of its parts relying on different datasets.

6.3   Solution 3

Solution 3 by Nuzzolese et al. [10] relied on the Article Content Miner (ACM) which extends the Metadata And Citations Jailbreaker (MACJa – IPA) [9], namely their approach which was submitted to the 2015 edition of the challenge. The tool integrates (i) the PDFMiner, a Python library[12], to extra the information from PDF; (ii) hybrid techniques based on Natural Language Processing (NLP), for instance Combinatory Categorial Grammar, Discourse Representation Theory (DRT), or Linguistic Frames and heuristics that exploit existing tools lexical resources and gazetteers to generate representation structures according to the DRT; (iii) FRED[13], a novel machine reader that produces RDF/OWL ontologies having classes depending on the lexicon used in the text; and (iv) modules to query external services to enhance and validate data.

---

[10] http://pdfx.cs.man.ac.uk.
[11] https://pdfbox.apache.org/.
[12] https://github.com/euske/pdfminer/.
[13] http://wit.istc.cnr.it/stlab-tools/fred.

6.4 Solution 4

Solution 4 by Sateli and Witte [14] relied on LODeXporter[14], a system composed from two modules: (i) a text mining pipeline based on the GATE framework to extract structural (syntactic processing) and semantic entities (semantic processing), leveraging existing NER tools; and (ii) a LOD exporter, to translate the document annotations into RDF according to custom rules. The text preprocessing occurs relying on PDFX to transform the PDF documents into XML documents[15], which are subsequently used by the GATE framework. The GATE framework then tokenises and lemmatises the text, detects sentence boundaries and performs gazetteering on the text, while the DBpedia Spotlight service is used for entity tagging. They also relied on their solution for the 2015 edition of the challenge [13]. In 2016, the PDF extraction tool used was changed and a number of additional or new conditional heuristics were added.

6.5 Solution 5

Solution 5 by Ramesh et al. [11] proposed an approach based on a three-level Conditional Random Fields (CRF) supervised learning approach. Their approach follows the same feature list as [5]. However, they extract PDF to an XML document that conforms to NLM JATS DTD, and generate RDF using an XSLT transformation tool dedicated for JATS. The Apache PDFBox library[16] is used to extract a stream of characters, their bounding boxes and information about their fonts. The aforementioned are fed into the three-level CRF model, namely (i) formatting, (ii) vocabulary, (iii) heuristic and (iv) language modeling features.

7 Evaluation

The different solutions which were submitted, were evaluated using the SemPub Evaluator[17] on a set of forty papers, as described in Sect. 4, and relying on a set of eight queries, as described in Sect. 5. The overall evaluation results for each solution are summarized in Table 2. The table presents the precision, recall and F-score for each solution, and the same values for those who participated in the 2015 edition as well. The best performing tool is the one with the highest F-score. For the 2016 edition of the challenge, the Solution 1 by Ahmad et al. was the winner of the best performing tool award.

Unfortunately, the best performing tool of the 2015 edition did not participate again for the 2016 edition. Nevertheless, the most innovative solution for the 2015 edition, namely current Solution 4 by Sateli and Witte, participated again and it was ranked second. For the 2016 edition, Solution 2 by Klampfl and Kern won the most innovative solution award.

---

[14] http://www.semanticsoftware.info/lodexporter.
[15] http://pdfx.cs.man.ac.uk.
[16] https://pdfbox.apache.org/.
[17] https://github.com/angelobo/SemPubEvaluator.

Table 2. Precision, recall and F-score for each Task 2 solution (2016 and 2015, where available).

	Authors	Precision	Recall	F-score	Precision 2015	Recall 2015	F-score 2015
#1	Ahmad et al.	0.775	0.778	0.771	–	–	–
#4	Sateli et al.	0.640	0.629	0.632	0.3	0.252	0.247
#2	Klampfl et al.	0.593	0.606	0.592	0.388	0.285	0.292
#3	Nuzzolese et al.	0.412	0.43	0.416	0.274	0.251	0.257
#5	Ramesh et al.	0.393	0.428	0.389	–	–	–

8 Conclusions

The 2016 edition of the Semantic Publishing Challenge was built on top of the previous ones. This continuity was crucial for the success of the event: the participation and the quality of the output encourage us to organise further editions in the future. Note in fact that all solutions which participated in the 2015 edition showed significant improvement in respect to their precision, recall and F-score, while new solutions proposed equally competitive approaches.

There is still room for improvements though. In fact, this year we also took the opportunity to review our experience in organising the challenge and we investigated in more detail both the overall organisation (tasks, datasets, evaluation procedures, etc.) and the results produced by the participants (approaches, tools, adopted vocabularies, etc.). More details can be found in [18].

Our conclusion is that challenges are very good enablers for producing Linked Data and helping the community to refine practices, datasets and tools.

Acknowledgments. Part of this research has been funded by the European Union under grant agreement no. 643410 (OpenAIRE2020).

References

1. Ahmad, R., Afzal, M.T., Qadir, M.A.: Information extraction for PDF sources based on rule-based system using integrated formats. In: Sack et al. [12], pp. 293–308
2. Bryl, V., Birukou, A., Eckert, K., Kessler, M.: What's in the proceedings? Combining publisher's and researcher's perspectives. In: García Castro, A., Lange, C., Lord, P., Stevens, R. (eds.) 4th Workshop on Semantic Publishing (SePublica). CEUR Workshop Proceedings, vol. 1155, Aachen (2014)
3. Di Iorio, A., Lange, C., Dimou, A., Vahdati, S.: Semantic publishing challenge - assessing the quality of scientific output by information extraction and interlinking. In: Gandon et al. [4], pp. 65–80
4. Gandon, F., et al. (eds.): SemWebEval 2015. CCIS, vol. 548. Springer, Heidelberg (2015)

5. Klampfl, S., Kern, R.: Machine learning techniques for automatically extracting contextual information from scientific publications. In: Gandon et al. [4], pp. 105–116

6. Klampfl, S., Kern, R.: Reconstructing the logical structure of a scientific publication using machine learning. In: Sack et al. [12], pp. 255–268

7. Lange, C., Di Iorio, A.: Semantic publishing challenge – assessing the quality of scientific output. In: Presutti, V., Stankovic, M., Cambria, E., Cantador, I., Di Iorio, A., Di Noia, T., Lange, C., Reforgiato Recupero, D., Tordai, A. (eds.) SemWebEval 2014. CCIS, vol. 475, pp. 61–76. Springer, Heidelberg (2014)

8. Milicka, M., Burget, R.: Information extraction from web sources based on multi-aspect content analysis. In: Gandon et al. [4]

9. Nuzzolese, A.G., Peroni, S., Recupero, D.R.: MACJa: metadata and citations jailbreaker. In: Gandon et al. [4], pp. 117–128

10. Nuzzolese, A.G., Peroni, S., Recupero, D.R.: ACM: article content miner for assessing the quality of scientific output. In: Sack et al. [12], pp. 281–292

11. Ramesh, S.H., Dhar, A., Kumar, R.R., Anjaly, V., Sarath, K., Pearce, J., Sundaresan, K.: Automatically identify and label sections in scientific journals using conditional random fields. In: Sack et al. [12], pp. 269–280

12. Sack, H., Dietze, S., Tordai, A., Lange, C.: SemWebEval 2016. CCIS, vol. 641. Springer, Heidelberg (2016)

13. Sateli, B., Witte, R.: Automatic construction of a semantic knowledge base from CEUR workshop proceedings. In: Gandon et al. [4], pp. 129–141

14. Sateli, B., Witte, R.: An automatic workflow for the formalization of scholarly articles' structural and semantic elements. In: Sack et al. [12], pp. 309–320

15. Shotton, D.: Semantic publishing: the coming revolution in scientific journal publishing. Learn. Publish. **22**(2), 85–94 (2009)

16. Shotton, D., Portwin, K., Klyne, G., Miles, A.: Adventures in semantic publishing: exemplar semantic enhancements of a research article. PLoS Comput. Biol. **5**(4), e1000361 (2009)

17. Tkaczyk, D., Bolikowski, L.: Extracting contextual information from scientific literature using CERMINE system. In: Gandon et al. [4]

18. Vahdati, S., Dimou, A., Lange, C., Di Iorio, A.: Semantic publishing challenge: bootstrapping a value chain for scientific data. In: Gonzalez-Beltran, A., Osborne, F., Peroni, S. (eds.) Semantics, Analytics, Visualisation: Enhancing Scholarly Data. LNCS. Springer, Heidelberg (2016)

Reconstructing the Logical Structure of a Scientific Publication Using Machine Learning

Stefan Klampfl[(✉)] and Roman Kern

Know-Center GmbH, Inffeldgasse 13, 8010 Graz, Austria
{sklampfl,rkern}@know-center.at

Abstract. Semantic enrichment of scientific publications has an increasing impact on scholarly communication. This document describes our contribution to Semantic Publishing Challenge 2016, which aims at investigating novel approaches for improving scholarly publishing through semantic technologies. We participated in Task 2 of this challenge, which requires the extraction of information from the content of a paper given as PDF. The extracted information allows answering queries about the paper's internal organisation and the context in which it was written. We build upon our contribution to the previous edition of the challenge, where we categorised meta-data, such as authors and affiliations, and extracted funding information. Here we use unsupervised machine learning techniques in order to extend the analysis of the logical structure of the document as to identify section titles and captions of figures and tables. Furthermore, we employ clustering techniques to create the hierarchical table of contents of the article. Our system is modular in nature and allows a separate training of different stages on different training sets.

Keywords: PDF extraction · Machine learning · Semantic publishing

1 Introduction

Semantic technology is playing an increasingly important role in scientific publishing today. It supports researchers, but also other scientific stakeholders like librarians and funding agencies, in managing collections of scientific literature and in assessing the quality of scientific output. Concrete problems tackled by the semantic publishing community include the generation of machine-readable representations of publications, the linking of papers to research data, and alternative metrics for the quality and impact of a publication. The ongoing growth of the volume of scholarly publications poses the need for automated processing systems that semantically enrich documents with information that support these tasks.

© Springer International Publishing Switzerland 2016
H. Sack et al. (Eds.): SemWebEval 2016, CCIS 641, pp. 255–268, 2016.
DOI: 10.1007/978-3-319-46565-4_20

To that end, the Semantic Publishing Challenge 2016[1] (SemPub 2016) asked participants to automatically extract information from papers of workshop proceeding volumes. This paper describes our contribution to Task 2 of this challenge, which focused on the extraction of specific pieces of information from the textual content of the papers given as PDF files. In particular, this task required the generation of a dataset in a linked open data format which allows to answer a given set of queries about the affiliations of the paper's authors, the countries of affiliations, supplementary material, titles of sections, table and figure captions, and funding information. This information should describe the internal organisation of the publication and also provide a deeper understanding of the context in which it was written.

We build here upon our contribution to last year's edition of the Semantic Publishing Challenge (SemPub 2015) [3,9]. There we participated with our existing system that uses a variety of supervised and unsupervised machine learning techniques to extract text blocks, label meta-data blocks, as well as to identify referenced papers from a given scientific article in PDF. In addition, we extracted funding information using named entity recognition techniques. Parts of this system have been described in [5–8,10]. A demonstration of the system can be accessed online[2], and the source code is available under an open source license[3]. In our contribution to the 2016 edition of the challenge we partly reused our techniques from last year's challenge (for extracting affiliations and funding agencies), partly we employed other processing steps from our existing pipeline that we had not used last year (the identification of section headings and their hierarchy as well as the extraction of table and figure captions), and partly we introduced novel aspects (such as the extraction of links to supplementary material).

This report is structured as follows. In Sect. 2 we briefly recap those parts of our contribution to the 2015 edition of the Semantic Publishing challenge that are also relevant for our current contribution. Section 3 then describes our approach for analysing the logical structure of a scientific article, which is the categorisation of document elements with respect to their role in the paper, e.g., headings or captions. In the next section, Sect. 4, we outline our algorithm for extracting the full hierarchy of (previously identified) section headings of the given scientific article (i.e., the table of contents). We used this to retrieve the titles of the first level sections, as required by the challenge. Given the descriptions of our algorithms we then elaborate in Sect. 5 how we address each of the queries of the challenge in detail. In Sect. 6 we present the performance of our system assessed with the evaluation tool provided by the challenge organisers. Finally, we conclude in Sect. 7.

---

[1] https://github.com/ceurws/lod/wiki/SemPub2016.

[2] http://code-annotator.know-center.tugraz.at.

[3] https://svn.know-center.tugraz.at/opensource/projects/code/trunk User: Anonymous, empty password.

2 Our Contribution to the Previous Edition of the Semantic Publishing Challenge

We participated in the 2015 edition of the Semantic Publishing challenge [3]. Task 2 of this challenge focused on extracting contextual information in the form of author's affiliations, citations, and funding agencies from the full text of papers in PDF. Our contribution [9] used supervised and unsupervised learning techniques to extract text blocks, label meta-data blocks, as well as to identify referenced papers. In addition, we extracted funding information using named entity recognition techniques. In this section, we briefly summarise our approach, as parts of it are the basis for our current contribution.

2.1 Extraction of Contiguous Text Blocks

We consider contiguous text blocks as the basic building blocks of a scientific article. These text blocks consist of several lines of single-column text, each of which is composed of a number of words, which themselves consist of multiple characters. To obtain these logical units we have to process the low-level character stream of the PDF file, which is in this case obtained from the open source Apache PDFBox[4] library. This low-level character stream only consists of a list of characters, their bounding boxes, and information about their font. Due to the unreliability of the information in this low-level stream, we require algorithms which are flexible enough to deal at the same time with both this noisy data and the variety of layouts of scientific publications.

We use methods from unsupervised machine learning, in particular clustering, to iteratively combine individual characters to words, lines, and blocks of text in a bottom-up manner. We employ a sequence of alternating *Merge* and *Split* steps: Each *Merge* step performs agglomerative clustering on vertical or horizontal distances to combine characters to words, words to lines, and lines to blocks. Each *Split* step is responsible for removing spurious mergings such as the combination of lines across columns.

The result of this stage is a hierarchical data structure containing the geometrical information of blocks, lines, and words, as well as the reading order of blocks within the document [1]. We have presented a more detailed description of this algorithm as well as an evaluation of the block extraction in [7].

2.2 Extraction of Author and Affiliation Meta-Data

For the extraction of author and affiliation meta-data from scientific articles we employed supervised machine learning techniques. This directly builds upon the output of the text block extraction and uses labelled training examples to classify the text blocks extracted from the first page of the article into multiple meta-data categories. Apart from author related information (names, e-mail addresses, and affiliations) we also categorise the title (and optional subtitle) of the article, the

---

[4] http://pdfbox.apache.org/.

name of the journal, conference, or venue, abstract and keywords. For author-related blocks we then re-apply the classification to the tokens of these blocks in order to obtain given names, surnames, and affiliations.

As a supervised learning mechanism we use Maximum Entropy (ME) [2] combined with Beam Search [13], which incorporates sequential information by taking into account the classification results of preceding instances in order to avoid unlikely label sequences. As training data set we used a subset of the PubMed database, which consists of a wide variety of article layouts. The features used for classification are derived from the layout, the formatting, the words within and around a text block, and common name lists for detecting author names. For more information about the features and the set of categories, the interested reader is referred to [5], where we describe our TeamBeam algorithm and show that it achieves a satisfactory performance on a number of different data sets.

2.3 Extraction of Funding Information

We used basic techniques from named entity recognition to extract information about research grants, funding agencies, and EU projects. In particular, we relied on a set of manually selected trigger phrases (e.g., "funded by", "supported by", "financed by"). We combined the information of the presence of one of the trigger phrases with the information of the noun phrases of the sentence, identified by part-of-speech tagging. The concrete distinction between research grants, funding agencies, and EU projects was based on heuristics, as defined by the query definitions of last year's challenge edition.

This work is embedded within the larger goal of ontologically mapping the domain of computer science, for which, in contrast to other domains like bio-medicine, ontologies did not exist which help to describe the content of scientific articles. In [10] we devised an ontological structure for the computer science domain, which describes the main concepts in computer science literature and also models the relationships between them, including the information about grants and funding agencies.

3 Analysis of the Logical Structure of a Scientific Article

The extraction of information about the internal organisation of a paper requires the analysis of the *logical* document structure, i.e., which parts of the document are main text elements, section headings, or table or figure captions, for example. This logical structure analysis builds upon the analysis of the *physical* structure, which refers to a hierarchical organisation of the paper into pages, columns, and text blocks, and which in our case is performed by the extraction of text blocks described in Sect. 2.1. In Sect. 2.2 we already tackled the supervised extraction of meta-data. Here we address the remaining content of a scientific article, in particular, section headings as well as table and figure captions, as required by the definition of the challenge's queries.

Starting from the set of extracted contiguous text blocks, we implemented a sequential pipeline of *detectors* each of which labels a specific type of block. These detectors are completely model-free and unsupervised and derive the categories only from information provided by the current document: they only use the labels given by previous detectors, the geometric information of the text blocks, their content including font information, as well as their geometric relations to other blocks. This unsupervised document structure analysis was a topic of our previous work [7,8]. In the following we describe our approach for detecting decorations, captions, main text, and section headings. Figure 1 shows a number of example pages from a scientific document with the resulting categorisation.

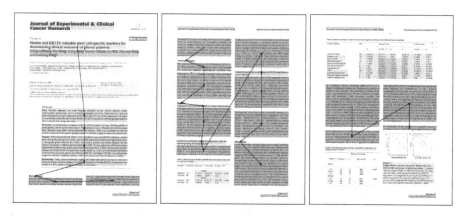

Fig. 1. Example pages with text blocks categorised into different classes (denoted by different colors): decorations (red), captions (cyan), main text (blue), and headings (green). The detection of table blocks (orange) is not described in this paper (see [7]). The first page contains meta-data blocks labelled with different colors. Arrows indicate the reading order of text blocks on each page. (Color figure online)

Decorations. Decorations contain information such as authors, titles, and page numbers and occur repeatedly at the border of each page, mostly inside headers and footers. The detection of decorations is helpful, since they can be removed from further consideration when identifying other parts of the document.

We detect decorations by associating top and bottom blocks across neighbouring pages based on both their content and their position on the page, a slightly modified variant of the approach in [11]. For each page, we sort all blocks on the page in four different orders: from top to bottom, from bottom to top, from left to right, and from right to left. In each ordering we consider the first 5 blocks, and for each block we calculate the maximum similarity to the 5 blocks on both the next and previous page. This similarity score is a value in the range $[0, 1]$ and given by the product between the content and the geometric similarity. The former is calculated from the normalised edit distance between the two content strings, where digits are replaced with "@" chars. The geometric similarity is the area of the intersection between the two bounding box rectangles divided by the area of the larger of the two bounding boxes.

A block is labelled as decoration if its score exceeds some predefined threshold (here, 0.25). The relatively low threshold allows for some noise in the block extraction stage, for example, on different pages of the same document headers might be extracted as a single or multiple blocks.

Captions. Captions are text blocks usually located directly above or below a figure or table explaining its contents. To detect a caption block we simply check whether its first word equals one of certain predefined keyword (viz., "Table", "Tab", "Tab.", "Figure", "Fig", "Fig.") and the second word contains a number (optionally followed by a punctuation, such as ":" or "."). This simple heuristic has been found sufficient for previous work [4,12].

Main Text. The main text of a scientific document is typically organised into one or multiple columns, structured into sections, and might be interleaved with tables or figures. In the context of the challenge, the detection of main text is a necessary prerequisite for detecting section headings.

We identified the following properties of text blocks containing the main text of most scientific articles: (i) they are left-aligned to a limited number of x coordinates (typically the number of columns), (ii) they have a similar width (if the text is justified, the width is virtually identical), (iii) the font of the majority of characters inside the block is the same for all main text blocks, and (iv) the majority of lines in the document belong to the main text.

In order to capture the similarities expressed by properties (i) and (ii), we applied hierarchical agglomerative clustering (HAC) on all blocks of a document in the two-dimensional feature space defined by the left x coordinate and the width of the blocks. As inter-cluster distance we used "single link"; as the distance between two blocks we used standard Euclidean distance, however, for two blocks with a different majority font we set the distance to positive infinity. This accounts for property (iii) and basically ensures that such pairs of blocks end up in different clusters, or, equivalently, all blocks inside one cluster share the same majority font. In the next step we sort the resulting clusters by their total size in *lines* in decreasing order, such that according to property (iv), the largest clusters should contain the main text blocks.

We have presented a more detailed description of this technique in [7]. The flexibility of the clustering algorithm deals with small disalignments, e.g., slightly indented blocks such as enumerations or lists, but some layouts might require a tuning of the threshold parameters.

Section Headings. The heading detection is based on the previous labelling of main text blocks and uses additional information from geometric relations between blocks, the reading order and the neighborhood between blocks. A necessary condition for a text block to be considered as a heading is that it occurs either immediately before a main text block in the reading order or is the top neighbour of a main text block. Furthermore, a candidate heading block has to be either left- or centre-aligned to the following main text block. Additionally, each of the following conditions must be met: (i) the text starts with either a number or an uppercase character, (ii) apart from an optional numbering it consists of at least one non-whitespace letter, (iii) it has a maximum number of lines (here: 3), (iv) the majority font size is at least as large as for the neighbouring main text block, (v) the distance to the neighbouring text block is lower than a threshold (here: 4 times the size of a line of the current heading block candidate). If these conditions are satisfied for unlabelled blocks, they are labelled as section headings [7].

Footnotes. Footnotes are text segments at the bottom of the page which contain the authors' comments to the main text or links to further information. We detected footnotes by looking for blocks at the bottom of the page that start with a superscripted number and which are not yet labelled as main text or decoration. We implemented this detector in the context of this challenge for the purpose of identifying supplementary material mentioned in footnotes in the form of URLs.

4   Extracting the Table of Contents of a Scientific Article

One part of the challenge required the identification of the titles of the first-level section headings of a scientific article. We build here on previous work that extracted the table of contents (ToC) of the paper by creating a tree of headings according to their hierarchy in the article [7,8]. The procedure of the ToC extraction is visualised in Fig. 2 for an example paper from the biomedical domain.

As the starting point for the ToC extraction we use the blocks labelled as headings (see Sect. 3). Our approach is divided into three stages: (i) grouping of headings with similar formatting, (ii) assigning a heading level to each of the groups, and (iii) use the sequence information within the article and the heading levels to recreate the complete hierarchy of section headings. The approach completely relies on unsupervised methods, therefore there is no need for training examples, and it also should work independently of domain.

In the first stage of the algorithm we group headings of the same level based on their formatting. We use the HAC clustering algorithm, where the distance function is a weighted sum of the differences of the mean character height and of the mean number of characters of two clusters. The distance between two clusters is set to infinity, i.e., headings are forced to belong to different groups, if one of the following criteria is met: (i) two headings are directly adjacent, (ii) either one of the clusters is made up exclusively upper-case characters, (iii) the difference in mean character heights differs by more than a threshold, or (iv) the level of the numbering that precedes the heading text differs.

The input for the second stage is the list of clusters that contain at least a single heading. These clusters are ordered according to their assumed heading

List of headings in the reading order:	Clusters with assigned heading level:	Resulting table of contents:
	Heading level 1:	
– Background		– Background
– Materials and methods	– Background	– Materials and methods
– Patients and Tissue Samples	– Materials and methods	• Patients and Tissue Samples
– Methods	– Methods	
– Immunohistochemical Staining	– Results	– Methods
	– Discussion	• Immunohistochemical Staining
– Statistical Analysis	– Abbreviations	
– Results	– Competing interests	• Statistical Analysis
– Expression of Nestin and CD133	– Authors' contributions	– Results
– Correlation of Nestin and CD133 expression	*Heading level 2:*	• Expression of Nestin and CD133
– Prognostic implications of Nestin and CD133		• Correlation of Nestin and CD133 expression
– Discussion	– Patients and Tissue Samples	
– Abbreviations	– Immunohistochemical Staining	• Prognostic implications of Nestin and CD133
– Competing interests	– Statistical Analysis	– Discussion
– Authors' contributions	– Expression of Nestin and CD133	– Abbreviations
	– Correlation of Nestin and CD133 expression	– Competing interests
	– Prognostic implications of Nestin and CD133	– Authors' contributions

Fig. 2. This example shows the workflow of our algorithm for extracting the table of contents of a specific scientific article. First, all headings are collected (left) and clustered based on their formatting. The resulting clusters are then sorted yielding a heading level for each cluster (middle). Finally, the table of contents are generated by processing the headings in the reading order (right) [7].

level by the following precedence rules: (i) number of prefix segments, (ii) difference in mean character height, (iii) preference of all upper-case clusters. The output then is a sequence of headings where the ranking defines the heading level.

The final stage takes the ranked list and produces a ToC tree by exploiting the sequence information. Starting from an empty tree with a single root node all headings are iterated in the sequence of how they appear within the article. The first heading is added as a child to the root node. If the level of the current heading is higher than that of the preceding heading it is added as a sibling to the last heading of the same level. If its level is lower it is added as a child of the preceding heading. By enforcing a valid tree hierarchy this approach corrects some errors that may have been introduced during the first two stages.

5   Addressing the Queries of the Semantic Publishing Challenge

In this section we describe how we addressed the queries of the Semantic Publishing Challenge 2016, based on the algorithms presented in this paper. For each query, we present how we extracted the relevant information. As our system is written in Java, we used the Apache Jena library[5] to convert this information into a linked data set in the form of an RDF file. We did not consider any existing

---

[5] https://jena.apache.org/.

ontologies to build the RDF output; rather we devised our own relations. The IRI structure and the desired output format were given in the task description. Figure 3 shows an example of a SPARQL query and the resulting output, which was stored as CSV for the evaluation. The system outputs were expected to consist of a single CSV file for each query and input document.

```
# Query Q2.2
PREFIX  ceur: <http://ceur-ws.org/#>
PREFIX  rdfs: <http://www.w3.org/2000/01/rdf-schema#>
PREFIX  rdf:  <http://www.w3.org/1999/02/22-rdf-syntax-ns#>
SELECT DISTINCT  ?country_iri ?country_fullname
WHERE
   { <http://ceur-ws.org/Vol-1518/#paper3> ceur:authors ?authors .
     ?authors rdfs:member ?author_iri .
     ?author_iri ceur:affiliations ?affiliations .
     ?affiliations rdfs:member ?affiliation_iri .
     ?affiliation_iri ceur:country ?country_iri .
     ?country_iri ceur:country-fullname ?country_fullname
   }


----------------------------------------------------------
| country_iri                          | country_fullname |
==========================================================
| <http://ceur-ws.org/country/germany> | "Germany"        |
----------------------------------------------------------
```

Fig. 3. Sample SPARQL query and output for Q2.2, which asked to derive the country of an affiliation. The evaluation was performed on the resulting tables stored in CSV format.

Q2.1: Retrieving the Affiliation String of an Author. This query required the extraction of one single string for each affiliation as it appears in the header of the paper. It was also necessary to extract the full name of each author.

In Sect. 2.2 we described how we used a supervised approach for classifying author names and affiliations. We reused this method from our contribution to last year's challenge, which had asked a similar query. We used a model that was pre-trained on a subset of the PubMed database, which consists of a wide variety of article layouts. To align author names to affiliations we took into account the index characters that are typically placed in a superscript font, similar to footnotes. These were primarily identifed by the *Index* label in the classification. In case no such label was assigned, we additionally implemented a heuristic approach that performed the matching based on digits following author names and preceding affiliation strings.

Q2.2: Deriving the Country of an Affiliation. For each affiliation it was also required to identify the country where the research institution is located. First, we looked for countries that are part of the affiliation string. For that we employed standard information extraction techniques using a gazetteer list of countries of the world, obtained by OpenStreetMap. In case no country name was explicitly mentioned in the affiliation, we queried an external service (Google Geocoding API[6]) with the affiliation string.

Q2.3: Identification of Links to Supplementary Material. In the context of this challenge, the term "supplementary material" refers to URLs that are mentioned in the paper, i.e., in the full text, in footnotes, or in appendices, and which link to additional material such as data sets, source code locations, or more detailed reports. In this work, we focused on supplementary materials in footnotes. In the spirit of Sect. 3, we implemented an additional detector for footnotes, extending our system for logical document structure analysis. If the text of the footnote consisted of an URL, we reported it as supplementary material.

Q2.4: Extracting the Titles of First-Level Section Headings. This query required the extraction of the titles of the first-level section headings of the paper in their correct ordering as they appear in the paper. To address this query we used our algorithm for extracting the table of contents of a scientific article (Sect. 4). In this algorithm, text blocks which are labelled as headings in the categorisation stage (Sect. 3) are organised in a tree using unsupervised clustering techniques that take into account the formatting of the headings. From the resulting tree, which represents the ordered table of contents, we take the top level headings and report them in the correct order. As required by the query description, we removed numberings from the heading title, if applicable, and manually included reference and acknowledgement sections, if not yet identified as first-level headings. In our system, the abstract is already detected by the earlier meta-data extraction stage; should the abstract heading be erroneously included in the extracted section headings, we removed it manually.

Q2.5: Identification of Table Captions. We identify table captions in the logical structure analysis stage (Sect. 3). We distinguished table captions from captions of other floating bodies by the starting keyword, e.g., "Table" or "Tab." Finally, we removed the labelling of the caption (e.g., "Table 1") and replaced it with an Arabic numbering as required by the query description.

Q2.6: Identification of Figure Captions. We identify figure captions in the logical structure analysis stage (Sect. 3). We distinguished figure captions from captions of other floating bodies by the starting keyword, e.g., "Figure" or "Fig." Unless they are labelled as figures, listings, pseudo code, and algorithms are not part of the resulting list of figures. Finally, we removed the labelling of the caption (e.g., "Figure 1") and replaced it with an Arabic numbering as required by the query description.

---

[6] https://developers.google.com/maps/documentation/geocoding/start.

Q2.7: Identification of Funding Agencies. This query required the extraction of the name, or the acronym, or both, of the funding agencies that are explicitly mentioned in the paper. We used named entity recognition techniques to extract funding information (Sect. 2.3). In this approach the funding agency is basically a noun phrase followed by a trigger phrase (e.g., "funded by"). We reused this method from our contribution to last year's challenge, which had asked a similar query. In order to distinguish funding agencies from EU projects (query Q2.8), we employed heuristics.

Q2.8: Identification of EU Projects. This query required the extraction of the name, or the number, or both, of the EU projects that are explicitly mentioned in the paper. We used named entity recognition techniques to extract funding information (Sect. 2.3). We reused this method from our contribution to last year's challenge, which had asked a similar query. In order to distinguish EU projects from funding agencies (query Q2.7), we employed heuristics.

6 Evaluation

The challenge organisers provided two non-overlapping sets of scientific articles: a training set (Training Dataset, TD) of 45 papers and an evaluation set (Evaluation Dataset, ED) of 40 papers. The training set was accompanied by a ground truth in the form of CSV files containing the outputs of each the queries for each of the document. With the evaluation tool, which was also given by the organisers, participants could evaluate their system output, which was expected in the same format, against this ground truth. The ground truth of the evaluation dataset was provided after the challenge. This evaluation tool works as follows. First, the true positives, false positives, and false negatives are collected for each query and document by comparing the system output with the gold standard output. During the evaluation process names, titles, and other strings are normalized. Then, from these numbers, precision, recall, and F-score are calculated, which are then averaged over all documents and queries, i.e., it calculates macro averages. It is important to note that precision, recall, and F-score are all 1, if both the true output and the system output are empty.

The results for both the training set and the evaluation set are shown in Tables 1 and 2, respectively, both in terms of the overall performance and the individual performances of each query. It can be seen that all the overall performance values are around 0.6, and slightly larger on the training set. Comparing the performance across different queries we observe that queries Q2.5 and Q2.6 (figure and table captions) achieve the best results (over 0.8), followed by queries Q2.3 and Q2.4 (supplementary material and section headings). It is interesting that the retrieval quality of authors, affiliations, and countries (queries Q2.1 and Q2.2) differs substantially between training and evaluation set, given that we used a supervised model trained on a separate dataset. We think that the relatively large performance values for queries Q2.7 and Q2.8 (funding agencies and EU projects) may result from the fact that in many cases the system output was (correctly) empty.

Table 1. Performance on the training dataset (TD) of 45 documents, both overall and per query. Precision, recall, and F-score are evaluated as macro averages over the documents.

Query	Precision	Recall	F-score
Q2.1	0.493	0.328	0.375
Q2.2	0.537	0.491	0.497
Q2.3	0.569	0.600	0.575
Q2.4	0.538	0.540	0.537
Q2.5	0.838	0.832	0.832
Q2.6	0.675	0.701	0.684
Q2.7	0.689	0.689	0.689
Q2.8	0.711	0.711	0.711
Overall	0.631	0.611	0.613

Table 2. Performance on the evaluation dataset (ED) of 40 documents, both overall and per query. Precision, recall, and F-score are evaluated as macro averages over the documents.

Query	Precision	Recall	F-score
Q2.1	0.238	0.201	0.204
Q2.2	0.229	0.263	0.237
Q2.3	0.548	0.600	0.560
Q2.4	0.585	0.597	0.589
Q2.5	0.771	0.790	0.778
Q2.6	0.856	0.874	0.853
Q2.7	0.663	0.675	0.667
Q2.8	0.850	0.850	0.850
Overall	0.593	0.606	0.592

7   Discussion

In this work we have presented our contribution to the Semantic Publishing Challenge 2016. We built upon our submission from the previous year, where we had used both unsupervised and supervised machine learning techniques as well as techniques from information extraction and named entity recognition to identify different meta-data categories, such as authors and affiliations, and to extract information about funding agencies and EU projects. We extended the analysis of a scientific document to extract deeper information about the internal organisation of the given publication. In particular, we explained how we identified titles of sections and captions of tables and figures. Furthermore, we recreated the table of contents as a hierarchy of section headings. Our system utilises the flexibility of unsupervised and supervised machine learning techniques.

The features for our algorithms are composed of layout information (e.g., the absolute and relative geometrical positioning of text blocks on a page), formatting information (e.g., the type, style, and size of fonts), and textual information.

One major problem with PDF and low-level parsing tools, such as PDF-Box, is that the information provided about individual characters in the PDF is inherently noisy, for example, height and width information might be wrong, or information about the font of some characters might be missing. This implicit noise affects every stage of our system and thus its overall performance.

Our system is flexible and modular in nature and allows a separate training of different stages on different training sets. In many cases we used a subset of the PubMed database as a training set, mainly because it provides a rigorous annotation of the complete content of each document, in particular, meta-data, main text, headings, and captions. The publications in this database are from a wide variety of journals the biomedical domain, which we consider as representative for the general domain of scientific articles. Still it might not perform well on a specific sub-domain, such as conference publications from computer science. This would have to be addressed by a different training set that is more representative for this type of publications, which to the best of our knowledge does not yet exist in a reasonable size.

One important aspect for potential future contributions to this challange is to actually take into account the information provided by the training set and the accompanying ground truth. Since our system is partly unsupervised and partly trained on a separate dataset, we ignored this information in both our previous and current contribution. While 45 documents might still be to few to build a representative supervised model, we believe that it might still provide very valuable information for certain subtasks, such as the identification of supplementary material or funding information.

On a related note regarding future work, in addition to our already existing approach for analysing the logical structure of a scientific article in an unsupervised manner, we plan to develop a supervised approach for categorising the logical elements of a scientific document. Instead of unsupervised detectors for individual types of document elements, we aim at training a separate classification model for headings, captions, etc. For each text block these models would output a real-valued score that reflects the likelihood of belonging to the respective class. A combination of scores across all classification models would then determine the final label of the text block. Alternative possibilities for a supervised analysis would be to employ a single classification model that considers multiple labels, or a sequence classifier that takes into account sequential information within the article. We expect a better performance with such a supervised technique, given that the training set is representative for the evaluation data set. It will be interesting to see if this is indeed the case, or if the unsupervised approach works better for the document set at hand.

Acknowledgements. The presented work was in part developed within the CODE project (grant no. 296150) and within the EEXCESS project (grant no. 600601) funded by the EU FP7, as well as the TEAM IAPP project (grant no. 251514) within the

FP7 People Programme. The Know-Center is funded within the Austrian COMET Program – Competence Centers for Excellent Technologies – under the auspices of the Austrian Federal Ministry of Transport, Innovation and Technology, the Austrian Federal Ministry of Economy, Family and Youth and by the State of Styria. COMET is managed by the Austrian Research Promotion Agency FFG.

References

1. Aiello, M., Monz, C., Todoran, L., Worring, M.: Document understanding for a broad class of documents. Int. J. Doc. Anal. Recogn. **5**(1), 1–16 (2002)
2. Berger, A.L., Pietra, V.J.D., Pietra, S.A.D.: A maximum entropy approach to natural language processing. Comput. Linguist. **22**(1), 39–71 (1996)
3. Iorio, A.D., Lange, C., Dimou, A., Vahdati, S.: Semantic publishing challenge – assessing the quality of scientific output by information extraction and interlinking. SemWebEval 2015. CCIS, vol. 548, pp. 65–80. Springer, Heidelberg (2015). doi:10.1007/978-3-319-25518-7_6
4. Gao, L., Tang, Z., Lin, X., Liu, Y., Qiu, R., Wang, Y.: Structure extraction from PDF-based book documents. In: Proceedings of the 11th Annual International ACM/IEEE Joint Conference on Digital Libraries, pp. 11–20 (2011)
5. Kern, R., Jack, K., Hristakeva, M., Granitzer, M.: TeamBeam - meta-data extraction from scientific literature. In: 1st International Workshop on Mining Scientific Publications (2012)
6. Kern, R., Klampfl, S.: Extraction of references using layout and formatting information from scientific articles. D-Lib Mag. **19**(9/10), 2 (2013)
7. Klampfl, S., Granitzer, M., Jack, K., Kern, R.: Unsupervised document structure analysis of digital scientific articles. Int. J. Digit. Libr. **14**(3–4), 83–99 (2014)
8. Klampfl, S., Kern, R.: An unsupervised machine learning approach to body text and table of contents extraction from digital scientific articles. In: Aalberg, T., Papatheodorou, C., Dobreva, M., Tsakonas, G., Farrugia, C.J. (eds.) TPDL 2013. LNCS, vol. 8092, pp. 144–155. Springer, Heidelberg (2013)
9. Klampfl, S., Kern, R.: Machine learning techniques for automatically extracting contextual information from scientific publications. In: Gandon, F., et al. (eds.) SemWebEval 2015. CCIS, vol. 548, pp. 105–116. Springer, Heidelberg (2015). doi:10.1007/978-3-319-25518-7_9
10. Kröll, M., Klampfl, S., Kern, R.: Towards a marketplace for the scientific community: accessing knowledge from the computer science domain. D-Lib Mag. **20**(11/12), 10 (2014)
11. Lin, X.: Header and footer extraction by page-association. In: Proceedings of SPIE vol. 5010, pp. 164–171 (2002)
12. Liu, Y., Mitra, P., Giles, C.L.: Identifying table boundaries in digital documents via sparse line detection. In: Proceeding of the 17th ACM Conference on Information and Knowledge Mining CIKM 2008, pp. 1311–1320. ACM Press (2008)
13. Ratnaparkhi, A.: Maximum entropy models for natural langual ambiguity resolution. Ph.D. thesis (1998)

Automatically Identify and Label Sections in Scientific Journals Using Conditional Random Fields

Sree Harsha Ramesh[1(✉)], Arnab Dhar[1], Raveena R. Kumar[1], Anjaly V.[1], Sarath K.S.[1], Jason Pearce[2], and Krishna R. Sundaresan[1]

[1] Surukam Analytics, Chennai, Tamil Nadu, India
{harsha,arnab,raveena,anjaly,sarath,krishna}@surukam.com
[2] Newgen KnowledgeWorks, Chennai, Tamil Nadu, India
jason@newgen.co

Abstract. In this paper, we describe a pipeline that automatically converts a journal article in the PDF format to an XML which conforms to NLM JATS DTD. First, the text and typographical features are extracted from the document using character level information. Then, we use a trickle down multi-level conditional random fields based classifier where at each level the pre-trained CRF model classifies a given line of text into one of the tags of DTD at a particular depth and feeds the resulting tag into the next level model as a feature. After identifying tags upto level three, we make use of separate supervised models for parsing authors, affiliations, references and citations. We employ heuristic based methods for matching affiliation to authors, and citation to references. The JATS XML thus generated, is converted into an RDF document. SPARQL queries are run on the RDF, to address the queries of Task 2 of the Semantic Publishing Challenge.

Keywords: Multi-level CRF · BIO encoding · NLM JATS · JATS2RDF

1 Introduction

Scientific journals have been typically published in the PDF format over the years and continue to do so ever-increasingly. However, the Portable Document Format (PDF) is optimized for presentation – it was created to be independent of application, hardware, and operating system [1] – but lacks structural information about sections within the documents and their labels and other such metadata, thus making indexing of documents for search and retrieval, very difficult. This necessitates systems which automatically parse journal content, annotate different sections and extract metadata.

Semantic Publishing Challenge (SemPub)[1] is a series of shared tasks, organised alongside the annual academic conference - Extended Semantic Web Conference (ESWC). The current edition of 2016 has three related tasks, with the

---

[1] Semantic Publishing Challenge 2016 - https://github.com/ceurws/lod/wiki/SemPub 2016.

© Springer International Publishing Switzerland 2016
H. Sack et al. (Eds.): SemWebEval 2016, CCIS 641, pp. 269–280, 2016.
DOI: 10.1007/978-3-319-46565-4_21

overarching theme of "annotating a set of multi-format input documents and to produce a Linked Open Data that fully describes these documents, their context, and relevant parts of their content".

This paper describes our submission to the Task 2 of Semantic Publishing Challenge 2016. As is the motivation behind this task, the approach described in this paper helps "provide a deeper understanding of the context in which a paper was written", by extracting contextual information from full text PDFs. The entities required to be extracted in this edition of the challenge were affiliations, countries in affiliations, supplementary material, titles of first-level sections, table and figure captions, funding agencies and EU projects that supported the research.

To this effect, we have created a system that predominantly employs Conditional Random Fields (CRF) [2]. CRF belongs to a class of probabilistic graphic models and is especially popular in sequence labelling, because of the context-aware predictions it can be trained to make, unlike ordinary classifiers. Each CRF module works on the output of the previous models and also includes a post-processing phase where we use heuristics to handle sparse edge cases which could not be learned by the algorithm. In Sect. 3, we describe the architecture of our system along with the tools which were used while implementing it.

2   Related Work

A rule-based approach to extract heading-based sectional hierarchies of HTML documents is proposed in [5]. HTML DOM (Document Object Model) tree analysis was used to identify the section and subsections of the documents. Some heuristics employed in this paper are that the section headings usually don't begin with punctuation symbols, that they are enclosed within line breaks, and that they are typographically[2] distinct from rest of the document. The DOM tree is processed to obtain blocks of elements in the document and it identifies the possible headings in the document. The obtained hierarchy is then rearranged based on the identified headings.

In [6] a method is described to identify section headers in Legal Briefs. To identify the section and type of section different machine learning approaches such as – naive Bayes, logistic regression, decision trees, support vector machines and neural networks were used. Informations about italics, bold are ignored and considering only analysis of the word and character sequences. Similar to [5], a header block of text is identified heuristically and is labelled using multi class classifiers.

The "Author and affiliation extraction and matching system" described in [7] consists of two steps – an extraction step and a matching step. Extraction step employs CRF, in which each token in the author and affiliation string is classified into one of the three classes, name (author name or affiliation tokens), symbol (correspondence marker) and separator (any separating token). Two models were

---

[2] Typographical features include information about typefaces, point size and line length.

learned for name and affiliation, using same feature set but, different training data. The feature set contains content features (related to textual content) and layout features (related to text layout). The second step, relational classification matches affiliations to authors using SVM.

The challenge of annotating content from PDF documents has been an open problem, with concerted efforts put forth by ESWC in solving it. SemPub has led to the development of tools [10,11] that automatically annotate elements within PDFs,thereby generating linked open data out of academic publications. The following papers, which were among the best submissions at the SemPub, define the state-of-the-art.

A processing pipeline was developed in [4] to analyze the structure of a PDF document incorporating machine learning techniques. The extraction of contiguous text blocks from raw data employs an unsupervised machine learning method. A supervised machine learning mechanism, combination of Maximum Entropy and Beam Search is employed for the extraction of metadata. The reference section was identified by using a set of heuristics and, reference string parsing process used a supervised machine learning approach based on CRF.

CERMINE [3], the state-of-the-art system for extraction of structured affiliations from scientific articles in PDF format, follows a modular approach towards metadata extraction. Document structure extraction and content classification are done using SVM, so as to decipher the reading order of a document in PDF. Author and affiliation extraction were done using heuristics and affiliation and reference parsing were done using CRF.

3   The Pipeline

This section explains each phase of the proposed architecture shown in Fig. 1. In each sub-section dedicated for explaining a particular phase of extraction, we also list the tools and resources used.

The first phase of the pipeline is the PDF extraction phase which extracts line and character level information from the journal articles in the PDF format. This serve as the basis for the subsequent models. An exhaustive list of features are extracted from this information. The feature file is fed into a sequence of three CRF models, where the model at each level classifies a given line of text into one of its respective categories and feeds the result into the next level as a feature.

We have followed the NLM Journal Archiving Tag Set (JATS) 3.0 DTD[3] specifications for generating an XML out of the text extracted from the PDF Extraction phase, and also defining the classes at each level of the CRF algorithms. Having identified the NLM tag three levels deep, there is an independent fine-grained classification phase for entities like affiliation, author, references and citation identification and mapping of citation to reference and mapping of

---

[3] NLM JATS DTD. http://dtd.nlm.nih.gov/archiving/tag-library/3.0/index.html.

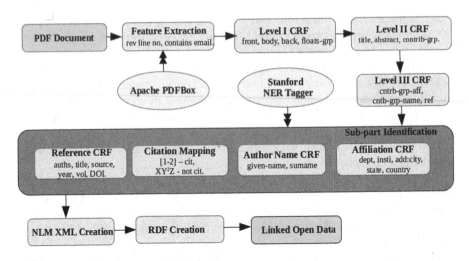

Fig. 1. A schematic diagram depicting the overall architecture of our system to automatically extract semantic information from PDF documents and convert it to NLM JATS XML and further convert it to RDF triples.

affiliation to author. At the end of this stage we have an XML which conforms to NLM JATS and then we generate RDF[4] using JATS2RDF XSLT transform.

3.1   PDF Extraction

We have used Apache PDFBox[5] library to extract the requisite information from the PDFs. PDFBox provides a stream of characters and their bounding boxes. A bounding box consists of four coordinates: the X and Y of the lower left corner, and the X and Y of the upper right corner of the character, where X is the width measured from the left margin of the page and Y is the height measured from the top of the page. It also provides information about the font of the character, such as the font family and font weight. We use the following heuristics to extract some metrics which will be further used in the feature extraction phase.

1. All the characters which lie at the same height belong to the same line of text.
2. Absence of any characters in particular region, across a span of rows indicates the presence of multiple columns.
3. If a character is of a smaller size than its predecessor but is roughly half its size above or below the Y coordinate of the lower left corner of the characters in the same line, indicates that it is superscript or subscript respectively.

---

4 Resource Description Framework (RDF), http://www.w3.org/RDF/.
5 Apache PDFBox. https://pdfbox.apache.org/.

3.2   Feature Extraction

The following features were generated from the raw features extracted in the PDF Extraction, phase and are subsequently fed into the Level I CRF, Level II CRF and Level III CRF models. In hindsight, we found that Klampfl and Kern [4] used a similar list of features and we have followed their method of categorizing them as follows:

Formatting Features. *fontSize* and *lineLength* are set[6] if the calculated value for a given line of text lies within 1 Standard Deviation from the feature value averaged over the entire list of sentences. *isCommonFontFamily* is set if the font family name of the current line is the same as the most occurring one in the document. Typical font family names are Times Roman and Helvetica. *lineNo* is the line number of the given line within the entire document. *fwdLineNoWithinPage* and *revLineNoWithinPage* are the forward line number and reverse line number within a given page respectively. *isBold* is set if the given line of text consists only of *bold* characters. As mentioned in PDF Extraction, PDFBox provides us with the font weight value of a given character – a value of 400 indicates normal weight whereas 700 indicates that it is bold.

Vocabulary Features. *containsIntroduction*, *containsAbstract*, *containsKeywords*, *containsReference*, *containsAcknowledgment*, *containsTOC*, *containsAssociatedContent*, *containsSupportingInformation*, *containsSynopsis*, *containsAuthorInfo*, *containsSection*, *containsFigure* and *containsTable* are set based on the presence of the respective word or its common misspellings, in the given line of text.

Heuristic Features. *containsTelephone*, *containsFax* and *containsEmail* are set based on respective regular expression matches. *containsSuperscript* is set, if the PDF Extraction phase found any superscript in the given line. *punctuationRatio* and *capsRatio* are encoded as categorical features based on the bucket – the bucketing ranges are heuristically determined – the punctuation count belongs to.

Language Modelling Features. We used Stanford 3-class NERTagger[7] [8] to identify the words which are Location, Person or Organization in a given sentence. Then we calculated features such as *departmentRatio* and *nameRatio* as the ratio of the length of the words tagged as "ORGANIZATION" or "PERSON" respectively, to the length of the given sentence.

---

[6] If it is a binary feature, then set would mean setting the value to 1. If it is a multi-categorical feature, then the values are discrete integers ranging from 0 to (number of buckets - 1).

[7] Stanford NER Tagger: http://nlp.stanford.edu/software/CRF-NER.shtml.

3.3 Level I CRF

To use the CRF algorithm described in [2], we have used the open-source implementation – CRF++[8]. The inputs to this tool are a feature file which has a whitespace-separated (spaces or tabular characters) list of features and a feature template file. The feature template file describes the semantics of these columns and also, enables the creation of the context windows. These windows allow the CRF++ package to look at the previous and next tokens to infer about the current state and decipher the context.

The first level CRF model is trained by feeding the feature file in CoNLL[9] format, generated in Sect. 3.1 along with the target classes such as front, body, back and floats-group. These classes correspond to the first level of the JATS XML.

1. *front*: It is a container element for article metadata.
2. *body*: It contains the main body of the article.
3. *back*: It contains references.
4. *floats-group*: It contains figures, tables and their captions and titles.

3.4 Level II CRF

The second level CRF model uses the same feature file as compiled in the Feature Extraction phase — Sect. 3.2, with one additional feature being the first level class predicted by the Level I CRF. The target classes for this model are *document-title, abstract-group, contrib-group, kwd-group, sec, ack, ref-list, fig, table-wrap*. These classes generally correspond to the nodes at a depth of 2 in the JATS XML.

1. *document-title*: It is a container element for article name.
2. *contrib-group*: It contains author details and affiliation information.
3. *abstract-group*: It is a container element for abstract, synopsis dek, and table of contents.
4. *kwd-group*: It contains a list – usually comma separated — of keywords that is associated with the whole document.
5. *sec*: A child of body element in Level I, this contains one or more paragraphs of main text and their section headers.
6. *ack*: This element contains acknowledgement.
7. *ref-list*: This element contains a list of references which are usually found at the *back* part of the document.
8. *fig*: This element consists of figure label and caption.
9. *table-wrap*: Within this tag we find table rows and columns and also table captions and footnotes.

---

[8] CRF++: https://taku910.github.io/crfpp/.
[9] CoNLL: http://www.cnts.ua.ac.be/conll2000/chunking/.

3.5   Level III CRF

The feature file used by the third level classifier is compiled by adding the second level class prediction as a new feature to the feature file used by the Level II CRF — Sect. 3.4. The target classes are mostly the same as those for the second level classifier, except *contrib-group* which is further classified into *contrib-group-contrib-name* and *contrib-group-aff*. These classes are not in the JATS DTD, but were chosen for our convenience. To identify first level section headings of a paper, i.e., Q4 of Task 2 - we have also added *p* and *head* classes to this model.

1. *contrib-group-contrib-name*: It is a container element for author name - given name and surname. The lines tagged as *contrib-group-contrib-name* are fed as inputs to the Author Name CRF in Sect. 3.6.
2. *contrib-group-aff*: It contains affiliation parts, namely, department name, institute name, street address, city, postal code, state and country. The lines tagged as *contrib-group-aff* are fed as inputs to the Affiliation CRF in Sect. 3.7.

3.6   Author Name CRF

This module uses CRF to identify given-name and surname regions of a line predicted as *contrib-group-contrib-name* by Level III CRF.

Here, we have assumed that these lines have names only, and that they do not have any affiliations. This is a heuristic we deduced from the CEUR dataset[10] released for the challenge.

Sentences are split on the tokens — comma and "and" — into individual names. Every name is assigned a unique author id, so as to map it to the corresponding affiliation, in a later step. The Author Name CRF model is trained on typographical features like *characterCase, tokenLength, isSingleCapitalLetter* and a keyword feature — *tokenAsFeature*, where the full token is fed back as a feature so as to help the model to memorize names.

BIO Tagging. The target class names, such as *given-name, surname* are encoded using BIO tagging scheme [9]. BIO encoding divides the tokens belonging to a certain tag, as either being begin-of-entity (B_X) or continuation-of-entity (I_X).

For e.g., *Sree Harsha Ramesh* where *Sree Harsha* is the GIVEN_NAME and *Ramesh* is the SURNAME could be tokenized and tagged as follows: *Sree* - B_GIVEN_NAME; *Harsha* - I_GIVEN_NAME; *Ramesh* - B_SURNAME. Once the text line containing author-name is tokenized and the textual features of each individual token are identified, it is passed through the learner. The learner will classify each token as either B_GIVEN_NAME, I_GIVEN_NAME, B_SURNAME or I_SURNAME.

---

[10] A subset of scientific journals published CEUR-WS.org - https://github.com/ceurws/lod/wiki/SemPub16_Task2#training-dataset-td2.

After classification, the tokens with consecutive B_X and I_X of the same tag are merged so as to form the complete token. For example, if we have a sequence of token-prediction tuples like (*Jean*, B_GIVEN), (-, I_GIVEN), (*Baptiste*, I_GIVEN), the tokens belonging to this list are merged to form the complete given name - *Jean-Baptiste*.

3.7   Affiliation CRF

This module uses CRF to identify entities such as department, institution, street-address, postal-code, city, state and country in lines predicted to be *contribgroup-aff* by Level III CRF.

In each sentence, superscripts and affiliation markers like *, †, ‡ and § are tagged as affiliation labels.

The Affiliation CRF model is trained on language features such as part-of-speech(POS) tag identified using Stanford Log-linear Part-Of-Speech[11] tagger and NER tag identified using Stanford Named Entity Recognizer. It also uses the following keyword-based features:

1. *isDepartmentMultiLang,isUniversityMultiLang*: simple heuristics to check if a token belongs to a list of words denoting university or department in languages like German, French, Italian, Spanish and English.
2. *isCity,isCountry*: The token is checked against a database of city-names[12] and Gazetteer list of countries.
3. *hasAddress*: The token is searched for words like 'road', 'street' and postal code patterns which denote address.

The target classes are BIO encoded. After decoding the tags like in Sect. 3.6, we obtain full classes such as department, institution, street-address, postal-code, city, state and country.

Also, affiliation markers[13] found in the Author Name CRF — Sect. 3.6, are mapped to their corresponding aff-ids identified in this section.

3.8   Citation Mapping

Citations are alphanumeric strings embedded in the document body that denote an entry in the bibliographic references section of a research article. Among citation systems, Vancouver system[14] uses sequential numbers either bracketed or superscript or both. The numbers refer to either footnotes, endnotes or references. Parenthetical referencing also known as Harvard referencing[15] has in-text

---

[11] Stanford Log-linear Part-Of-Speech Tagger - http://nlp.stanford.edu/software/tagger.shtml.

[12] Maxmind Free World Cities Database - https://www.maxmind.com/en/free-world-cities-database.

[13] Symbols like *, †, ‡ and §, or numbers 0–9.

[14] Vancouver System of Referencing - https://en.wikipedia.org/wiki/Vancouver_system.

[15] Harvard Referencing - https://en.wikipedia.org/wiki/Parenthetical_referencing.

citations enclosed within parentheses. Our model is trained to identify citations that follow the Vancouver system, which is predominantly the style followed by the papers published by CEUR-WS.org.

From the Level III CRF — Sect. 3.5, sentences which are classified as *sec* are tokenized and passed through the model created for citation identification.

Citation identification model is trained on POS tag, NER tag, punctuation features like *containsParentheses*, keyword features like *containsEtAl*, typographical features like *isBold* and *isSuperscript*. The target classes — CITATION and NOT_CITATION, are BIO Encoded.

3.9   Reference Parsing

A bibliographic list item usually contains one or more citations describing a referenced work. This module works best for atomic references, which contain a single citation.

From the Level III CRF — Sect. 3.5, text lines which are classified as *ref-list* are tokenized and passed through the model created for reference parsing. Each reference string is tokenized by splitting at the occurrence of every whitespace character. Punctuation symbols such as *comma(,)*, *semicolon(;)*, and *hyphen(-)* are considered to be individual tokens, because these symbols have an inherent meaning in the particular reference style used in a document.

To train a CRF model, each token in the tokenized ref line, is tagged with the textual and layout features such as *leadingCaps*, *hasCaps*, *allCaps*, *allNums*, *isYear*, *trailingDot*, *isHyphen* and *digitRatio*. The target classes are BIO encoded.

After resolving the BIO encoded tags like in Sect. 3.6, we obtain full classes such as Author, Article Title, Source, Year, Volume, Issue, DOI, First Page, Last Page, Note and URI.

3.10   NLM XML and RDF Creation

In the previous sections, author names and their corresponding affiliations are resolved into one xml node called *contrib-group*, and similarly the references module generates a separate xml node called *ref-list*. Likewise, the remaining annotations are compiled into a DOM conforming to the NLM JATS standard.

To generate an RDF from the consolidated XML, we have used an XSLT transform called JATS2RDF [12] that automates the creation of RDF metadata from a JATS-marked up document. The mappings defined in this transform function are based on various SPAR[16] ontologies and also other vocabularies such as the Dublin Core Metadata Initiative (DCMI) Metadata Terms and the Friend of a Friend (FOAF) Vocabulary.

Table 1 shows example RDF mappings for JATS elements such as country and affiliation. For the JATS elements not mapped in the JATS2RDF transform

---

[16] SPAR - the Semantic Publishing and Referencing Ontologies is an integrated ecosystem of various ontologies like DoCO and CiTO.

Table 1. JATS to RDF mapping

Element/attribute name	XML example	RDF translation
country (child of address and aff)	<aff> <country> XXX < /country> < /aff>	:this-agent-contact-info vcard:address [a vcard: Address; vcard:country-name "XXX"].
aff	<contrib> .... <aff>....< /aff> .... <contrib>	:this-agent pro:holdsRoleInTime [pro:withRole scoro:affilate ; pro:relatesToOrganization :this-organization pro:relatesToDocument :conceptual-work]

The first column gives the JATS element or attribute name, the second shows an example XML usage from the JATS specification, and the third describes the mapping of to RDF.

function, but which are relevant to this challenge, such as *fig* and *table-wrap*, we have defined mappings based on the DoCO[17] ontology.

4   Results and Evaluation

We trained our system on the 146 workshop papers published by CEUR-WS.org that were released as training datasets for 2015[18] and 2016[19] challenges. The results shown here would be updated with the performance scores on the evaluation dataset which is yet to be released.

Table 2 shows the training level precision, recall and f-scores for individual tags described in the Level I, Level II and Level III CRF modules. The values seen in the table were computed using the perl script - conlleval.pl[20].

The consolidated NLM XMLs were generated and transformed into RDF documents as explained in Sect. 3.10. Then we ran the SPARQL queries written specifically to address the aforementioned eight queries of Task 2. When the 320 queries (40 different queries for each of the challenge's 8 queries) were posed against the populated knowledge base, our system yielded an average F-measure of 0.612 (Precision: 0.629, Recall: 0.62), when evaluated against the gold data, as calculated by the tool - SemPubEvaluator[21] released for this challenge. These figures are different from the reported F-measure of 0.389 (Precision: 0.393, Recall: 0.428)[22] due to the presence of empty lines in the output files, which were misconstrued for query output.

---

[17] Document Components Ontology (DoCO), http://purl.org/spar/doco.

[18] https://github.com/ceurws/lod/wiki/SemPub15_Task2#training-dataset-td2.

[19] https://github.com/ceurws/lod/wiki/SemPub16_Task2#training-dataset-td2.

[20] http://www.cnts.ua.ac.be/conll2000/chunking/conlleval.txt.

[21] https://github.com/angelobo/SemPubEvaluator.

[22] https://github.com/ceurws/lod/wiki/SemPub2016#winners.

Table 2. CRF training accuracy for sentence level annotation

	Tags	Precision	Recall	F score
First level	front	93.61	80.38	86.49
	body	96.71	98.09	97.39
	back	91.9	96	93.91
	floats-group	58.06	30	39.56
Second level	abstract-group	95.24	99.55	97.35
	ack	93.10	48.21	63.53
	contrib-group	98.94	96.88	97.89
	document-title	96.77	96.77	96.77
	fig	93.62	100.00	96.7
	sec	98.35	99.11	98.73
	kwd-group	100.00	80.00	88.89
	ref-list	96.47	99.75	98.08
	table-wrap	100.00	81.25	89.66
Third level	contrib-group aff	94.23	97.35	95.77
	contrib-group contrib name	91.43	78.05	84.21
	p	98.51	99.21	98.86
	head	69.08	54.12	60.69

In every row, the tag and its corresponding precision, recall and F score values are shown.

5 Conclusion and Future Work

In this paper, we proposed an approach to automatically identify sections and their labels by using a sequence of multiple CRF models where each model builds on the predictions of the preceding models. A JATS XML was then generated which was transformed into an RDF document, thereby contributing towards building a knowledge base. As part of SemPub-2016, we have addressed all of the eight queries of Task 2. We have shown that our post-challenge evaluation scores are comparable against those of the pattern matching approaches presented at SemPub-2016, thereby underscoring the value of integrating machine learning and natural language processing techniques into information retrieval systems. Future work would focus on improving the overall accuracy of the meta-data extraction module of our system by training on a bigger, annotated dataset.

References

1. Rosenthol, L.: Developing with PDF: Dive Into the Portable Document Format. O'Reilly Media Inc., Sebastopol (2013)
2. Lafferty, J., McCallum, A., Pereira, F.: Conditional random fields: probabilistic models for segmenting and labeling sequence data. In: Proceedings of ICML, pp. 282–289 (2001)

3. Tkaczyk, D., Szostek, P., Fedoryszak, M., Dendek, P.J., Bolikowski, L.: CERMINE: automatic extraction of structured metadata from scientific literature. Int. J. Doc. Anal. Recogn. (IJDAR) **18**, 317–335 (2015). Springer

4. Klampfl, S., Kern, R.: Machine learning techniques for automatically extracting contextual information from Scientific Publications. In: Gandon, F., Cabrio, E., Stankovic, M., Zimmermann, A. (eds.) SemWebEval 2015. CCIS, vol. 548, pp. 105–116. Springer, Heidelberg (2015). doi:10.1007/978-3-319-25518-7_9

5. Pembe, F.C., Güngör, T.: Heading-based sectional hierarchy identification for HTML documents. In: 22nd International Symposium on Computer and Information Sciences, ISCIS, pp. 1–6. IEEE (2007)

6. Vanderbeck, S., Bockhorst, J., Oldfather, C.: A machine learning approach to identifying sections in legal briefs. In: MAICS, pp. 16–22 (2011)

7. Do, H.H.N., Chandrasekaran, M.K., Cho, P.S., Kan, M.Y.: Extracting and matching authors and affiliations in scholarly documents. In: Proceedings of the 13th ACM/IEEE-CS Joint Conference on Digital Libraries, pp. 219–228. ACM (2013)

8. Finkel, J.R., Grenager, T., Manning, C.: Incorporating non-local information into information extraction systems by Gibbs sampling. In: Proceedings of the 43rd Annual Meeting on Association for Computational Linguistics. Association for Computational Linguistics (2005)

9. Ramshaw, L.A., Mitchell, P.M.: Text chunking using transformation-based learning (1995). arXiv preprint: arXiv:cmp-lg/9505040

10. Iorio, A.D., Lange, C., Dimou, A., Vahdati, S.: Semantic publishing challenge – assessing the quality of scientific output by information extraction and interlinking. In: Gandon, F., Cabrio, E., Stankovic, M., Zimmermann, A. (eds.) SemWebEval 2015. CCIS, vol. 548, pp. 65–80. Springer, Heidelberg (2015). doi:10.1007/978-3-319-25518-7_6

11. Lange, C., Di Iorio, A.: Semantic publishing challenge – assessing the quality of scientific output. In: Presutti, V., et al. (eds.) SemWebEval 2014. CCIS, vol. 475, pp. 61–76. Springer, Heidelberg (2014)

12. Peroni, S., Lapeyre, D.A., Shotton, D.: From markup to linked data: mapping NISO JATS v1.0 to RDF using the SPAR (Semantic Publishing and Referencing) ontologies. In: Journal Article Tag Suite Conference (JATS-Con) Proceedings 2012 [Internet]. National Center for Biotechnology Information (US), Bethesda (MD) (2012). http://www.ncbi.nlm.nih.gov/books/NBK100491/

ACM: Article Content Miner for Assessing the Quality of Scientific Output

Andrea Giovanni Nuzzolese[1], Silvio Peroni[2],
and Diego Reforgiato Recupero[3(✉)]

[1] Semantic Technology Laboratory, ISTC-CNR, Rome, Italy
andrea.nuzzolese@istc.cnr.it
[2] Digital and Semantic Publishing Laboratory,
Department of Computer Science and Engineering,
University of Bologna, Bologna, Italy
silvio.peroni@unibo.it
[3] Department of Mathematics and Computer Science,
University of Cagliari, Cagliari, Italy
diego.reforgiato@unica.it

Abstract. This paper presents the *Article Content Miner* (a.k.a. *ACM*), i.e., a method for processing the research papers in PDF format available for the 2016 edition of the Semantic Publishing Challenge in order to extract relevant semantic data and publish them in a RDF triplestore according to the *Semantic Publishing And Referencing (SPAR) Ontologies* (http://www.sparontologies.net). In particular, the extraction of all the information needed for addressing the queries of the second task of the challenge (https://github.com/ceurws/lod/wiki/SemPub16_Task2) is guaranteed by ACM by using techniques based on Natural Language Processing (i.e., Combinatory Categorial Grammar, Discourse Representation Theory, Linguistic Frames), Semantic Web technologies and good Ontology Design practices (i.e., Content Analysis, Ontology Design Patterns, Discourse Referent Extraction and Linking, Topic Extraction).

Keywords: Machine reading · SPAR Ontologies · Semantic Publishing

1 Introduction

The knowledge management of scholarly products is an emerging research area in the Semantic Web field known as Semantic Publishing [38]. The Semantic Publishing refers to the publication of information on the web as documents accompanied by semantic markup. Here semantics is employed to provide a way to computers to understand the structure and the meaning of the published information, making more efficient search and integration of data. The Semantic Publishing is aimed at contributing to the realisation of the Web of Data by providing access to semantic enhanced scholarly products in order to enable a variety of tasks focused on the exploitation of scholarly data, such as knowledge discovery, knowledge exploration and data integration. However, most of the

H. Sack et al. (Eds.): SemWebEval 2016, CCIS 641, pp. 281–292, 2016.
DOI: 10.1007/978-3-319-46565-4_22

research outcomes are still locked up in flat PDF documents that do not provide any machine-readable data and prevent publishing and accessing scholarly data as Linked Data.

Here we propose the *Article Content Miner* (a.k.a. *ACM*) as a solution for addressing such a problem. ACM is a method and a tool for processing research papers available as PDF documents in order to extract relevant semantic data and publish them as Linked Data. ACM uses the *Semantic Publishing And Referencing (SPAR) Ontologies*[1] [30] as the reference model for organising scholarly knowledge extracted from PDFs.

ACM implements a novel solution for dealing with natural language and extracting relevant metadata from scholarly articles by hybridising techniques based on Natural Language Processing (i.e., Combinatory Categorial Grammar, Discourse Representation Theory, Linguistic Frames) with Semantic Web technologies and good Ontology Design practices (i.e., Content Analysis, Ontology Design Patterns, Discourse Referent Extraction and Linking, Topic Extraction). Additionally, ACM employs FRED[2] [16], a novel machine reader that is quickly spreading [8,17,18,34,36,37] along the Semantic Web community and that we have deployed, for some of the queries of the Semantic Publishing Challenge 2016 (task 2)[3]. We remind that the queries to be answered for such a task include the following items:

- Affiliations in a paper;
- Countries in affiliations;
- Supplementary material;
- Sections;
- Tables;
- Figures;
- Funding agencies;
- EU projects.

ACM has been developed on top of [28], our first prototype for the processing of research papers available in CEUR-WS.org and stored as PDF files.

ACM extracts the following kind of information: the name of the authors and their related affiliations including the countries, supplementary information such as downloads links, sections, tables and figures present within the paper, funding agencies and EU projects information.

More in detail, the paper is organised as follows: Sect. 2 presents the related work; Sect. 3 presents materials and methods related to our project, while Sect. 4 introduces our contribution, i.e., ACM. In Sect. 5 we will show the results we obtained with ACM and a comparison with [28]. Finally, in Sect. 6 we conclude the paper with remarks and sketch out some future works and directions where we are headed.

---

[1] http://www.sparontologies.net.

[2] http://wit.istc.cnr.it/stlab-tools/fred.

[3] https://github.com/ceurws/lod/wiki/SemPub16_Task2.

2   Related Work

Most of the literature about the extraction of metadata and citations from scholarly articles in the research area of Semantic Publishing has converged, during last years, into the *Jailbreaking the PDF* initiative [19]. This initiative is aimed at creating a formal flexible infrastructure to extract semantic information from PDF documents by combining existing solutions and tools for extracting data and annotations, and for identifying the argumentative discourse of scholarly papers.

Cermine [39] provides a Java library and a web service for extracting metadata and content from PDF files containing academic publications. It does not include any OCR phase, but it analyses only the PDF text stream found in the input documents. The workflow inspects the entire content of the document and produces two kinds of output in NLM format [27]: the document's metadata and parsed bibliographic references.

PDFMiner [29] is a Python tool for extracting information from PDF documents, which focuses entirely on getting and analysing text data and allows to extract the outline of a paper and its tagged content, to reconstruct the original layout by grouping text chunks and to convert the PDF to an HTML.

PDFX [5] is a rule-based system designed to reconstruct the logical structure of scholarly articles stored in PDF, regardless of their formatting style. The system's output is an XML document that describes the input article's logical structure in terms of title, sections, tables, references, etc. and also links it to geometrical typesetting markers in the original PDF, such as paragraph and column breaks.

ParseCit+SectLabel [23] is an open source system to solve two related subtasks in logical structure discovery: (i) logical structure classification, and (ii) generic section classification. ParseCit uses the machine learning methodology of conditional random fields (CRF) [21] - i.e., a model that blends sequential labeling techniques with pointwise entropy-based classification. ParseCit+SectLabel is an open source system that extends ParsCit in order to provide logical structure discovery and classification of a scholarly article.

CiTalO [8] is an algorithm and a tool that allows inferring the rhetorical function of citations linking scholarly articles. CiTalO relies on (i) FRED [35] for generating a logical representation (expressed as RDF/OWL) of a sentence containing a citation and on (ii) a set of rules for interpreting such a logical representation in order to map the rethorical function of a citation to the properties of the *Citation Typing Ontology (CiTO)* [32].

All the approaches described so far extract the high level structure of the PDF document and only some focus on the extraction of fine-grained information (Cermine [39] extracts affiliations, countries and section titles or PDFX [5] that extracts section titles and ParsCit [23] that extracts section titles, table/figure captions and affiliations) as required by the Semantic Publishing Challenge 2016 (task 2)[4].

---

[4] https://github.com/ceurws/lod/wiki/Task2.

Approaches that tried to solve the same problem of ours, such as [2,12], were presented at the Semantic Publishing challenge [22] held at the Extended Semantic Web Conference 2014[5], that consisted of similar tasks. [2] proposes an hybrid method for the extraction and characterization of citations in scientific papers by combining machine learning with rule-based techniques. The solution consists of extraction of metadata, bibliography parsing, section titles processing, and fine-grained semantic annotation on the sentence level of texts. [12] presents a solution to extract and map data of workshop proceedings published from HTML to RDF. The solution exploits RML [13], which is an extension of the R2RML mapping language [7] for defining customized mapping rules from data expressed in heterogeneous formats to the RDF data model.

To the best of our knowledge none of the previously mentioned works (except CiTalO to some extent) use Combinatory Categorial Grammar, Discourse Representation Theory or Linguistic Frames. Therefore, *ACM* is the first method of its kind that leverages the results of many NLP components by reengineering and unifying them in a unique RDF/OWL graph designed by following semantic web ontology design practices. ACM is flexible and domain-independent: it can easily be adapted to answer queries in several other domains.

3   Article Content Miner

In this section we introduce the materials and methods used by our *Article Content Miner (ACM)* project.

3.1   Materials and Ontologies

All the data that are extracted through the scripts introduced in Sect. 4 are stored according to the *Semantic Publishing and Referencing (SPAR) Ontologies*[6] [30]. Such ontologies form a suite of orthogonal and complementary ontology modules for creating comprehensive machine-readable RDF metadata for all aspects of semantic publishing and referencing. In particular, they allow researchers to describe far more than simply bibliographic entities such as books and journal articles, by enabling RDF metadata to include information related to citations, bibliographic records, specific sections of documents, and various aspects of the scholarly publication process. In the context of the ACM project, four of them are relevant for the data related to the questions of the challenge:

- the *FRBR-aligned Bibliographic Ontology (FaBiO)*[7] [32] is an ontology for describing entities that are published or potentially publishable (e.g., journal articles, conference papers, books), and that contain or are referred to by bibliographic references;

---

[5] http://2014.eswc-conferences.org/.

[6] http://www.sparontologies.net.

[7] http://purl.org/spar/fabio.

- the *Document Components Ontology (DoCO)*[8] [4] is an ontology for describing the various components of an article, such as figures, tables, sections, etc.;
- the *Publishing Roles Ontology (PRO)*[9] [33] is an ontology for the characterisation of the roles of agents – people, corporate bodies and computational agents in the publication process. These agents can be authors, editors, reviewers, publishers or librarians;
- the *Funding, Research Administration and Projects OntologyFRAPO)*[10] [31] is an ontology for describing the administrative information of research projects, e.g., grant applications, funding bodies, project partners, etc.

3.2   Methods

ACM employs and integrates several tools and techniques. Natural Language Processing (NLP) techniques have been used to pre-process the text, to break it down in sections and sentences, and to extract specific sub-sections.

First of all, for extracting the plain text of the articles, we use the PDFMiner[11] Python library, which is a tool for extracting information from PDF documents that focuses entirely on getting and analyzing text data.

Once we extracted the text, we have developed on top of the Stanford CoreNLP [24] and the Natural Language Toolkit (NLTK, http://www.nltk.org). CoreNLP and NLTK are two of the leading platforms for building programs to work with human language data and that includes several corpora and lexical resources, along with a suite of text processing libraries for classification, tokenization, stemming, tagging, parsing, and semantic reasoning.

One more tool that we have included in ACM is FRED[12] [35]. FRED automatically produces RDF/OWL ontologies and linked data from text. FRED was successfully applied in the past to several semantic web applications [8,15–18,34,36,37]. FRED formally represents, integrates, improves, and links the output of several NLP tools. The backbone deep semantic parsing is currently provided by Boxer [3], which uses a statistical parser (C&C) producing Combinatory Categorial Grammar trees, and thousands of heuristics that exploit existing lexical resources and gazetteers to generate representation structures according to Discourse Representation Theory (DRT) [20]. The basic NLP tasks performed by Boxer, and reused by FRED, include: (mostly) verbal event detection, semantic role labeling with VerbNet and FrameNet roles, first-order logic representation of predicate-argument structures, logical operators scoping (called boxing), modality detection, and tense representation. FRED produces RDF/OWL ontologies having classes (and related taxonomies) depending on the lexicon used in the text. In order to provide a public identity to such classes, FRED exploits Word-Sense Disambiguation (WSD) to resolve classes into WordNet or BabelNet

---

[8] http://purl.org/spar/doco.

[9] http://purl.org/spar/pro.

[10] http://purl.org/cerif/frapo.

[11] https://github.com/euske/pdfminer/.

[12] http://wit.istc.cnr.it/stlab-tools/fred.

[26]. FRED can use any WSD system, such as UKB [1] or Babelfy [25]. WSD also enables FRED to generate alignments to two top-level ontologies: WordNet supersenses and a subset of DOLCE+DnS Ultra Lite (DUL) classes.

Figure 1 shows a RDF graph produced by FRED for the example sentence "This work has been funded by the Federal Ministry of Education and Research, Germany (BMBF), within the SMART project". The example sentence has been taken from the acknowledgment section of one of the paper of the training set and gives some hints on the *patient* and *agent* roles for the instance of the verb *fund*. Patient would be the object funded whereas the agent would be the funding entity.

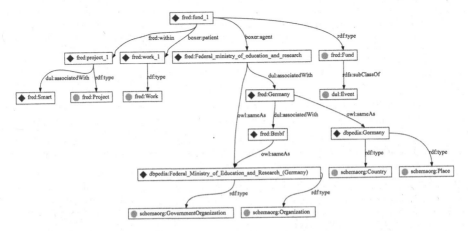

Fig. 1. RDF graph for the sentence "This work has been funded by the Federal Ministry of Education and Research, Germany (BMBF), within the SMART project".

4   Implementation Details

The workflow implemented in ACM for extracting all the data of interest is actually organised as a sequential execution of several scripts and existing tools written in different languages – mainly Python and Java. The first script executed, called *text extractor* (based on PDFMiner), is responsible for extracting the full text of the article from PDF files. All the other scripts start from the outcome of the *text extractor* and return appropriate JSON objects that contain the data needed for answering all the queries of the challenge and reported below. Then, all these JSON objects are converted into RDF according to the SPAR Ontologies[13] [30] and published on the ACM triplestore, available at http://six. eelst.cs.unibo.it:8080 (dataset *ACM*).

---

[13] http://www.sparontologies.net.

4.1   Queries Q2.1 and Q2.2: Affiliations

The main requirement of question 2.1 is to identify in a PDF article the information about the affiliations of the authors. Affiliations are complex pieces of information, as they contain data of different nature. For example, an affiliation string provides references to entities that can be typed as organisations, such as research units/laboratories (e.g., STLab) or institutions (e.g., National Research Council). Moreover, an affiliation string typically contains geographical information about the organisation that can be general (the city and the country where the organisation is based) or more detailed (addresses, postal codes, etc.). The first step performed by ACM is the identification of the chunk of words that provides authors' names and their affiliations. ACM uses a predefined set of heuristics for identifying authors and affiliation syntactically. The basic intuition is that these kinds of information typically occur in the front-matter of scholarly articles. Then, ACM relies on Named Entity Recognition (NER) based on Stanford CoreNLP [24] plus a set of statistical rules for identifying organizations, sub-organizations and units.

The rules have been defined by recording frequent occurring patterns used for describing organisations in author's affiliations introduced in scholarly articles. Additionally, ACM keeps track of extracted data in order to use already parsed affiliation strings as background knowledge. Once the data about affiliations have been extracted, ACM maps each affiliation to its corresponding author (even in case of multiple affiliations), by using (i) the order in which affiliations and authors' names appear in a paper, (ii) special marker used in a source article for coupling authors and affiliations (e.g., †, ‡, *), and (iii) the background knowledge consisting of previously parsed articles.

Question 2.2 requires to extract data about affiliations and to identify the country where each research institution is located in. ACM answers this question by completing the parsing step presented in Q2.1 with the recognition of Named Entities that identify countries.

4.2   Queries Q2.3, Q2.4, Q2.5 and 2.6: Supplementary Materials and Components

Four of the task 2 queries concern the identification of specific aspects of the content of the article in the testbed, namely, supplementary materials (Q2.3), sections (Q2.4), tables (Q2.5) and figures (Q2.6). In order to be able to provide a correct identification of these components, we approach the identification of these parts by applying several algorithms that allow us to automatically identify them starting from their structural organisation. In particular, we aim at applying our algorithm developed in past works, that analyses each input document separately, and for each one it performs the recognition of low-level structural patterns as presented in [10]. Then, starting from the outcomes of the previous algorithm, another mechanism we presented in [11] applies some heuristics and association rules in order to associate one of the classes describing document components [4] (among which those requested by the questions of the challenge) to each portion of text of the document.

4.3   Queries Q2.7 and Q2.8: Funding Agencies and EU Projects

The challenge's instructions state that the analysis of research grant, funding and EU projects must be restricted to those whose number or identifier is explicitly mentioned in the underlying paper. This means that it is not possible to look for other information in external data sources. For example, the EU system http://cordis.europa.eu/projects/home_en.html would have helped us a lot with the extraction of detailed information of EU projects. This requirement drove us to focus on the analysis of when a grant, funding agency or EU project can be mentioned in a given paper. The first consideration we had is that those information are often mentioned within the acknowledgement section of a paper and only in rare cases as footnote in other sections (usually in the first page, within the introduction). As an example, in the training set, only 7 papers out of 109 included information related to queries Q2.7 and Q2.8 as footnote in the first page. Also, 57 papers contained a dedicated acknowledgement section and 45 did not provide any data about it. This consideration helped us to limit the search for grants, funding agencies and EU projects to the acknowledgement section of a given paper or to footnotes present in the first page and containing identified keywords.

Query Q2.7: Funding Agencies. As a funding agency is usually explicitly mentioned (if present) using verb such as *fund, promote, support* we decided to apply FRED to each sentence of the acknowledgement section/footnote. We want to underline that FRED does not use any external ontology related to info about funding agencies but only lexical resources such as Verbnet or Wordnet. Once we obtained the RDF graph from FRED for each sentence of the acknowledgement section of each processed paper, we looked for any of the verbs above, and, in particular, if any node had a *boxer:agent* or *vn.role:Agent* role property connected to the verb. That node would correspond to the agency itself. For instance, in the example sentence reported in Fig. 1, the reader may notice the presence of the instance node *fund_1*, verb of type *Fund*. The role *boxer:agent* in such an example reports the name of the mentioned funding agency. After we extracted the information tied to the agent role we augmented the result using some regular expressions in case of FRED's errors and in presence of compound expressions involving projects. The reason is that it may happen that a given paper is supported by a project and not an agency, and we would obtain that information from FRED's graph. With this step we removed such information from our result. A further step was needed to remove punctuations from our final results.

Query Q2.8: EU Projects. We analysed the possible forms to mention EU projects in a research paper. There several ways a EU project can be mentioned: either explicitly mentioned using verbs such as *support, fund, acknowledge,* or by referencing to it using the acronym or the EU call identifier. Therefore we decided to not use any NLP tool nor FRED and, in order to extract EU projects names, we identified a set of regular expressions that included terms such as *fp6, fp7, 7th programme, european project, etc.*. The extracted information were then cleaned to remove punctuations, common expressions and words included into a stop-word list we defined.

5   Results

ACM was built on top of MACJa, a method we have introduced in [28]. ACM was improved taking into account some feedbacks and experience from the Semantic Publishing Challenge 2015[14] where MACJa was first proposed. ACM was able to outperform the results of MACJa as shown in Table 1. The reader is suggested to check this page[15] to see the performances of all the competitors of ACM participating at the Task 2[16] of the Semantic Publishing Challenge 2016[17].

Table 1. Comparison between ACM and MACJa

Method	Precision	Recall	F-score
Macja	0.274	0.251	0.257
ACM	0.412	0.43	0.416

6   Conclusions

We have presented ACM, a framework for processing the research papers stored as PDF files. Its goal is to extract relevant semantic information and create a RDF triplestore according to the Semantic Publishing And Referencing (SPAR) Ontologies. ACM employees Discourse Representation Theory, Combinational Categorial Grammar, Linguistic Frames, NLP, Semantic Web and the most prominent Ontology Design practices to achieve its goal. The information that ACM can extract are limited to those identified within the Task 2[18] of the Semantic Publishing Challenge 2016[19]. Future directions we are already headed include the exploitation of the Abstract Meaning Representation (AMR) [14] and its property to deal with semantic interpretation of text. We are extending FRED to represent information in AMR with the aim of extracting relevant information from research papers.

---

[14] https://github.com/ceurws/lod/wiki/SemPub2015.

[15] https://github.com/ceurws/lod/wiki/SemPub2016.

[16] https://github.com/ceurws/lod/wiki/SemPub16_Task2.

[17] https://github.com/ceurws/lod/wiki/SemPub2016.

[18] https://github.com/ceurws/lod/wiki/SemPub16_Task2.

[19] https://github.com/ceurws/lod/wiki/SemPub2016.

References

1. Agirre, E., Soroa, A.: Personalizing pagerank for word sense disambiguation. In: EACL 2009, Athens, Greece. The Association for Computer Linguistics (2009)
2. Bertin, M., Atanassova, I.: Hybrid approach for the semantic processing of scientific papers. In: Semantic Publishing Challenge (2014)
3. Bos, J.: Wide-coverage semantic analysis with boxer. In: Bos, J., Delmonte, R. (eds.) Semantics in Text Processing, pp. 277–286. College Publications, London (2008)
4. Constantin, A., Peroni, S., Pettifer, S., Shotton, D., Vitali, F.: The document components ontology (DoCO). In: Semantic Web - Interoperability, Usability, Applicability. IOS Press, Amsterdam (2016). doi:10.3233/SW-150177
5. Constantin, A., Steve, P., Andrei, V.: Fully-automated PDF-to-XML conversion of scientific literature. In: Proceedings of the ACM Symposium on Document Engineering, pp. 177–180. ACM, New York (2013). doi:10.1145/2494266.2494271
6. d'Aquin, M., Baldassare, C., Gridinoc, L., Sabou, M., Angeletou, S., Motta, E.: Supporting next generation semantic web applications. In: Proceedings of WWW/Internet Conference (2007)
7. Das, S., Sundara, S., Cyganiak, R.: R2RML: RDB to RDF Mapping Language. W3C recommendation (2012). http://www.w3.org/TR/r2rml/
8. Di Iorio, A., Nuzzolese, A.G., Peroni, S.: Towards the automatic identification of the nature of citations. In: Castro, A.G., Lange, C., Lord, P.W., Stevens, R. (eds.) SePublica, CEUR Workshop Proceedings, vol. 994, pp. 63–74 (2013). CEUR-WS.org
9. Di Iorio, A., Nuzzolese, A.G., Peroni, S., Shotton, D., Vitali, F.: Describing bibliographic references in RDF. In: Garcia Castro, A., Lange, C., Lord, P., Stevens, R. (eds.) Proceedings of 4th Workshop on Semantic Publishing (SePublica 2014) (2014). http://ceur-ws.org/Vol-1155/paper-05.pdf
10. Di Iorio, A., Peroni, S., Poggi, F., Vitali, F.: Dealing with structural patterns of XML documents. J. Am. Soc. Inf. Sci. Technol. **65**(9), 1884–1900 (2014). doi:10. 1002/asi.23088. Wiley, Hoboken
11. Di Iorio, A., Peroni, S., Poggi, F., Vitali, F., Shotton, D.: Recognising document components in XML-based academic articles. In: Proceedings of the 2013 ACM symposium on Document Engineering (DocEng 2013), pp. 181–184. ACM, New York (2013). doi:10.1145/2494266.2494319
12. Dimou, A., Vander Sande, M., Colpaert, P., De Vocht, L., Verborgh, R., Mannens, E., Van de Walle, R.: Extraction and semantic annotation of workshop proceedings in HTML using RML. In: Semantic Publishing Challenge (2014)
13. Dimou, A., Vander Sande, M., Colpaert, P., Mannens, E., Van de Walle, R.: RML: a generic language for integrated RDF Mappings of Heterogeneous Data. In: Workshop on Linked Data on the Web (2014)
14. Flanigan, J., Dyer, C., Smith, A.N., Carbonell, J.: Generation from Abstract Meaning Representation using Tree Transducers (Accepted to NAACL HTL, 2016)
15. Gangemi, A.: A comparison of knowledge extraction tools for the semantic web. In: Cimiano, P., Corcho, O., Presutti, V., Hollink, L., Rudolph, S. (eds.) ESWC 2013. LNCS, vol. 7882, pp. 351–366. Springer, Heidelberg (2013)
16. Gangemi, A., Draicchio, F., Presutti, V., Nuzzolese, A.G., Reforgiato Recupero, D.: A machine reader for the semantic web. In: Blomqvist, E., Groza, T. (eds.) International Semantic Web Conference (Posters & Demos), CEUR Workshop Proceedings, vol. 1035, pp. 149–152 (2013). CEUR-WS.org

17. Gangemi, A., Nuzzolese, A.G., Presutti, V., Draicchio, F., Musetti, A., Ciancarini, P.: Automatic typing of DBpedia entities. In: Cudré-Mauroux, P., et al. (eds.) ISWC 2012, Part I. LNCS, vol. 7649, pp. 65–81. Springer, Heidelberg (2012)

18. Gangemi, A., Presutti, V., Reforgiato Recupero, D.: Frame-based detection of opinion holders and topics: a model and a tool. IEEE Comp. Int. Mag. 9(1), 20–30 (2014)

19. Garcia, A., Murray-Rust, P., Burns, G.A., Stevens, R., Tkaczyk, D., McLaughlin, C., Belin, A., Iorio, A., García, L., Gruson-Daniel, C., Mounce, R., Nuzzolese, A.G., Peroni, S., Spinks, J., Villazon-Terrazas, B., Corcho, O., Giraldo, O., Wabiszewski, M.: PDFJailbreak-a communal architecture for making biomedical PDFs semantic. In: Proceedings of BioLINK SIG (2013)

20. Kamp, H.: A theory of truth and semantic representation. In: Groenendijk, J.A.G., Janssen, T.M.V., Stokhof, M.B.J. (eds.) Formal Methods in the Study of Language, vol. 1, pp. 277–322. Mathematisch Centrum, Amsterdam (1981)

21. Lafferty, J., McCallum, A., Pereira, F.C.N.: Conditional random fields: probabilistic models for segmenting and labeling sequence data. In: Proceedings of 18th International Conference on Machine Learning, pp. 282–289. Morgan Kaufmann, San Francisco (2001)

22. Lange, C., Di Iorio, A.: Semantic publishing challenge – assessing the quality of scientific output. In: Presutti, V., et al. (eds.) SemWebEval 2014. CCIS, vol. 475, pp. 61–76. Springer International Publishing, Heidelberg (2014)

23. Luong, M.T., Dung Nguyen, T., Kan, M.Y.: Logical structure recovery in scholarly articles with rich document features. Int. J. Digit. Libr. Syst. (IJDLS) 1(4), 1–23 (2010)

24. Manning, C.D., Surdeanu, M., Bauer, J., Finkel, J., Bethard, S.J., McClosky, D.: The stanford CoreNLP natural language processing toolkit. In: Proceedings of 52nd Annual Meeting of the Association for Computational Linguistics: System Demonstrations, pp. 55–60 (2014)

25. Moro, A., Raganato, A., Navigli, R.: Entity linking meets word sense disambiguation: a unified approach. Trans. Assoc. Comput. Linguist. 2, 231–244 (2014)

26. Navigli, R., Ponzetto, S.P.: BabelNet: the automatic construction, evaluation and application of a wide-coverage multilingual semantic network. Artif. Intell. 193, 217–250 (2012)

27. NLM. http://dtd.nlm.nih.gov/archiving/

28. Nuzzolese, A., Peroni, S., Reforgiato Recupero, D.: MACJa: Metadata and Citations Jailbreaker (2015). doi:10.1007/978-3-319-25518-7

29. PDFMiner: Python PDF parser and analyzer (2010)

30. Peroni, S.: Semantic Web Technologies and Legal Scholarly Publishing (2014). ISBN 978-3-319-04776-8

31. Peroni, S.: Example of use of FRAPO #1. figshare (2015). http://dx.doi.org/10.6084/m9.figshare.1549721

32. Peroni, S., Shotton, D.: FaBiO and CiTO: ontologies for describing bibliographic resources and citations. Web Seman. Sci. Serv. Agents World Wide Web 17, 33–43 (2012). doi:10.1016/j.websem.2012.08.001

33. Peroni, S., Shotton, D., Vitali, F.: Scholarly publishing and linked data: describing roles, statuses, temporal and contextual extents. In: Sack, H., Pellegrini, T. (eds.) Proceedings of the 8th International Conference on Semantic Systems (i-Semantics 2012), pp. 9–16. ACM Press, New York. doi:10.1145/2362499.2362502

34. Presutti, V., Consoli, S., Nuzzolese, A.G., Reforgiato Recupero, D., Gangemi, A., Bannour, I., Zargayouna, H.: Uncovering the semantics of wikipedia wikilinks. In: 19th International Conference on Knowledge Engineering and Knowledge Management (EKAW 2014) (2014)

35. Presutti, V., Draicchio, F., Gangemi, A.: Knowledge extraction based on discourse representation theory and linguistic frames. In: ten Teije, A., Volker, J., Handschuh, S., Stuckenschmidt, H., d'Acquin, M., Nikolov, A., Aussenac-Gilles, N., Hernandez, N. (eds.) EKAW 2012. LNCS, vol. 7603, pp. 114–129. Springer, Berlin (2012)

36. Reforgiato Recupero, D., Consoli, S., Gangemi, A., Nuzzolese, A.G., Spampinato, D.: A semantic web based core engine to efficiently perform sentiment analysis. In: Presutti, V., Blomqvist, E., Troncy, R., Sack, H., Papadakis, I., Tordai, A. (eds.) ESWC Satellite Events 2014. LNCS, vol. 8798, pp. 245–248. Springer, Heidelberg (2014)

37. Reforgiato Recupero, D., Presutti, V., Consoli, S., Gangemi, A., Nuzzolese, A.G.: Sentilo: frame-based sentiment analysis. Cogn. Comput. **7**, 211–225 (2014)

38. Shotton, D.: Semantic publishing: the coming revolution in scientific journal publishing. Learn. Publ. **22**(2), 85–94 (2009)

39. Tkaczyk, D., Szostek, P., Jan Dendek, P., Fedoryszak, M., Bolikowski, L.: CERMINE - automatic extraction of metadata and references from scientific literature. In: Proceedings of the 11th IAPR International Workshop on Document Analysis Systems, pp. 217–221 (2014)

Information Extraction from PDF Sources Based on Rule-Based System Using Integrated Formats

Riaz Ahmad[✉], Muhammad Tanvir Afzal,
and Muhammad Abdul Qadir

Department of Computer Science, Capital University of Science & Technology,
Islamabad, Pakistan
r.ahmadafridi@gmail.com, {mafzal,aqadir}@cust.edu.pk

Abstract. Information extraction from the PDF sources is a tedious task. Most of the existing approaches use either tag-based format such as HTML and XML, or Plain-text format for the extraction of information. In this paper, we present an information extraction technique for research papers which exploits both XML and text formats intelligently. The various patterns and rules are prepared from integrated formats. Furthermore, the intelligent processing of XML and Plain-text for various situations compliments the approach to achieve high accuracy. The proposed approach is a heuristic based approach that extracts the information about logical structure and supportive materials of research papers.

Keywords: Information extraction · Research papers · Ontology · PDF parser · Regular expression · XML and plain-text formats

1 Introduction

In the scientific community, most of the published research work is represented and published in the form of text file such as PDF documents. However, the automatic retrieval of desired information from PDF documents is not a direct and simple way due to the lack of formal computer understanding structure. Therefore, it is required to understand the structure of documents and to build the algorithms to reduce the variance between text documents and computer recognizable representation.

In this research area, many different approaches have been developed to extract information either from XML documents [1–5] or from Plain-text document [6, 8–10]. None of these approaches utilize the patterns in both XML & Text formats to detect the desired information of research papers. For the simple and straight forward solutions, using only one of these formats is not enough. Therefore, we propose an approach that uses the patterns of both formats i.e. XML & Plain-Text, of the PDF documents. The XML documents are obtained by the PDFx online tool[1] for our experiment.

The dataset is provided in ESWC (European Semantic Web Conference) 2016 conference for the task2 in Semantic Publishing challenging 2016[2]. This challenge

---

[1] http://pdfx.cs.man.ac.uk.

[2] http://2016.eswc-conferences.org/assessing-quality-scientific-output-its-ecosystem.

© Springer International Publishing Switzerland 2016
H. Sack et al. (Eds.): SemWebEval 2016, CCIS 641, pp. 293–308, 2016.
DOI: 10.1007/978-3-319-46565-4_23

consists of three tasks (1) Extraction and assessment of workshop proceedings information (2) Extracting information from the PDF full text of the papers, and (3) Interlinking. This paper is focusing the second task of the challenge.

The best way of extracting information from PDF documents is the direct interaction with the content of the document either in XML or Plain-text format. The content of the document consists of different patterns along with some additional characters. Therefore, we need to pre-process the content of the documents to remove the additional characters from the desired structured elements such as 'Author', 'Sections', 'Figure', 'Table', 'Citations', and 'Supplementary materials' etc.

In this paper, we apply our proposed approach on the CEUR (Central European) Workshop training set. The proposed model of PDF source documents is prepared by using different heuristic and rules. Based on different formats of research document, we propose the integrated way of extracting the logical structure of the document.

2   Lesson Learned: Pattern and Rules Analysis

In pattern and rules analysis task, we have critically analyzed the content of both 'XML' and 'Text' formats for the information extraction about the research paper elements. These elements are 'Authors with affiliation and country', 'First level Section heading', 'Figure', 'Table', 'Supplementary Materials links', 'Funding agency', and 'Funded project'. The extracted rules of this analysis are stored in knowledge base. In the following part of current section, each research paper element is discussed with real snapshots.

In the investigation of 'Authors' with their additional information such as 'Affiliation' and 'Country', we have analyzed different formats and patterns of such information in both XML and Plain-text formats of 45 research papers in ESWC training dataset. In below part, one of such formats is highlighted in Fig. 1. The snapshot shows the 'authors list', 'affiliation & country context'. The authors list contains one or more than one author. Each author is represented with affiliation such as represented by triangle shape in Fig. 1. The affiliation & country context consists of one or more affiliation lines. The affiliation line represents two information 'Affiliation Name' and 'Country name' as shown in Fig. 1 by line and circle respectively.

In Fig. 1, the analysis of author related information such as author, affiliation, and country is shown. In this scenario, we present the start and end identifiers in both input formats. In Fig. 2(a), the snapshot of author information has been shown in XML

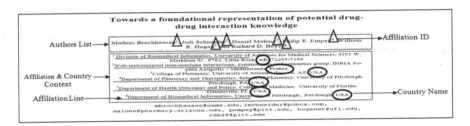

Fig. 1. Author related information

document with the <region> tag. The region tag is not used specifically for author information. In XML format in Fig. 2(a), the author list and affiliation & country information is indicated by the <region> tag. This tag is widely used to represent some other information such as image, content etc. Hence, the region tag is not suitable to detect the author information from XML documents. The XML document has no proper start and end identifier to extract the author relevant information.

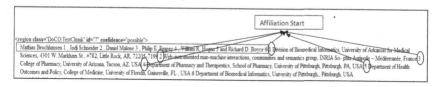

(a) Author information in XML format

While in the Plain-text format, the required information is denoted by content pre-processing, start and end identifiers such as numeric, newline (\n), and carriage return (\n) characters as shown in Figure (b). This analysis shows that the extraction of author related information might be extracted in better way from Plain-text format of PDF document of research paper.

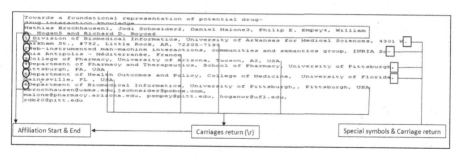

(b) Author information in Plain-text format

Fig. 2. Analysis of author information in different formats

In the analysis of 'first level section heading', we have observed the three formats of section heading in PDF documents as shown in Table 1. The first format 'Numeric Capital Case' of section heading consists of section 'heading number (1)' and 'heading name (INTROUDCTION)' in capital case. The second format 'Roman Capital Case' is denoted by 'roman heading number (II)' and 'uppercase name (RELATED WORK)'.

Table 1. Various formats of section headings

Section headings	Examples
Numeric capital case	1. INTRODUCTION
Roman capital case	II. RELATED WORK
Numeric title case	1.1. Related Work

The third format of section heading is represented by 'numeric heading number (1)' along with 'heading name (Related Work)' in title case.

In Fig. 3(a), the various formats of first level section headings are highlighted in XML formats of PDF documents. The PDFx tool properly assigns the '<h1>' tag to 'section heading' in both cases 'Numeric Title Case' and 'Numeric Capital Case'. While in the roman capital case, the PDFx tool does not assign any tag. This analysis shows that the XML format is better for Numeric 'Title' and 'Capital' cases of section headings. The roman title case might not be detected from the XML format. Therefore, in this case, we are using the Plain-text format of PDF documents in our analysis. The snapshot of Plain-text format of section headings is given in Fig. 3(b). The Plain-text format is also suitable for the extraction of section headings in numeric and roman with capital cases. The numeric with title case is not properly extracted due to the commonality in content.

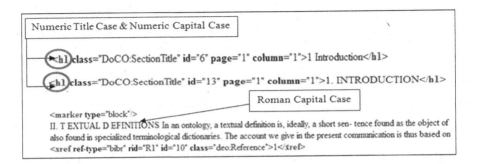

(a)   Snapshots of first level section headings in XML format

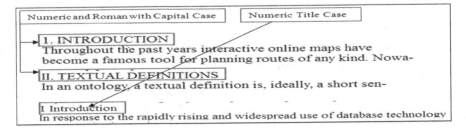

(b)   Snapshots of first level section headings in Plain-text format

Fig. 3. Analysis of section headings in both XML and plain-text format

In the Fig. 4, both the XML and Plain-text formats of same PDF document are critically analyzed for the pattern detection of figure caption. For example, the four figures such 'Fig. 1.', 'Fig. 2', 'Fig. 4.', and 'Fig. 5.' might be extracted correctly from XML format by using '<caption>' tag and 'Figure' or 'Fig' keyword as highlighted by red rectangle in Fig. 4. The green rectangle shows the figure pattern which is wrongly tagged by the PDFx tool. In this case, our proposed system will identify the missing

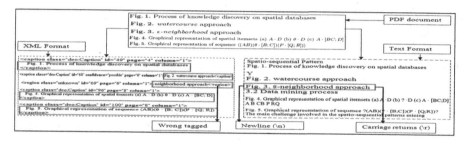

Fig. 4. Analysis of figure caption patterns in both XML and plain-text formats (Color figure online)

pattern of figure caption such as 'Fig. 3' in the content of Plain-text format using newline, Fig or Figure keyword, and carriage return information. In this way the frequency of figure caption can be improved by the integration of both formats of PDF documents. Similarly, the table caption is analyzed with '<caption>' tag and 'Table' keyword from XML document, while the 'newline', 'Table' keyword, and carriage return '\n' information can be used to retrieve table caption from 'Plaintext' format.

In Fig. 5(a) we have shown the snapshot of supportive material links in research paper document. The links in research papers are represented in the form of URL. The PDFx parser converts the links into external link '<ext-link>' tags. While the PDFbox[3] java library converts the supportive materials links into plain text. In XML format, the links can be extracted using 'http' and '<ext-link>' patterns. Similarly, the links can be indicated in plain text by using 'Newline' character (\n), 'http' keyword, and 'Carriage Return (\r)' as shown n Fig. 5(b).

```
¹ http://bioportal.bioontology.org/ontologies/SNOMEDCT, as of September 2014
² http://mimic.physionet.org/database.html
³ http://code.google.com/p/topic-modeling-tool/
⁵ Bone Dysplasia Ontology - http://bioportal.bioontology.org/ontologies/BDO
```

(a)   Snapshot of supportive materials links from research paper

(b)   Supportive materials links in XML and Plain-text format

Fig. 5. Analysis of URL links

In the analysis of funded 'project name' in the research papers, we have observed '11' research papers in training dataset that contains project name. The analysis shows that most of the time the project name is used in the 'Acknowledgments' section than

---

[3] https://pdfbox.apache.org/.

the 'footnote' part. We have critically analyzed the content of footnote and acknowledgments parts. The snapshots of these parts are taken from some of '11' research papers as shown in Fig. 6. We analyzed that most of time the project name exist between 'the' and 'project' words shown in red color in Fig. 6. Based on this assumption, the 'Rule1' is prepared. The green color patterns also follow the 'Rule1'. However, due to the extra words such as double quotes, keywords like 'finalized' or running', and semantically related words ("project" or "that is funded") in these patterns, the 'Rule1' does not work. Therefore we need to process these patterns before applying 'Rule1'. The second rule 'Rule2' is prepared based on assumption that the project-name start with only 'project' keyword. Both rules can be suffering due to the carriage return (\r) and newline (\n) characters. The pre-processing is performed in content processing component shown in Fig. 1.

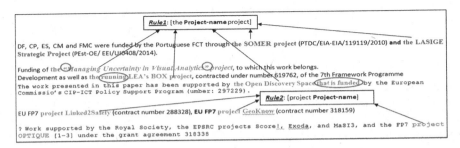

Fig. 6. Pattern analysis of project name in the content of research papers (Color figure online)

Similarly, the extraction of funding agencies names and acronyms from the text of scientific document is very crucial task due to absent of well-defined patterns. Therefore, we analyzed deeply the content of the research papers. Initially, the ESWC training dataset is exploited and it was identified that most of the time, the name of funding agencies along with their acronyms exists in the acknowledgements part. The training dataset consists of 45 papers, in which 17 papers contain either both project name and acronym or one of them. In the content analysis of these 17 papers, we have found different types of patterns along with various start and end identifier. Figure 7 shows the patterns of funding agencies names and acronyms in the content snapshots of some research papers. These patterns are identified by various start and end identifiers. The start and end identifiers are denoted by rectangle and circle shapes respectively. The analysis shows that start and end identifiers are helpful in the detection of funding agency names and acronyms. Sometimes, the research work are funded or supported by more than one funding agencies. Usually, the additional agencies names are started with combination of words such as 'and by the', 'and by a', 'and the', and 'and by' etc. These words will help to identify the end of first agency name and the start of second agency name as shown in Fig. 7.

Fig. 7. Pattern analysis of funding agency name in the content of research papers

3 Proposed System Architecture

The proposed system architecture consists of several modules such as PDF file conversion, content pre-processing, Rule Identifier, ruled-based research paper information extraction and triplification of structured elements. The detailed architectural layout is shown in Fig. 8.

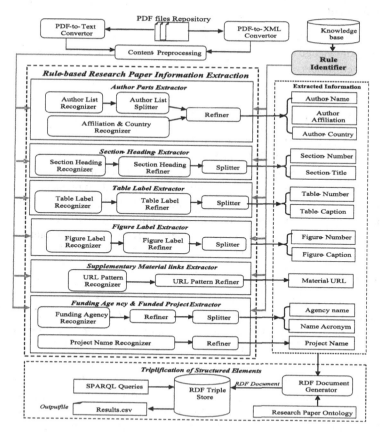

Fig. 8. Research paper information extraction system architecture

3.1   PDF Parsers

In this research work, information extraction is performed from PDF documents using both XML and Plain-text formats. Therefore, we have selected two parsers such as PDF-to-XML and PDF-to-Text for conversion of PDF file into XML and Plain-text formats respectively. Furthermore, the parsed formats use as input for 'content pre-processing' and 'pattern and rule analysis' tasks of proposed architecture as shown in Fig. 8.

3.2   Content Processing

In the content processing step, all the discussed preprocessing steps in the pattern and rule analysis part about each elements of research paper are performed. It is very important part in information retrieval task that provides the normalized form of content to extract the desired information.

3.3   Rule Identifier

The rule identifier is searching component of proposed architecture. It is used to search rules in the knowledge based repository to identify the logical and supportive elements in the research papers.

3.4   Rule-Based Research Paper Information Extraction

After detailed analysis of pattern identification of various elements in both XML and Text formats, we have developed and store various rules in knowledge base. These rules will be identified by rule identifier to extract various elements from the content of testing scientific documents. The extraction of all required components was a challenging task. The critical analysis of all training papers helped us to craft generic heuristics and patterns to identify required information with high accuracy. However, due to space limitations, we were only able to discuss few important cases with examples in the following sections.

(a) **Author Parts Extractor**

In our proposed architecture, the extraction of author name, affiliation, and country information from the research paper is done by author parts extractor. This part consists of (1) author list recognizer, (2) affiliation & country recognizer (3) author list splitter, and (4) refiner. The author list recognizer is used to identify and extract the list of authors. The affiliation & country recognizer is exploited to identify the affiliation and country parts. The splitter divides the author list into number of authors. The refiner clean some additional characters such as carriage returns, comma etc. The authors with their affiliation extraction is very challenging task due to the various formats in research papers. Therefore, these formats need different heuristics and rules. One of these formats is highlighted with its complexity in the Fig. 1. The heuristic to handle the format in Fig. 1 consists of several steps. In the step1, the title of research paper is detected using XML format. In the step2, the author & affiliation context is identified from the Plain-text format by using the "Title" and "Abstract" keyword. In the step3, the author names with affiliation id "Mathias Brochhausen[1]", "Catarina Martins[2,3] "are collected by using regular expression "[A-Za-z]{1}[A-Za-z\s\.\u00C1\u00E9\u00F3\u00D3 \u00D1\u00FA-]*\s?\d\s?(,\d)*". Then in the same step, we have separated the author names and affiliation id. In step4, each affiliation part along with affiliation id is extracted by using the pattern "\n?\d\s?[A-Za-z\s,#\.-]*". In the same step, the 'affiliation id' and 'affiliation country' information is separated. The country name verification is done by using the countries name repository. Finally, in this step the affiliation name is detected using affiliation id and country name information. In the step5, each author name is integrated with the concerned affiliation names based on affiliation id. These experimental analyses shown that each format require a strong heuristic and pattern for the required information retrieval.

(b) **Section Heading Extractor**

In the Fig. 8, the section heading extractor is used to extract the first level headings of sections from research paper document. It consists of section heading recognizer, section heading refiner, and splitter. The section heading recognizer has the ability to identify the section headings in the processed content of XML or Text format. The section heading refiner is used to remove the wrong patterns and additional characters with the output of the recognizer. Finally, the splitter will separate the structured elements such as 'section number' and 'section title' from the first level section heading. The section heading extractor exploits the heuristic and rules which exist in the form of regular expressions. First all the regular expressions are verified over the content of both XML and Plain-text formats in "EDITpad Pro 7[4]" tool and then it is used in java code.

The functionality of section heading recognizer, section heading refiner, and splitter is explained in below part. In the Fig. 9, the PDF document is parsed into XML document by the PDFx tool. This tool represents the section heading by tags '<h1>' and '<h2>' due to different formats of section heading represented by red rectangle in

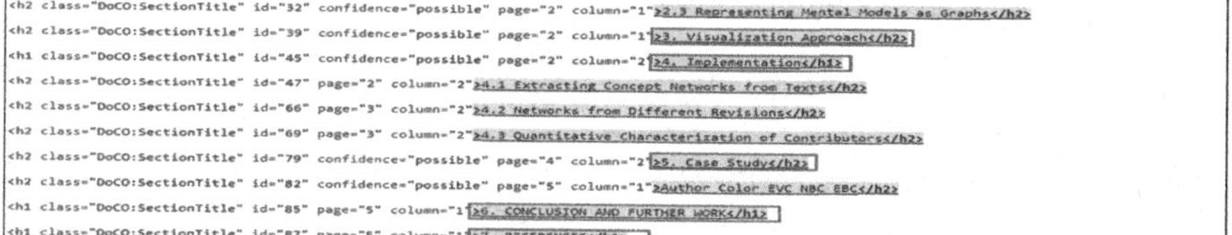
```
<h2 class="DoCO:SectionTitle" id="6" confidence="possible" page="1" column="1">General Terms Keywords</h2>
<h1 class="DoCO:SectionTitle" id="8" page="1" column="1">1. INTRODUCTION</h1>
<h2 class="DoCO:SectionTitle" id="17" confidence="possible" page="1" column="2">2. Background 2.1 Collaborative Writing Activities in Education</h2>
<h2 class="DoCO:SectionTitle" id="25" confidence="possible" page="1" column="2">2.2 Visualization Approaches for Collaborative Writing</h2>
<h2 class="DoCO:SectionTitle" id="32" confidence="possible" page="2" column="1">2.3 Representing Mental Models as Graphs</h2>
<h2 class="DoCO:SectionTitle" id="39" confidence="possible" page="2" column="1">3. Visualization Approach</h2>
<h1 class="DoCO:SectionTitle" id="45" confidence="possible" page="2" column="2">4. Implementation</h1>
<h2 class="DoCO:SectionTitle" id="47" page="2" column="2">4.1 Extracting Concept Networks from Texts</h2>
<h2 class="DoCO:SectionTitle" id="66" page="3" column="2">4.2 Networks from Different Revisions</h2>
<h2 class="DoCO:SectionTitle" id="69" page="3" column="2">4.3 Quantitative Characterization of Contributors</h2>
<h2 class="DoCO:SectionTitle" id="79" confidence="possible" page="4" column="2">5. Case Study</h2>
<h2 class="DoCO:SectionTitle" id="82" confidence="possible" page="5" column="1">Author Color EVC NBC EBC</h2>
<h1 class="DoCO:SectionTitle" id="85" page="5" column="1">6. CONCLUSION AND FURTHER WORK</h1>
<h1 class="DoCO:SectionTitle" id="87" page="5" column="1">7. REFERENCES</h1>
```

Fig. 9. Section heading recognition in XML document by section heading recognizer (Color figure online)

4 https://www.editpadpro.com/.

PDF document such as '1. INTRODUCTION', '2. Background', '3. Visualization Approach', '4. Implementation', '5. Case Study', '6 CONCLUSION AND FURTHER WORK', and '7. REFERENCES'. However, most of the time the '<h2>' tag is used to represent the second level heading such as '2.2', '2.3', '4.1', '4.2', '4.3' etc. To solve the various formats problem with section heading, first we extracted all patterns of '<h1>' and '<h2>' by section heading recognizer. Second, the output of section heading recognizer is transferred to section heading refiner. The refiner removes the second level headings by using 'RegEx2' and also removes some additional characters such as '>', '</h1>' or '</h2>' with section headings.

Finally, the refiner generates the accurate section heading. The output of refiner is further processed by the splitter to produce the structured elements such as 'section Number' and 'section Title' of each section heading in research paper as shown in Fig. 10.

Section Heading		Section-Number	Section-Title
1.	INTRODUCTION	1	INTRODUCTION
2.	Background	2	Background
3.	Visualization Approach	3	Visualization Approach
4.	Implementation	4	Implementation
5.	Case Study	5	Case Study
6.	CONCLUSION AND FURTHER WORK	6	CONCLUSION AND FURTHER WORK
7.	REFERENCES	7	REFERENCES

Fig. 10. Section heading conversion into structured elements

Similarly, the roman capital cases are identified from the Plain-text format by using regular expression. The roman numbers are replaced with numeric numbers such as 'I', 'II' with '1', '2' etc. Some sections are represented in research papers without section numbers such as 'Acknowledgments' or 'References' as shown in Fig. 11. In this approach, the section numbers are also assigned to such sections.

```
I.   INTRODUCTION
Ontologies have on the one hand axioms that form parts of
II.  TEXTUAL DEFINITIONS
In an ontology, a textual definition is, ideally, a short sen-
III. AXIOMS IN ONTOLOGIES
Axioms in ontologies restrict the intended meaning of a
IV.  CORRESPONDENCES BETWEEN TEXTUAL AND
LOGICAL DEFINITIONS
As we have seen, axioms and textual definitions have
V.   USING THE CORRESPONDENCES TO HELP IN
DEFINITION CHECKING
In ontologies that use semi-automated systems to create
VI.  OTHER USEFUL WAYS OF CHECKING THE CONTENTS
OF TEXTUAL DEFINITIONS
In ontologies, definitions should include only necessary
VII. CONCLUSION
In this communication, we showed through examples that
VIII. ACKNOWLEDGMENT
Work on this paper was supported by the Swiss National
Science Foundation (SNSF).
REFERENCES
```

Fig. 11. Roman capital cases in plain-text document

(c) Table Label Extractor

The table label extractor is used to identify the table label inside the content of research document. It consists of three parts such as 'Table label recognizer', 'Table label refiner', and 'Table label splitter'. The table label recognizer highlight the table labels in the XML document by using the regular expression "> (Table|TABLE) [A-Za-z0-9\s\.:,\(\)*%/-]{4,} </caption>" as shown in Fig. 12. The output of table label recognizer is further refined by removing the additional characters such as '>' and '</caption>'.

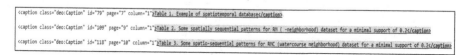

Fig. 12. Table label recognition in XML document

The result of refiner is further divided into structural elements such as table number and table caption by splitter is shown in Fig. 13.

Table Label
Table 1. Example of spatiotemporal database
Table 2. Some spatially sequential patterns for RM (-neighborhood) dataset for a minimal support of 0.2
Table 3. Some spatio-sequential patterns for RMC (watercourse neighborhood) dataset for a minimal support of 0.2

Table-Number	Table-Caption
1	Example of spatiotemporal database
2	Some spatially sequential patterns for RM (-neighborhood) dataset for a minimal support of 0.2
3	Some spatio-sequential patterns for RMC (watercourse neighborhood) dataset for a minimal support of 0.2

Fig. 13. Table label conversion into structured elements

(d) Figure Label Extractor

The figure label extractor is used to extract the figure label from research paper. It consists of three parts. The first part is figure label recognizer which identifies the figure label in both XML and Plain-text formats as shown in red rectangle in Fig. 4. The figure recognizer extracts the patterns from XML document by using the regular expression ">(Figure|Fig\.)   [A-Za-z0-9\s\.:',   \u2018\u2019\[\]\{\}\(\)& = ;-]* </caption>". If a pattern does not found in the XML format, then the figure recognizer finds the missed pattern and then use the figure number of the missed pattern in the regular expression "\n(Fig.|Figure)\s3(\.|:) [A-Za-z0-9:! +\.,\s\(\)/?;\[\]-]*\r", to find the missed pattern in the Plain-text document. The second part'figure label refiner' remove the extra characters from the output of recognizer such as '>', '</caption>', '\r', '\n' etc. The result of refiner is processed by third part 'splitter' to produce structured elements such as figure number and figure caption. In the Fig. 14 the output of figure caption extractor has been shown.

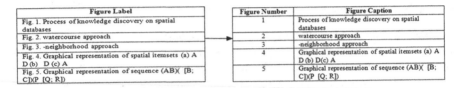

Fig. 14. Figure label conversion into structured elements

(e) Supplementary Material Links Extractor

The extractor of supplementary material consists of two parts (1) URL pattern recognizer, and (2) URL pattern refiner. Both XML and Plain-text formats are suitable for the extraction of supplementary materials links. We observed that the accuracy of links or URL extraction from Plain-text format is better than the XML format. Therefore, the link extractor is built based on Plain-text format. The URL pattern recognizer uses the regular expression "http[A-Za-z0-9\\.#%,:/_-]*" to find out the links from the Plain-text format of PDF document. The regular expression wrongly identifies some of the links due to the carriage return and newline characters which occurred with colon (:\r\n), forward slash (/\r\n), and hyphen sign (-\r\n) as shown in Fig. 15. Therefore, we need to replace these characters with ":", "/", and "-". The URL pattern refiner removes some additional characters from the output of the recognizer to produce the final output as links of supportive materials.

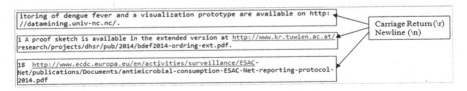

Fig. 15. Carriage return problem with links or URL

(f) Funding Agency and Funded Project Extractor

In this part, we have extracted the information about funding agency and funded project in a research paper.

The extraction of funding agencies name and acronyms as shown in Fig. 8 is done by the content processor, extractor, refiner and splitter. In the content processing step, start identifiers such as 'contribution of', 'supported from', 'supported by the', 'funded by', 'grant from', 'funding from', 'support from' etc. have been replaced with 'supported by' common identifier. The end identifiers such as 'double quotes', 'comma', 'colon', 'closed bracket' etc. have been replaced with 'dot' common identifier. The regular expression is prepared based on 'supported by' and 'dot' identifiers such as "support[A-Za-z]*\s(by|of)[A-Za-z0-9:,\s/\(\)\u2018\u2019\u2013-]*\.". The recognizer uses this pattern to search the name of funding agency along with acronyms. The refiner removes the additional words from the recognizer output by using such as pattern "[A-Za-z\s]*support[A-Za-z]*\s(of|by)(\s(the|a))?\s|\(|\.". In the last step, the splitter split the funding agency name and acronyms in two parts. The snapshot of the output is given in Fig. 16.

fund-agency-name		fund-agency-acronym
"Center for Service Innovation"	31B	"CSI"
"Austrian Science Fund"	23B	"FWF"
"National Library of Medicine"	30B	
"National Center For Advancing Translation al Sciences of...	92B	
"Agency for Healthcare Research and Quality"	45B	"AHRQ"
"Swiss National Science Foundation"	35B	"SNSF"
"Future and Emerging Technologies"	34B	"FET"

Fig. 16. The output of funding agency extractor

The project name extraction is performed by using project name recognizer & refiner. The project name recognizer extracts the area of content in research paper which consists of project name by using the regular expressions "(\bthe\b|\bThe\b) [A-Za-z0-9\s'\u201D\u201C\u2018\u2019-]{6,}\b(p|P)roject\b|(p|P)roject\s [A-Za-z0-9u201D\u201C\u2018\u2019'-]{4,}|(\u201C|")[A-Z]*(\u201D)". These regular expressions are prepared based on the deep analysis of project name recognition as shown in Fig. 6. The refiner part of 'project name extractor' find the position of last occurrence of the word 'the' in the highlighted text 'the portuguese FCT through the SOMER project' and gets the substring such as 'the SOMER project' and finally remove the additional words from substrings like 'the' and 'project'.

4   Triplification of Structured Elements

In this section, the extracted structured elements are converted into the RDF triples. It consists of two core components (1) Research Paper Ontology, and (2) RDF Document Generator. The research paper ontology is used to capture all the extracted structured elements of research papers by our extraction tool. The RDF document generator produces the RDF document based on the research paper ontology and then it store the RDF document in the RDF Triple store. Initially, the SPARQL queries are designed over the RDF document which consists of 45 research papers of training dataset. Then the same queries are tested over the RDF of evaluation dataset of 40 research papers. The RDF of 45 documents contains 2,488 triples while the RDF of evaluation data contains 1,815 triples. Jena Java Library[5] was used to implement the designed queries. The RDF document, SPARQL queries, and triples related information are available on this link[6]. In Fig. 17 we have mentioned the research paper ontology to represent the entities and their properties.

---

[5] http://Jena.apache.org.

[6] http://www.cdsc-cust.org/downloads.

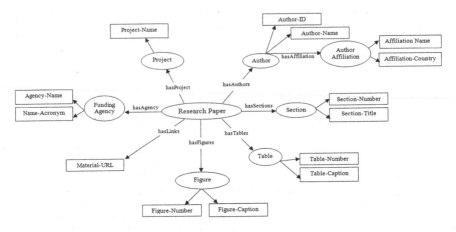

Fig. 17. Research paper ontology

5 Experiments

In this section, we present the brief details of dataset and experimental results. For the experimental study, two datasets are selected from the ESWC challenge. The first training dataset consists of 45 research papers while the second evaluation dataset consists of 40 research papers. The RDF of each training and evaluation datasets contains of 2,488 and 1,815 triples respectively. The execution of eight designed queries against each research paper in both datasets generates 360 and 320 (CSV extension) files respectively. The generated CSV files are evaluated with the gold standard dataset using evaluation tool. The gold standard dataset and evaluation tool is provided in the ESWC challenge available at link[7]. The score of our proposed technique over both datasets is given in Table 2. The average F-score over both dataset is 0.783.

Table 2. Experimental analysis of our proposed technique

Datasets	Precision	Recall	F-score
Training	0.798	0.802	0.795
Evaluation	0.775	0.778	0.771

6 Conclusions

This paper presents a comprehensive system for the extraction of informational components from PDF format of research papers of CEUR workshop. The proposed system utilizes powerful capabilities of converted XML and crafted rules from plain text formats. The in depth analysis has identified different situations in which XML

---

[7] https://github.com/angelobo/SemPubEvaluator.

document or plain text document may help to extract the required structured information more accurately. The extracted information includes: authors, their affiliation & country, first level section heading, table caption, figure caption, funding agency name, and project name. The challenges for extraction of each component have been presented using real snapshots followed by a comprehensive proposed methodology to extract the required information. After the extraction of required information, triplification of data was done using paper's ontology. This triplified data will help semantic agents to answer complex queries. The proposed approach has been developed over 45 training dataset which is not sufficient data to check the performance of suggested technique. Although the rules identified were made generic as much as possible, however, the approach can further benefit from other diversified publication datasets. In some typical cases, the approach may not work due to limitations of crafted rules built for CEUR dataset.

Acknowledgments. We would like to thank the computer science department of CUST (Capital University of Science & Technology) to provide us research Lab and knowledgeable support to conduct this research activity. We are also thankful to the members of CDSC (Center for Distributed & Semantic Computing) research group for supporting us in different occasion to complete this research work.

References

1. Do, H.H.N., Chandrasekaran, M.K., Cho, P.S., Kan, M.Y.: Extracting and matching authors and affiliations in scholarly documents. In: Proceedings of the 13th ACM/IEEE-CS Joint Conference on Digital Libraries, pp. 219–228. ACM (2013)
2. Di Iorio, A., Peroni, S., Poggi, F., Vitali, F., Shotton, D.: Recognising document components in XML-based academic articles. In: Proceedings of the 2013 ACM Symposium on Document Engineering, pp. 181–184. ACM (2013)
3. Kim, S., Cho, Y., Ahn, K.: Semi-automatic metadata extraction from scientific journal article for full-text XML conversion. In: Proceedings of the International Conference on Data Mining (DMIN), p. 1 (2014). The Steering Committee of the World Congress in Computer Science, Computer Engineering and Applied Computing (WorldComp)
4. Luong, M.T., Nguyen, T.D., Kan, M.Y.: Logical structure recovery in scholarly articles with rich document features. In: Multimedia Storage and Retrieval Innovations for Digital Library Systems, vol. 270 (2012)
5. Milicka, M., Burget, R.: Information extraction from web sources based on multi-aspect content analysis. In: Gandon, F., et al. (eds.) SemWebEval 2015. CCIS, vol. 548, pp. 81–92. Springer, Heidelberg (2015). doi:10.1007/978-3-319-25518-7_7
6. Mohemad, R., Hamdan, A.R., Othman, Z.A., Noor, N.M.: Automatic document structure analysis of structured PDF files. Int. J. New Comput. Architect. Appl. (IJNCAA) 1(2), 404–411 (2011)
7. Manabe, T., Tajima, K.: Extracting logical hierarchical structure of HTML documents based on headings. Proc. VLDB Endow. 8(12), 1606–1617 (2015)
8. Nuno, M., Fátima, R.: Extracting structure, text and entities from PDF documents of the portuguese legislation. Institute of Engineering, Polytechnic of Porto, Portugal (2012)

9. Ramakrishnan, C., Patnia, A., Hovy, E., Burns, G.A.: Layout-aware text extraction from full-text PDF of scientific articles. Source Code Biol. Med. 7(1), 1 (2012)
10. Saleem, O., Latif, S.: Information extraction from research papers by data integration and data validation from multiple header extraction sources. In: Proceedings of the World Congress on Engineering and Computer Science, vol. 1 (2012)
11. Constantin, A., Pettifer, S., Voronkov, A.: PDFX: fully-automated PDF-to-XML conversion of scientific literature. In: Proceedings of the 2013 ACM Symposium on Document Engineering, pp. 177–180. ACM (2013)

An Automatic Workflow for the Formalization of Scholarly Articles' Structural and Semantic Elements

Bahar Sateli and René Witte[(✉)]

Semantic Software Lab,
Department of Computer Science and Software Engineering,
Concordia University, Montréal, Canada
{sateli,witte}@semanticsoftware.info

Abstract. We present a workflow for the automatic transformation of scholarly literature to a Linked Open Data (LOD) compliant knowledge base to address Task 2 of the Semantic Publishing Challenge 2016. In this year's task, we aim to extract various contextual information from full-text papers using a text mining pipeline that integrates LOD-based Named Entity Recognition (NER) and *triplification* of the detected entities. In our proposed approach, we leverage an existing NER tool to ground named entities, such as geographical locations, to their LOD resources. Combined with a rule-based approach, we demonstrate how we can extract both the structural (e.g., floats and sections) and semantic elements (e.g., authors and their respective affiliations) of the provided dataset's documents. Finally, we integrate the LODeXporter, our flexible exporting module to represent the results as semantic triples in RDF format. As the result, we generate a scalable, TDB-based knowledge base that is interlinked with the LOD cloud, and a public SPARQL endpoint for the task's queries. Our submission won the second place at the SemPub2016 challenge Task 2 with an average 0.63 F-score.

1 Introduction

The *Semantic Publishing Challenge*, which started in 2014, is a recent series of competitive efforts to produce linked open datasets from multi-format and multi-source input documents. The 2016 edition of the challenge[1] pursues the goal of assessing the quality of scientific output through an automatic analysis of scholarly documents for fine-grained contextual information. The dataset under study consists of 85 papers (45 training and 40 evaluation) taken from computer science workshop proceedings, published by CEUR-WS.org. The goal is to extract authors, affiliations, funding bodies, as well as a document's sections and floating elements and ultimately, populate a knowledge base with the

---

[1] Semantic Publishing Challenge 2016, https://github.com/ceurws/lod/wiki/SemPub 2016.

© Springer International Publishing Switzerland 2016
H. Sack et al. (Eds.): SemWebEval 2016, CCIS 641, pp. 309–320, 2016.
DOI: 10.1007/978-3-319-46565-4_24

results. Such information, when extracted, can be used to automatically provide an overview of the authors' scientific output, the universities and research institutions involved, as well as the active funding bodies in a domain.

In this paper, we present our automatic workflow we developed to address Task 2 of the challenge that can extract structural and semantic elements from the full-text of a document and transform the detected entities into RDF triples, which are eventually made persistent in a scalable, TDB-based knowledge base with a public SPARQL endpoint. The generated knowledge base is evaluated against a set of eight pre-defined queries for its correctness and completeness and exploited as a means of assessing the quality of scientific production in the respective workshops.

The challenge queries are concerned with searching for entities, categorized as follows:

- Authors, their Affiliations (**Q2.1**) and the country where the affiliation is located in (**Q2.2**);
- Supplementary Material (**Q2.3**), mentioned as part of the authors' contributions;
- Structural elements, in particular first-level Sections (**Q2.4**), and floating elements like Tables (**Q2.5**) and Figures (**Q2.6**); and
- Names of Funding Agencies (**Q2.7**) and European Projects (**Q2.8**) supporting the research presented in the paper.

Note that you can find supplementary material, such as the populated knowledge base and the text mining pipeline resources at http://www.semanticsoftware.info/sempub-challenge-2016.

2   Design

We designed an automatic workflow (Fig. 1) that starts from a set of documents (called a *corpus*), which go through multiple processing phases, and produces semantic triples as output. The *Syntactic Processing* phase breaks down the full-text of the documents into smaller segments and pre-processes the text for further semantic entities. The *Semantic Processing* phase takes the results of the syntactic analysis and attempts to annotate various entities in a text. Finally, each document's annotations are translated into semantic triples, according to a series of custom *mapping rules*, and stored in a knowledge base.

2.1   Syntactic Processing

In order to remove the various formatting and typesetting styles of the input documents, rather than working directly with the PDF files, we scrape the text from them and then process the extracted plain text. The full-text of a document is segmented into *tokens* – smaller, linguistically meaningful parts, like words, numbers and symbols. Subsequent syntactical processing components process

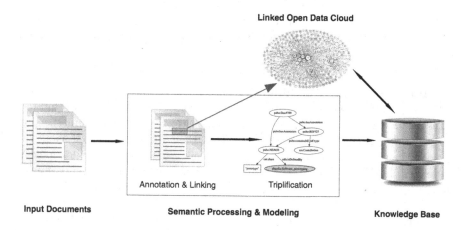

Fig. 1. Overall workflow of our text mining pipeline

the tokenized text into sentences and all sentence constituents are tagged with a Part-of-Speech category, like noun, verb or adjective. Finally, we lemmatize the tokens in a text to store their root format in order to ignore their morphological variations (e.g., "university" and "universities" will have the same canonical form).

2.2   Semantic Processing

The semantic analysis of the pre-processed text is conducted in an iterative fashion. In each iteration the semantic analysis method uses the output of the upstream processing steps in order to generate new *annotations* used incrementally to generate the desired entities.

Gazetteering. In text mining applications, gazetteers play the role of carefully curated, readily available knowledge resources. Essentially, gazetteers are dictionaries of entities with pre-defined semantics, e.g., the list of all countries or months of a year. In our approach, we use several gazetteers to annotate tokens, like "*Department*" or "*Institute*" that the downstream processing algorithms use, for example, to detect names of universities or research institutions in the authors' affiliations.

Named Entity Linking. In this year's challenge, we integrate a LOD-based Named Entity Recognition (NER) tool that allows us to ground various types of named entities in the documents to their corresponding resources on the LOD cloud, using a Universal Resource Identifier (URI). This approach not only alleviates the burden of manually updating and curating our gazetteers, e.g., of country names, but also by virtue of traversing the semantic links of a semantic named entity, we will are able to find additional, machine-readable information where needed. In our approach, we rely on an existing NER tool that can link the surface form of a document's terms to LOD resource, which we further filter to retain only nouns and noun phrases.

Rule-Based Pattern Matching. We developed a set of hand-crafted rules that capture various syntactical (grammatical) structure of the entities of interest in Task 2. Most of the rules rely on the annotations generated from the pre-processing phase: the tokens, their root forms, their POS tag, as well as all the entities in a text that were matched against our gazetteers. Our rules look at pre-defined, hand-crafted sequences of annotations and when a match is found, generate an annotation with a semantic type, e.g., an Affiliation annotation.

Authors and Affiliations. We focused the authors and affiliations extraction to the region of text between a document's title and abstract section (which we call the metadata body), since both were present in all of the training set documents. The detection of authors largely reuses the rules from our last year's submission to the challenge [1]. We start from our gazetteers of common first names. For instance, all first name tokens followed by an upper initial token in the metadata body are annotated as Authors in a text.

For this year's challenge, we developed a new approach for detecting affiliations: Using a NER tool, we first annotate all tokens in the metadata body that can be mapped to a resource in the LOD cloud with either a location, city or country semantic type. Subsequently, we apply our pattern matching rules against the sequence of word and symbol tokens in text adjacent to the grounded location entities. The rules below show exemplary patterns, where the semantic types are shown in capital letters:

(1) (UPPER INITIAL) (University | Universität | Universidad) (of | de) (TOWN | CITY | COUNTRY)
↪ matched strings: *"Universidad de Chile"*, *"Technische Universität Wien"*

(2) (COUNTRY ADJECTIVE) (ORGANIZATION) (of | de) (UPPER INITIAL)
↪ matched strings: *"Indian Institute of Technology"*, *"National Institute of Aging"*

In addition to institutions names, our pipeline also detect various organizational units like research centres, departments and research group naming patterns. If an organizational unit is found in a text in adjacency of an institution name (e.g., a university name), then the longest span of text covering both entities will be annotated as the Affiliation annotation.

Affiliation Locations. This year, we employed a new approach in detecting where an affiliation is located. The goal is to find the country name of the affiliations in the metadata body. To this end, our new component applies a set of heuristics with a fallback strategy that looks at the Affiliation annotations and the country named entities generated by the NER tool. While investigating the training set documents, we found out that detecting such a relation is a complex task: First, we observed that the line-by-line scraping of the documents' text often mixes up the order of the metadata body entities. This way, e.g., in a two-column affiliation information, the university names are scraped in one line next to each other and the next line has both of their country information. Second, there exists a number of documents in the training set, which do not have the country name in the metadata body. Our *Affiliation–Location* Inferrer component, optimistically applies a set of rules on the metadata body:

- In the case that there is only one affiliation and one country entity in the metadata body, it matches the two entities together with a high confidence.
- If there is more than one annotation of each type (i.e., Affiliation and Location), it makes two lists from the annotations sorted by their start offset in a text. It then iterates through the lists and matches each affiliation with a location that has a greater offset, but is located in the shortest distance from the affiliation annotation.
- Finally, if no Location annotation is available in the metadata body, it tries to find the country name from the DBpedia ontology. To do so, it first performs a DBpedia Lookup[2] operation using the affiliation label in the document in order to find a DBpedia resource URI. Subsequently, it executes a federated query against the public DBpedia SPARQL endpoint,[3] looking for triples where the subject matches the affiliation URI, and the predicate is one of 'country' or 'state' properties from the DBpedia ontology. If such a triple is found, it returns the English label of the object (country) and infers a relation between the affiliation and the country name.

Sections and Floats. We curated a set of *trigger* words in the documents that are used in figures and tables' captions, such as *"Fig."*, *"Figure"* and *"Table"*. The detection of sections and floats are dependant on whether the text scraping process was also successful in segmenting the document. In the case where the segmentation is properly done, we merely check for the existence of our trigger words in the candidate segments, i.e., text boundaries detected by the scraping tool. If the boundary truly represents a caption, we further analyze it for symbols and numbers and classify the caption as either a Figure or a Table. In case we cannot find a number in the caption, a counter incrementally numbers the generated annotations, ordered by their starting offset.

We use a similar approach in detecting document Sections: If the text scraping tool cannot successfully extract the section headers, we check them against a gazetteer of conventional section headers (e.g., "Introduction", "Results and Discussion", or "Conclusion") to find matching strings. If no match is found, the pipeline will simply skip the section detection phase.

Funding Agencies. The challenge in finding the names of funding agencies is to distinguish the agency or organization names that funded the work presented by the authors of a paper. We previously investigated in [2] that so-called *deictic* phrases are used by authors in the context of scholarly discourse to refer the

---

[2] DBpedia Lookup, https://github.com/dbpedia/lookup.
[3] DBpedia SPARQL endpoint, http://dbpedia.org/sparql.

readers to the document under study, such as *"In this work"* or *"The presented paper"*. We use the approach described in [2] to find deictic phrases in a document and look at the tokens and annotations following such phrases to elect a candidate sentences for funding agency name detection. Similar to our approach from last year, the agency name is detected as either *(i)* one or more upper-initial word tokens, or *(ii)* an annotated organization name.

(3) (DEIXIS) (VERB) (funded) (by | within) (the) (UPPER INITIAL | ORGANIZATION)
    ↪ matched string: *"This research is funded by the DebugIT project."*

We plan to further improve the detection by incorporating a dependency parser in our text mining pipeline so that the funding agency names can be extracted from the noun phrase following the *"funded by"* verb phrase in the sentence's dependency tree.

2.3   Knowledge Base Construction

Using the rule-based text mining approach described above, we can extract the entities we are interested in to answer the challenge queries. The next step is to transform the extracted annotations into semantic triples and store them in a LOD-compliant knowledge base.

Semantic Vocabularies. In our semantic model, we reuse the vocabularies from the Document Components Ontology (DoCO) [3] to describe the structural elements of documents, namely, the sentences, sections and floats. For the authors and affiliations we use the FOAF[4] and Bibliographic Ontology (BIBO).[5] All other semantic classes and relationships, such as modeling annotations in a document, embedded annotations and the relations between annotations are described using our PUBlication Ontology (PUBO).[6]

LODeXporter. We use our *LODeXporter*[7] component in our text mining workflow to populate a knowledge base. It accepts a set of custom mapping rules as input and transforms the designated document's annotations into their equivalent RDF triples. For each annotation type that is to be exported, the mapping rules have an entry that describes: *(i)* the annotation type in the document and its corresponding semantic type, *(ii)* the annotation's features and their corresponding semantic type, and *(iii)* the relations between exported triples and the type of their relation. Given the mapping rules, the mapper component then iterates over the document's entities and exports each designated annotation as the subject of a triple, with a custom predicate and its attributes, such as its features, as the object.

---

[4] FOAF, http://xmlns.com/foaf/spec/.
[5] BIBO, http://purl.org/ontology/bibo/.
[6] PUBO, http://lod.semanticsoftware.info/pubo/pubo.rdf.
[7] LODeXporter is currently considered to be in pre-release and available at http://www.semanticsoftware.info/lodexporter.

3   Implementation

We implemented our text mining pipeline described in Sect. 2 based on the *General Architecture for Text Engineering* (GATE) framework [4], shown in Fig. 2. The pipeline accepts scientific literature in PDF or XML format from local or remote URLs as input and stores the extracted entities in form of an RDF document in a knowledge base as output. In this section, we provide the details of our GATE pipeline's processing resources.

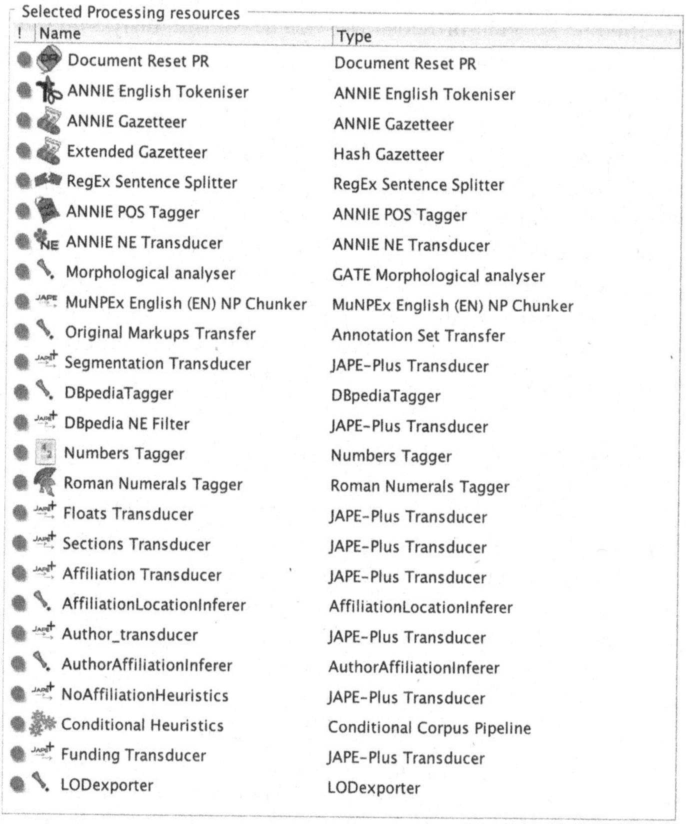

!	Name	Type
	Document Reset PR	Document Reset PR
	ANNIE English Tokeniser	ANNIE English Tokeniser
	ANNIE Gazetteer	ANNIE Gazetteer
	Extended Gazetteer	Hash Gazetteer
	RegEx Sentence Splitter	RegEx Sentence Splitter
	ANNIE POS Tagger	ANNIE POS Tagger
	ANNIE NE Transducer	ANNIE NE Transducer
	Morphological analyser	GATE Morphological analyser
	MuNPEx English (EN) NP Chunker	MuNPEx English (EN) NP Chunker
	Original Markups Transfer	Annotation Set Transfer
	Segmentation Transducer	JAPE–Plus Transducer
	DBpediaTagger	DBpediaTagger
	DBpedia NE Filter	JAPE–Plus Transducer
	Numbers Tagger	Numbers Tagger
	Roman Numerals Tagger	Roman Numerals Tagger
	Floats Transducer	JAPE–Plus Transducer
	Sections Transducer	JAPE–Plus Transducer
	Affiliation Transducer	JAPE–Plus Transducer
	AffiliationLocationInferer	AffiliationLocationInferer
	Author_transducer	JAPE–Plus Transducer
	AuthorAffiliationInferer	AuthorAffiliationInferer
	NoAffiliationHeuristics	JAPE–Plus Transducer
	Conditional Heuristics	Conditional Corpus Pipeline
	Funding Transducer	JAPE–Plus Transducer
	LODexporter	LODexporter

Selected Processing resources

Fig. 2. The sequence of processing resources of our text mining pipeline

3.1   Text Pre-processing

Different from our last year's approach [1], we use PDFX[8] [5] to transform PDF articles to XML documents. The GATE framework itself relies on the Apache

---

[8] PDFX, http://pdfx.cs.man.ac.uk.

Tika library[9] for extracting the textual content of the XML files, while preserving the original XML elements. For the scope of this paper, we exclude a discussion on text extraction from PDF documents; a comprehensive overview can be found in [5]. We also re-use GATE's ANNIE and Tools plugins [6] to tokenize and lemmatize the text, detect sentence boundaries, and perform gazetteering on the text.

3.2   Named Entity Detection with Spotlight

To detect domain-specific entities in the documents, we rely on the LOD cloud, in particular DBpedia [7]. This provides for a rich, continuously-updated resource in a standard semantic format. By linking entities detected in documents to LOD URIs, we can semantically query a knowledge base for all papers on a specific entity (URI). For the actual entity tagging, we use an external tool, DBpedia Spotlight [8] version 0.7 with a statistical model for English. To integrate this web service into a GATE text mining pipeline, we use our *LODtagger* plugin.[10] This component sends the entire UTF-8 formatted text of a document as a RESTful POST request to a given Spotlight endpoint and receives the results in JSON format, which are subsequently parsed and transformed to GATE annotations. For the challenge queries, we annotate the metadata body of each document for city and country named entities, which are used in rules for affiliation detection in our pipeline.

3.3   Rule-Based Extraction of Contextual Entities

The rules described in Sect. 2.2 are implemented using GATE's JAPE language that provides for defining regular expressions over a document's annotations (by internally transforming them into finite-state transducers). Our text mining pipeline contains several *transducers* that apply our JAPE rules on the documents' text. Each JAPE rule has two main parts: a Left-Hand Side (LHS) pattern that essentially describes a regular expression over a document's annotations; When matched, the Right-Hand Side (RHS) of the rule adds a new semantic, typed annotation to the document. Additional information relevant to annotations are stored as their *features*. Figure 3 shows an example JAPE rule used in our pipeline to extract an Affiliation annotation in a document. The rule shown matches a sequence of organizational base entries in our gazetteer (e.g., 'University') with an adverb (e.g., 'of'), followed by a city named entity.

Our submission this year also features a number of heuristics to automatically re-try detecting or correcting the authors and affiliations in a fuzzy manner, if the JAPE rules do not strictly match any patterns in a document. As shown in Fig. 2, if there is no Affiliation annotation is found in the document after all JAPE rules are executed, a set of conditional heuristics will then be applied

---

[9] Apache Tika, https://tika.apache.org/.

[10] LODtagger, http://www.semanticsoftware.info/lodtagger.

on the document, for example, to blindly annotate spans of text between the last author name and the first mention of a country named entity, which may represent an Affiliation.

```
Rule: BaseINCity(
  ({Lookup.majorType=="org_pre"})?
  {Lookup.majorType == "org_base"}
  ({Token.orth == "upperInitial",
   !Lookup.majorType == "org_base"})?
  ({Lookup.majorType == "org_in"})+
  ({Affiliation_location.loc_type == "city"}
  | {Lookup.majorType == "location"})
):mention
-->
:mention.Affiliation_univ = {
  rule = "BaseINCity",
  content = :mention@cleanString }
```

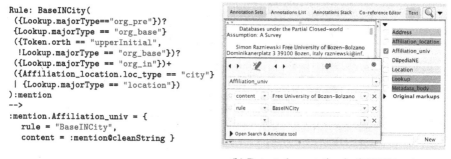

(a) Example JAPE rule   (b) Detected annotation in GATE Developer

Fig. 3. Example JAPE rule (left) to extract an Affiliation entity and the generated annotation in GATE's graphical user interface

3.4 Knowledge Base Population

The mapping rules shown in Fig. 4 describe the specifications of exporting GATE annotations into several inter-connected semantic triples: Each Author annotation in the document should be exported with `<foaf:Person>` as its type, and its verbatim content in a text using the `<cnt:chars>` predicate. Similarly, Affiliation annotations are exported with their 'locatedIn' feature describing the country they are located in from the GeoNames ontology (`<gn:locatedIn>`).[11] Subsequently, the value of the 'employedBy' feature of each Author annotation is used to construct a `<rel:employedBy>` relation between an author instance and its corresponding affiliation instance in the knowledge base. We used vocabularies from our PUBO ontology wherever no equivalent term was available in existing Linked Open Vocabularies. For example, we use the `<pubo:containsNE>` property to build a relation between the metadata body and the entities that appear within its start and end offsets in a document. There exists several other mapping rules that we custom-tailored to model the extracted entities, so that the exported triples can be queried for the challenge Task 2. Ultimately, the LODeXporter processing resource generates all of the desired RDF triples from the document's annotations, and stores them in a scalable, TDB-based[12] triplestore.

---

[11] GeoNames Ontology, http://www.geonames.org/ontology/documentation.html.
[12] Apache TDB, http://jena.apache.org/documentation/tdb/.

```
@prefix map: <http://semanticsoftware.info/mapping#> .
@prefix rdf: <http://www.w3.org/1999/02/22-rdf-syntax-ns#> .
@prefix cnt: <http://www.w3.org/2011/content#> .
@prefix rel: <http://purl.org/vocab/relationship/> .
@prefix foaf: <http://xmlns.com/foaf/0.1/> .
@prefix gn: <http://www.geonames.org/ontology#> .

### Annotation Mapping ###
map:GATEAuthor a map:Mapping ;
        map:type    foaf:Person ;
        map:GATEtype  "Author" ;
        map:hasMapping map:GATEContentMapping .

map:GATEAffiliation a map:Mapping ;
        map:type    foaf:Organization ;
        map:GATEtype  "Affiliation" ;
        map:hasMapping map:GATEContentMapping ;
        map:hasMapping map:GATELocatedInFeatureMapping .

### Feature Mapping ###
map:GATEContentMapping a map:Mapping ;
        map:type    cnt:chars ;
        GATEattribute "content" .

map:GATELocatedInFeatureMapping a map:Mapping ;
        map:type    gn:LocatedIn ;
        GATEfeature "locatedIn" .

### Relation Mapping ###
map:AuthorAffiliationRelationMapping a map:Mapping ;
        map:type rel:employedBy ;
        map:domain map:GATEAuthor ;
        map:range map:GATEAffiliation ;
        GATEattribute "employedBy" .
```

Fig. 4. Excerpt of the mapping rules for exporting Authors, Affiliations and their relations

3.5   Query Results Export

We published the populated knowledge base described in the previous section through a Jena Fuseki[13] server. In order to conform to the output format required for the automatic challenge evaluation tool,[14] we implemented a Java command-line tool that executes a set of hand-crafted queries against the Fuseki HTTP endpoint and transforms the results into Comma Separated Value (CSV) files. The command-line tool accepts a list of document URIs (e.g., http://ceur-ws.org/Vol-1006/#paper2) and a template for each of the challenge Task 2 queries. Each query template is designed to answer one of the Task 2 queries using same vocabularies explained in Sect. 3.4 and contains a placeholder for the document URI. Our query export tool parameterizes each query template with the correct URI form and performs a web service request against the knowledge base interface. Subsequently, the generated CSV files can be directly fed into the evaluator tool for generating our tool's performance report against the gold standard.

---

[13] Jena Fuseki, https://jena.apache.org/documentation/serving_data/.

[14] SemPubEvaluator, https://github.com/angelobo/SemPubEvaluator.

4   Results and Discussion

We analyzed the challenge training and evaluation sets of 45 and 40 documents, respectively, with our automatic workflow. The input documents are in PDF format and range from 3 to 16 pages for full-papers with various publisher-specific formatting. On average, each document contained 3331 and 5306 tokens in the training and evaluation sets, respectively. Table 1 shows the statistics of the document sets and the number of generated triples. The number of triples includes all the exported entities, their properties and inter-relationships, as well as the mapping rules themselves. On average, the workflow execution time is around 6 s per document (on a late 2013 MacBook Pro with 2.3 GHz Intel Core i7 and 16 GB memory), where a majority of time is consumed by DBpedia Spotlight NER web service.

Table 1. Quantitative analysis of Task 2 training and evaluation sets processing

Document set	#Documents		#Pages			#Tokens			#Triples
	Full	Abstract	Min	Max	Avg	Min	Max	Avg	
Training	41	4	1	16	8.5	262	14401	3331	3880
Evaluation	38	2	1	15	6.9	293	13059	5306	3044

We evaluated the performance of our approach on the training set against the provided gold standard. Note that in our submission we only analyzed the documents to address queries Q2.1, Q2.2, and Q2.4–7. The SemPubEvaluator tool, however, computes the average metrics over all entries in the gold standard. On the training set, we obtained an average of 0.619, 0.598 and 0.605 for precision, recall and F-score, respectively. On the evaluation set, our approach achieved an average of 0.64, 0.629 and 0.632 for precision, recall and F-score, respectively, which placed us as the runner-up in the challenge.[15]

An error analysis of our results showed that many of Type I errors (i.e., false positives) are the results of fuzzy detection of affiliation names, where the span of annotated text also contained characters that were not in the gold standard, or when the boundary of the author names had mistakenly overlapped with the affiliations, e.g., in the case of Hasso Plattner Institute. Type II errors (i.e., false negatives), on the other hand, were mostly due to affiliation names in languages other than English, or when the organizational unit of the affiliation was missing from the annotation spans.

---

[15] The best performing tool reported 0.775, 0.778, and 0.771, for precision, recall and F-score, respectively. The complete scoreboard is available on https://github.com/ceurws/lod/wiki/SemPub2016.

5 Conclusions

Semantic publishing research aims at making scientific publications readable and semantically understandable to computers. The long-term vision is to enable automated discovery and consumption of scholarly artifacts, like articles and datasets, by humans and machines alike. In this paper, we provided the details of our automatic workflow for the extraction of contextual information from full-text of computer science articles to address Task 2 of the Semantic Publishing Challenge 2016. We described our text mining pipeline which uses a rule-based pattern matching approach, combined with a named entity recognition tool to annotate the challenge datasets with various semantic entities. We then explained how we integrated our LODeXporter plugin to export the annotations to RDF triples, thereby populating a LOD-compliant knowledge base, in which the entities are inter-linked with resources on the linked open data cloud.

References

1. Sateli, B., Witte, R.: Automatic construction of a semantic knowledge base from CEUR workshop proceedings. In: Gandon, F., et al. (eds.) SemWebEval 2015. CCIS, vol. 548, pp. 129–141. Springer, Heidelberg (2015). doi:10.1007/978-3-319-25518-7_11

2. Sateli, B., Witte, R.: Semantic representation of scientific literature: bringing claims, contributions and named entities onto the linked open data cloud. PeerJ Comput. Sci. **1**, e37 (2015). doi:10.7717/peerj-cs.37

3. Shotton, D., Peroni, S.: DoCO, the Document Components Ontology (2011)

4. Cunningham, H., Maynard, D., Bontcheva, K., Tablan, V., Aswani, N., Roberts, I., Gorrell, G., Funk, A., Roberts, A., Damljanovic, D., Heitz, T., Greenwood, M.A., Saggion, H., Petrak, J., Li, Y., Peters, W.: Text Processing with GATE (Version 6). University of Sheffield, Department of Computer Science, Sheffield (2011)

5. Constantin, A., Pettifer, S., Voronkov, A.: PDFX: fully-automated PDF-to-XML conversion of scientific literature. In: Proceedings of the 2013 ACM Symposium on Document Engineering (DocEng 2013), pp. 177–180. ACM, New York (2013)

6. Cunningham, H., Maynard, D., Bontcheva, K., Tablan, V.: GATE: a framework and graphical development environment for robust NLP tools and applications. In: Proceedings of the 40th Anniversary Meeting of the Association for Computational Linguistics (ACL 2002) (2002)

7. Auer, S., Bizer, C., Kobilarov, G., Lehmann, J., Cyganiak, R., Ives, Z.G.: DBpedia: a nucleus for a web of open data. In: Aberer, K., et al. (eds.) ASWC 2007 and ISWC 2007. LNCS, vol. 4825, pp. 722–735. Springer, Heidelberg (2007)

8. Mendes, P.N., Jakob, M., García-Silva, A., Bizer, C.: DBpedia spotlight: shedding light on the web of documents. In: Proceedings of the 7th International Conference on Semantic Systems, pp. 1–8. ACM (2011)

Author Index

Printed in the United States
By Bookmasters